JOHN R. DIXON

*Department
of Mechanical Engineering
University of Massachusetts*

Thermodynamics I:
an introduction to energy

Printed in the United States of America

Library of Congress Cataloging in Publication Data

DIXON, JOHN R

Thermodynamics I: an introduction to energy.

1. Thermodynamics. I. Title.

QC311.D49 536′.7 74–9711

ISBN 0–13–914887–6

10 9 8 7 6 5 4 3 2 1

PRENTICE-HALL INTERNATIONAL, INC., *London*
PRENTICE-HALL OF AUSTRALIA, PTY. LTD., *Sydney*
PRENTICE-HALL OF CANADA, LTD., *Toronto*
PRENTICE-HALL OF INDIA PRIVATE LIMITED, *New Delhi*
PRENTICE-HALL OF JAPAN, INC., *Tokyo*

To Randal John Dixon

Contents

PART III

An introduction to the kinetic theory of ideal gases

Appendices

Acknowledgments

I would like to take this opportunity to acknowledge my teachers J. Fletcher Osterle (Carnegie-Mellon) and J. B. Jones (V.P.I.). I took no courses from them; these men teach with their lives. To the extent the book is correct and valuable, they have undoubtably helped; to the extent it is not perfect I am responsible.

Warren Rohsenow has been a helpful commentor and critic throughout. It's a book because of his encouragement, and a better book because of his comments.

I wish also to acknowledge the careful and efficient typing of the manuscript by Monigue Barnett, Alis Glazier and Helga Ragle. Finally, Hans Goettler's thorough checking of the manuscript, indexing, and preparation of the solutions manual were of tremendous value.

"I had a lover's quarrel with the world"

Robert Frost

Preface

I have a lover's quarrel with the thermodynamics books for engineering students. The general quality of these books is excellent, and each has its own outstanding, unique features. I love them, and there are so many of them that it boggles the mind to think of adding yet another one.

However, my quarrels are four in number. In this book I have tried to deal with three of them. The fourth, I believe, can at this time only be dealt with by the human teacher. First, on the level of philosophy of science, I believe our engineering science books should stress the inductive nature of physical principles and develop the operational viewpoint articulated so well by P. W. Bridgman. Thus I have attempted to develop the laws of thermodynamics inductively out of the students' experience instead of deductively from axiomatic statements of so-called "first" principles.

Second, I don't believe that entropy is the most significant or useful aspect of the Second law for engineers. The Second law has to do with the concept of degradation of energy; that is, with loss of useful *work* potential. Entropy is simply the available measure of degradation. It is useful as a property, but abstract. Degradation, on the other hand, because it is a work term, is an easily grasped concept. Thus I have attempted to introduce the Second law through the concept of degradation of energy. The result, I believe, is a clearer physical meaning for entropy.

Third, it has always seemed to me that thermo texts have left the student too much on his own in learning to solve problems and, at the same time, have denied him the opportunity to make the typical assumptions necessary to apply the laws. Thus I have tried to develop a problem-solving approach in the book, and have attempted to show the student how and why typical physical assumptions are made. I wish the text could be supplemented by a programmed text on problem-solving in thermodynamics, but I have so far been unable to make this wish a practical reality.

As a lover of thermodynamics books, my fourth quarrel is with the fact that it is so difficult in a book to give students a really sensitive physical intuition for the concepts of energy conservation and energy degradation. I would like my students to sense and be able to describe the energy flows in the devices, systems, and processes with which they work. Of course, I want them also to be skilled in the details of analyzing energy systems, but I do wish that such skill were complemented by what I can only at this juncture call an articulate conceptual understanding and awareness of the First and Second laws at work in nature all the time. Providing such a sensitivity is something I can leave only to the human teacher.

My unique contributions to the evolution of pedagogical engineering thermodynamics are restricted to Part I of the book; that is, to development and exposition of the principles. Part II contains the traditional applications to gas and vapor power and refrigeration cycles, to mixtures, air conditioning, and combustion. I have tried to do these sections as well and as capably as I can but they are included for completeness, not because they are unique. I have no lover's quarrel with the way these topics are treated in the many other texts.

I have included a chapter on introductory kinetic theory and arranged the book so that it can be woven into the classical treatment, left to the end, or ignored completely, at the discretion of the teacher. That is, it seems to me, as it should be.

John R. Dixon

Thermodynamics I

PART I

Principles

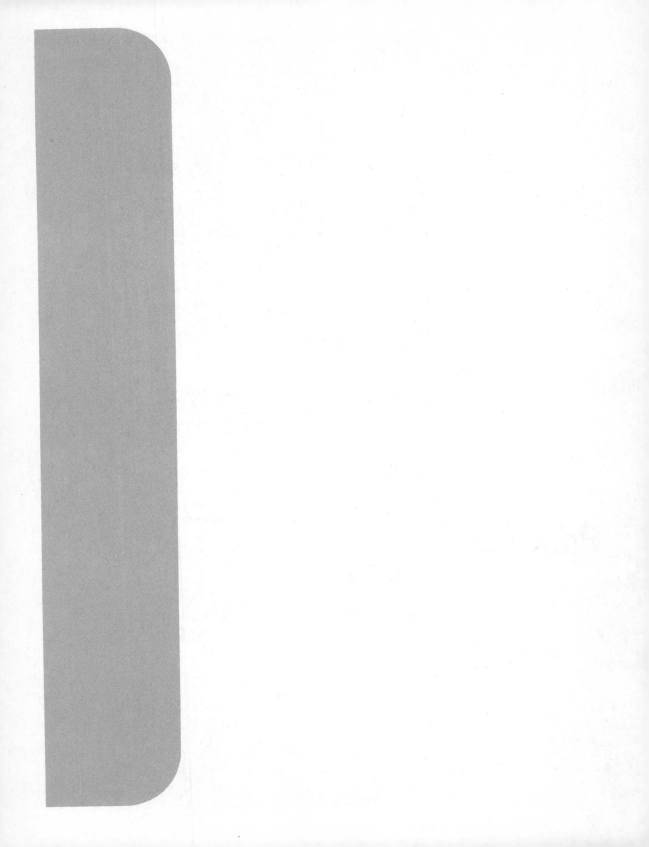

1

Energy, man, and environment

Gerry Smith had the time and the cash for his long awaited trip around the country. His old car was in good shape and his parents had given their approval, but still Gerry could not go. He just hadn't been able to save enough *E* (for energy) rationing stamps for the gasoline for the trip, and he hadn't been able to borrow enough from his friends and family against his future allotment. With more money, he might have been able to buy the necessary stamps at the Energy Exchange but that would have left him short of cash for the trip. This year again he would have to settle for a low energy vacation, hiking or biking or sailing, and save his *E* stamps for the trip next year.

Reflecting on this state of affairs, Gerry recalled that all the national presidential candidates in the 1984 election campaign had vowed not to institute energy rationing. But then the price of energy became so high that rationing and price control was the only way to keep the country going. The cost of scarce energy for mining, transportation, manufacturing, etc., was becoming so high that the price of everything was going up too rapidly. The *E* stamp system created a mess of bureaucracy, counterfeiting, black markets (largely solved by the establishment of an exchange where those who didn't need their allotment could sell stamps through a broker to those who wanted more), and confusion: "That'll be $6.50, sir, and 112 *E* stamps." But at least since rationing began, most people have the energy required for necessities; though it is still very expensive, the new inverted rate schedule enables the average citizen to get by. The government has given assurances that rationing will be dropped as soon as the crisis passes.

This is a book about energy. It will attempt to help you understand the principles and practices of energy in a way that will enable you not only to deal professionally with energy but also to appreciate the associated societal implications. The scenario above is technological fiction of course. Whether energy will develop into a crisis or remain merely a very serious problem is not yet predictable. The answer will depend on complex interactions and timing of developments in science, politics, culture, economics, and technology.

What is clearly predictable, however, is that society and its engineers will be very concerned with energy for a good many years to come. How much energy will be needed? Of what type? How long will the fossil fuels last? Nuclear fuels? What other resources are available? How can new energy be distributed? What are the effects of cost on demand for energy? Which research and development projects should receive priority? What tradeoff between energy and pollution should or will be tolerated? How can present systems be made more efficient? What new and better systems can be proposed? If a crisis comes, is rationing feasible?

In this first chapter we shall present background information on energy and man: the need and demand for energy, the resources available, the environmental factors, and the present and possible future techniques for converting energy from one form to another. Man's interest in energy and energy conversion is not new, however, so let us begin at the beginning....

1-1 Energy in the past

In order to build civilizations, man has required more than his own personal human energy. At first some men were able to multiply their own energy by forcing or hiring other men to work for them. Thus the pyramids were built. Animals provided man with even more energy leverage. By domesticating them, man could travel farther, farm better, etc. The plough, for example, dates back to at least 3000 B.C. Waterwheels were used by about 100 B.C. Wind was purposefully used: first for sailing and then with windmills about 900 A.D. for grinding, pumping water, etc. The steam engine entered energy history about 1700 A.D.

Compared with what we are used to, the preceding developments seem dreadfully slow progress. After 1700 A.D., however, things began to happen more rapidly, probably because a more systematic and quantitative approach was made possible by the increase of scientific inquiry. It would be wrong, however, to assert that the science of energy preceded the technology. It didn't. The laws of energy were not understood and articulated until the 1850's; though by that time steam engines were well advanced. (Watt's engine came in 1775). But an atmosphere of rational, purposeful development and experimentation motivated and led to more efficient and rapid change. By 1800, steam engines were a keystone of the industrial revolution. Fulton's "folly" powered itself on the Hudson River in 1807.

Electric generators and motors arrived in 1831, the gasoline engine in 1876 (cars a few years later), electric lamps in 1879, steam turbines in 1884, Einstein's $E = mc^2$ in 1905, gas turbines in 1930, rockets in 1940, the atomic bomb in 1945, and jets in 1950.

It would be hard to exaggerate the importance of the relationship between energy and the development of any kind of civilization. Once an energy use level is achieved by a people, whatever that level, they become very nearly totally dependent on maintaining it. Some nations today have by now achieved quite high levels indeed as we shall see. And these nations are very dependent on their energy, too.

In terms of resources, the description above shows the progression from man himself to animals, to water and wind, and to the fossil fuels (coal, oil, gas); now of course we are seeing the advent of nuclear energy. We shall soon take a longer look at the question of energy resources but first let's find out what is known about how much energy man uses.

1-2 Energy usage[1]

Each person in the United States consumes an average of 180 *million* Btu annually.[2] That's the energy equivalent of 13,000 gallons of fuel oil.

Table 1-1 Approximate United States energy usage in 1970

	Annual amount	*Percent of total*	*Annual rate of growth (%)*
Household and commercial (about half of this for home heating)	15.8×10^{15} Btu	30	?
Transportation (about two-thirds for cars, trucks, buses)	16.3×10^{15} Btu	31	6
Industrial	20.7×10^{15} Btu	39	?
Total	52.8×10^{15} Btu	100	4
Total per capita	180×10^{6} Btu		1
Used as electricity	5.2×10^{15} Btu	10	9–10
Used to *generate* electricity	17.0×10^{15} Btu	32	9–10

Table 1-1 shows how the energy used in the United States is divided among various applications and forms. More important perhaps, the table shows annual rates of growth for the rapidly growing areas. To help you understand the significance of the rate of growth figures, Table 1-2 has been included.

[1] The data in this chapter are taken from several papers in the September, 1971, issue of *Scientific American* entitled "Energy and Power."

[2] To convert Btu to joules, multiply by 1055. To convert Btu to kilowatt-hours, divide by 3413.

Table 1-2 shows that the annual growth rate of our total energy usage of 4%, for example, results in a doubling about every 18 years—*every* 18 years! That means quadrupling in 36 years, and so on. Even a 1% growth rate means doubling every 70 years, quadrupling in 140 years, and so on.

Table 1-2 Annual rate of increase, percent versus approximate years required to double

Annual rate of increase (%)	*Approximate years required to double usage*
1	70
1.25	60
1.5	47
2.0	35
2.5	28
3.0	23
4.0	18
5.0	14
6.0	12
7.0	10
8.0	9
9.0	8
10.0	7

Note: Students who want to remember a quick rule of thumb can try this (it applies to bank accounts, too):

$$\text{Approximate years to double} \simeq \frac{70}{\text{annual rate of increase (\%)}}$$

Going back to Table 1-1, it can also be noted that our most inefficient energy usages are the fastest growing. Depending a bit on how you define efficiency, our cars, buses, and trucks give us about 25% of our energy input as useful output. Table 1-1 makes it possible to calculate the overall efficiency of electrical generation. The total energy used, mostly fossil fuels, to produce electricity is 17.2×10^{15} Btu and the electricity produced for use is 5.2×10^{15} Btu. Hence, the overall efficiency of generation and distribution is 5.2/17.0 or about 31%. That means, for example, that a home which heats with electricity must use more than twice as much total energy as one which heats directly with fuel oil or gas because the home fuel furnace is about 55 to 70% efficient over a heating season.

When the worldwide use of energy is considered, we find that the United States, with 6% of the world's population, uses 35% of its energy. Table 1-3 shows some per capita comparisons for various nations and areas of the world.

The rate of growth of energy usage per capita in the world as a whole is about 1.3%, producing a doubling about every 50 years or so. Note, this is

Table 1-3 1970 Per capita energy use

United States	180×10^6 Btu	$= 190 \times 10^9$ joules
Canada, United Kingdom	125×10^6 Btu	$= 132 \times 10^9$ joules
Western Europe	$60\text{–}100 \times 10^6$ Btu	$= 63 - 105 \times 10^9$ joules
U.S.S.R.	70×10^6 Btu	$= 74 \times 10^9$ joules
Japan, Italy, Argentina, Spain	30×10^6 Btu	$= 32 \times 10^9$ joules
Eastern Europe	$20\text{–}75 \times 10^6$ Btu	$= 21 - 79 \times 10^9$ joules
South America	$10\text{–}25 \times 10^6$ Btu	$= 11 - 26 \times 10^9$ joules
India, Africa	$\sim 5 \times 10^6$ Btu	$= 5 \times 10^9$ joules

per capita. Population increases must be superimposed to find the total energy usage growth rate.

One interesting feature of energy usage is the almost direct way that it correlates with such things as degree of industrialization, gross national product, and "standard of living." With this in mind, look at Table 1-3 and note how the more industrial nations use more energy. The significance of this is the strong desire of almost all nonindustrial nations today to industrialize as rapidly as possible. The possible impact on the world's energy usage of the industrialization of large populations like those in China, India, Africa, and South America is tremendous. Of course, they won't be *able* to industrialize if the energy is not available but if they industrialize and reach a high per capita energy use rate and then populations grow or energy resources deplete, ...

1-3 Energy resources

We are living during a brief period of history when fossil fuels (coal, oil, gas) are the primary energy resource. We live between the wood and nuclear (probably) or solar periods. See Fig. 1.1, which applies to the United States. The figure indicates that if present trends and predictions are fulfilled, fossil fuels will peak in importance as an energy source before the year 2000 A.D., though they will still be supplying over two-thirds of the total energy to the nation at that time. But nuclear fuel will apparently be taking over. As we shall soon see, something must take over because there isn't that much fossil fuel available.

Figure 1.1 deals with the percentage of energy coming from various sources. Another factor has to do with the actual quantities of energy being supplied. These are shown in Table 1-4. Though fossil fuels will peak and begin to decline on a percentage basis, they will still be growing on an absolute basis as the United States' source of energy well beyond the year 2000 A.D. This is important because coal, oil, and gas are *capital*-type energy resources, as distinguished from *income*-type resources. That is, coal-oil-gas fuels are stored as capital much like money in a savings account (but not drawing interest!). They have been put there over eons by the sun and nature. They are not replenishable. When they are gone, they are gone.

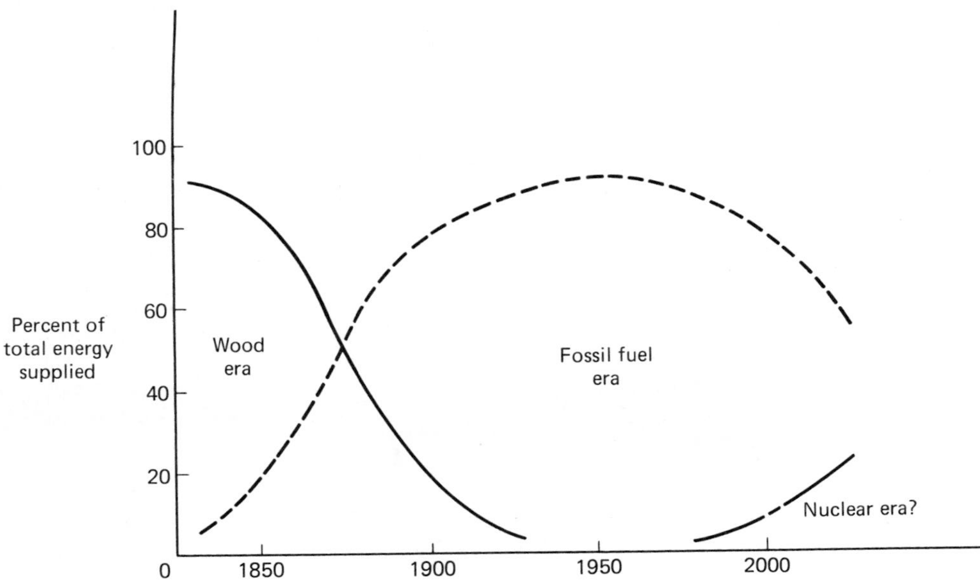

Figure 1.1 Energy eras in the United States.

Table 1-4 United States energy use per year (10^{15} Btu)

Year	*Water*	*Wood*	*Fossil fuels*				*Nuclear*	*Total*
			Coal +	*Gas* +	*Oil* =	*Total*		
1850	0	2	0.2 +	0 +	0 =	0.2	0	2
1900	0.2	2	7 +	0.2 +	0.2 =	7.4	0	10
1950	1	1	13 +	7 +	13 =	33	0	35
1975 (estimated)	2	0	17 +	26 +	35 =	78	2	82
2000 (guess)	2	0	30 +	55 +	46 =	131	27	160

Note: You'll find discrepancies in the data and estimates between Tables 1-3 and 1-4. That's the way it is. Different individuals using different raw information obtain different numbers but the *general* conclusions and trends are the same.

How much of these fuels is in the bank? Not all that much. Of course, how much fuel there is depends to some extent on how much you are willing to pay to find it, get it out of the ground, transport it, and refine it. (Oil from Alaska is harder to get and more expensive to use in the United States than oil from Texas, and offshore oil drilling is more expensive than onshore.) Using "guesstimates" based on an allowed doubling of costs, United States fossil fuel resources are as follows:

Coal	$5000–7000 \times 10^{15}$ Btu	$= 1.5 - 2.0 \times 10^{15}$ kw-hr
Oil	$600 - 1100 \times 10^{15}$ Btu	$= 0.2 - 0.3 \times 10^{15}$ kw-hr
Gas	$600 - 1100 \times 10^{15}$ Btu	$= 0.2 - 0.3 \times 10^{15}$ kw-hr
Total	$6200–9200 \times 10^{15}$ Btu	$1.9 - 2.6 \times 10^{15}$ kw-hr

It should be noted that coal represents more than three-fourths of our estimated fossil fuel reserves. Efforts to produce gas or oil economically from coal are underway. One major concern, however, is that much of the coal is only available through strip mining and much of it occurs in our plains states, which are now used for growing grain and corn. Half of Iowa and 40% of Illinois is underlain with strippable coal. If we mine this coal, there will be serious environmental effects—and some plans will have to be made to replace the food lost. There is also a related problem in the fact much of our coal is high in sulfur. If we do not want this sulfur in our atmosphere, current research and development efforts to learn how to remove the sulfur economically must be successful.

Now if we go to the projected annual usages for the various types of energy and calculate the cumulative requirements for the United States for the 50-year period from 1970 to 2020, we find some frightening results:

Coal	1350×10^{15} Btu	$= 0.4 \times 10^{15}$ kw-hr
Oil	2500×10^{15} Btu	$= 0.7 \times 10^{15}$ kw-hr
Gas	2000×10^{15} Btu	$= 0.6 \times 10^{15}$ kw-hr
Total	5850×10^{15} Btu	1.7×10^{15} kw-hr

Comparison of the resource estimates with the estimated requirements indicates a very short life expectancy indeed for the fossil fuel era. The United States' own oil and gas can be expected to be either gone or very expensive long before 2020, making the country almost wholly dependent on imports for these fuels.

When one looks at nuclear fuel, the situation is better, but only a little. The uranium-235 used in ordinary reactors is not a plentiful resource. Based on an estimated doubling of cost, there are about $10{,}000 \times 10^{15}$ Btu available in the United States. If by the year 2000 we are using 160×10^{15} Btu per year and that annual figure is growing rapidly, that fuel will serve for less than 50 years.

It appears that persons living now are likely to deplete the supply of fossil fuel in the United States and that their children are likely to deplete the supply of nuclear fuel! Of course, all these estimates are *only* estimates. No one knows for sure how fast population will continue to grow, how much the per capita energy demand will grow, where new fossil or nuclear fuels are likely to be discovered, and so on. But such factors affect the timing, not the overall conclusion. Population and energy use per person simply cannot continue to grow indefinitely. Until they are stabilized, there is going to be an energy "problem." Or do you now wish to join those who call it a "crisis"?

It should be noted that the temporary fuel oil and/or gasoline shortages that are beginning to occur are not properly called an *energy* crisis. There is, as yet, no shortage of energy resources, even of fuel oil. The supply and demand problems are caused by a mixture of economic, bureaucratic and political factors and may be seen as symptomatic of things to come but they are not an *energy* crisis.

When one looks at the world's resources, the situation is a little but not a whole lot better. Based again on a doubled cost, here are the estimated non-United States world resources:

Fossil fuels		
Coal	20,000–30,000 $\times 10^{15}$ Btu	$= 5.8 - 8.8 \times 10^{15}$ kw-hr
Oil	3,000– 6,000 $\times 10^{15}$ Btu	$= 0.9 - 1.8 \times 10^{15}$ kw-hr
Gas	2,000– 5,000 $\times 10^{15}$ Btu	$= 0.6 - 1.5 \times 10^{15}$ kw-hr
Total	25,000–41,000 $\times 10^{15}$ Btu	$= 7.3 - 12.1 \times 10^{15}$ kw-hr
U-235	100,000 $\times 10^{15}$ Btu	29.3×10^{15} kw-hr

The world's energy requirements are hard to project into the year 2020 A.D. because of outstanding questions about population and industrialization but the range is probably 15,000 to 40,000 $\times 10^{15}$ Btu. Thus, the situation is more like that in the United States than it is different.

What are some ways out of this situation for the United States? The country might plan to import more fuel, especially oil and gas but this has tremendous economic and political implications. We already import about 25% of our oil and the imports are an important factor in heating homes along the East coast. (*How* important depends on whether you are a legislator from the Southwest or from New England!) If we are forced to import huge quantities of oil just to keep going, the economic drain is bound to be significant. Perhaps even more dangerous, however, are the political problems, even wars, that could develop from a nation like the United States becoming dependent on importing essential oil from all over the world. Threats to that oil supply anywhere in the world would be threats to our national security. One does not need to be a political scientist to see that there is danger in that situation. We shall, of course, import even more oil as time goes on despite the danger of involvement but the possibility of "crisis" also grows as we do.

Another way out is to level or reduce population and/or per capita energy consumption. Given the problems of rationing energy, think of the problems of rationing children! We may do one or both, by fiat or perhaps it will happen culturally,[2] but one would hardly place large bets on either happening in the *near* future. At least not without a real survival crisis for motivation.

We might hope to develop some new sources of energy, a possibility to be discussed a bit later. First, however, we must look at another dimension of the problem (or crisis): pollution.

[2] The United States birth rate in 1971 equaled 2.1 children per female in the childbearing age bracket, a rate that would give us zero population growth if it is maintained for 70 years!

1-4 Energy and pollution

Energy and pollution are intimately related. We shall discuss pollution from energy generation and conversion here in four major categories: air pollution, water pollution, radiation pollution, and thermal pollution.

Air pollution derives mostly from the combustion of fuels, in a power generating plant, factory, home, vehicle, etc. Air pollution is a complex subject and there is not space here to discuss all its aspects in detail so we shall touch on only a few of the most important factors.

Of the gaseous pollutants, sulfur dioxide (SO_2) is receiving a great deal of attention. It is poisonous and noxious and may be linked to respiratory diseases. In 1966, an estimated 28 million tons of SO_2 were discharged into the air in the United States, about half from power generating plants and the rest from industry, homes, vehicles, etc. Thus, there is considerable interest in low sulfur fuels (rare and hence expensive) and in techniques for removing sulfur from fuels before burning or from the exhaust gases. Certainly the projected increases in power generation and in total fossil fuel consumption are of great concern when one considers SO_2 levels in the atmosphere. No doubt something must be done soon.

Other air pollutants from combustion include oxides of nitrogen (NO, NO_2), carbon dioxide (CO_2), carbon monoxide (CO), hydrocarbons, and particulates. There is concern for what CO_2 levels may do to the incident solar radiation and hence to the earth's climate as a whole. Any who come from cities or towns where coal is burned in quantity know about the particulate problem. The link between the automobile and air pollution has been felt and observed by nearly all of us.

The automobile deserves special mention in any discussion of energy and pollution. It is a major source of pollution. Figure 1.2 shows the pollution standards set by the Environmental Protection Agency (EPA) for cleaning up automobile engines. These standards are going to be hard to meet and will cost money. They may also reduce engine efficiency, thereby further increasing fuel consumption but it must be done. Whether the effects of high energy costs will change us from an automobile society or not remains to be seen—probably not. There is always the possibility of the electric car ....

All power generating plants, conventional and nuclear, also create some water pollution with the chemicals they use for corrosion and slime control, etc.

When nuclear plants are considered, there is not much air pollution in the conventional sense but there is radioactivity discharged into both the air and water from the plant and there is the very serious problem of disposal of radioactive wastes. Whether these problems are solvable in an economical and safe way is a matter of dispute.

Of course, air and water pollutants *can* be controlled—for a price! Pollution control systems and equipment cost money to buy and maintain and usually

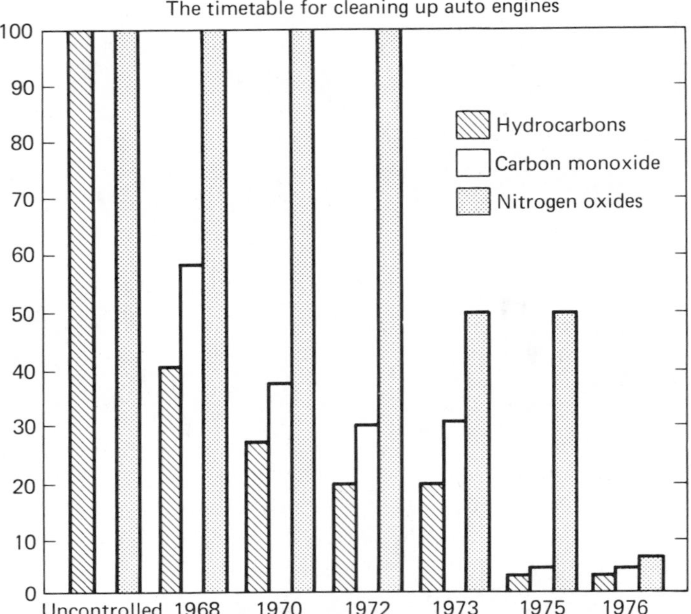

Figure 1.2 Percent emissions from automobiles permitted by the Environmental Protection Agency (EPA). Data from the National Academy of Sciences—National Research Council.

also adversely affect the performance of the energy conversion system. The question of how to account for the cost of such equipment and who to charge is intimately involved in the energy problem or crisis. But more energy will mean more pollutants of one kind or another and if the atmosphere is to be kept clean and livable, then these pollutants will have to be controlled to some extent. Thus, not only is fuel likely to become more expensive but so is pollution control and hence energy.

Of all the types of pollution, thermal pollution is the most inherently intractable. *All* energy used is ultimately discharged into the earth's atmosphere (air, land, water) and thence radiated back to space. Eventually we may have to worry about this global problem but, for the present, the problem is more with local effects on air and water temperatures near cities and/or power plants. Because the generation of electricity is only about one-third efficient, two-thirds of the total energy input is converted to waste heat directly at the generating plant. (There are also other losses in transmission lines, transformers, etc.) Boilers are about 80 to 85% efficient so 15 to 20% of the input goes to increase the local air temperature, not usually a significant problem now. The rest usually goes to increase the temperature of the cooling water that is taken from a nearby lake, river, or ocean. The water used for cooling is itself raised in temperature by 15 to 20°F. Depending on the size of the river or lake, or on ocean location, effects on the marine ecology can be important. Already about 10% of the total river flow in the United States is used in power plants. Nuclear plants use more cooling water and so as time goes on, a lot more cooling water will be needed. If cooling towers, wet or dry, are used, the heat is transferred to the air instead of to water so this sometimes helps reduce local water thermal pollution. The possibility of doing something constructive with the "waste" heat is discussed in the next section.

The level of efficiency of electrical generation, about one-third, is the reason electrical home heating is not as pollution free as one might suspect by looking only at the all-electric home—where admittedly no air pollution occurs. A gas or oil heating system is typically 55 to 70% efficient, compared with the 31% efficiency of getting the electrical power to the house (not counting land use for transmission lines or gas lines, roads for oil trucks, etc.). Thus, the electrically heated home uses about twice the energy resources and potentially creates twice the air pollution, though there are differences between small home systems and large central systems that *may* help to alleviate this difference.

Let us look now at some of the various kinds of energy converters that man has devised or at least conceived.

1-5 Energy conversion

Man mostly wants and needs energy to do work for him. Except for the heat required to keep his homes and buildings comfortable and for some industrial processes, man uses energy mostly as mechanical work to propel his various vehicles and power his many machines. Unfortunately, it is not mechanical energy that man finds available to him in quantity in nature. Instead he finds hydro, thermal, nuclear, and chemical energy resources and he must convert these resource energy forms into the forms he wants.

Table 1-5 illustrates some energy conversion chains from natural resource to application. Heat for buildings and industrial processes is not included for the sake of clarity. Man has found electricity to be a most convenient intermediate form of energy between his natural resources and his applications, except for transportation (column 6), and he may yet require electricity somewhere in the transportation chain before long as oil reserves are depleted and nuclear power becomes available. The table emphasizes the many conversion steps required for our primary sources (chemical and nuclear) to reach their final application.

The conversion of energy along these chains from one form to another is neither easy nor cheap and it cannot be done with an efficiency of 100%.

Table 1-5 Some energy conversion chains—resource to application

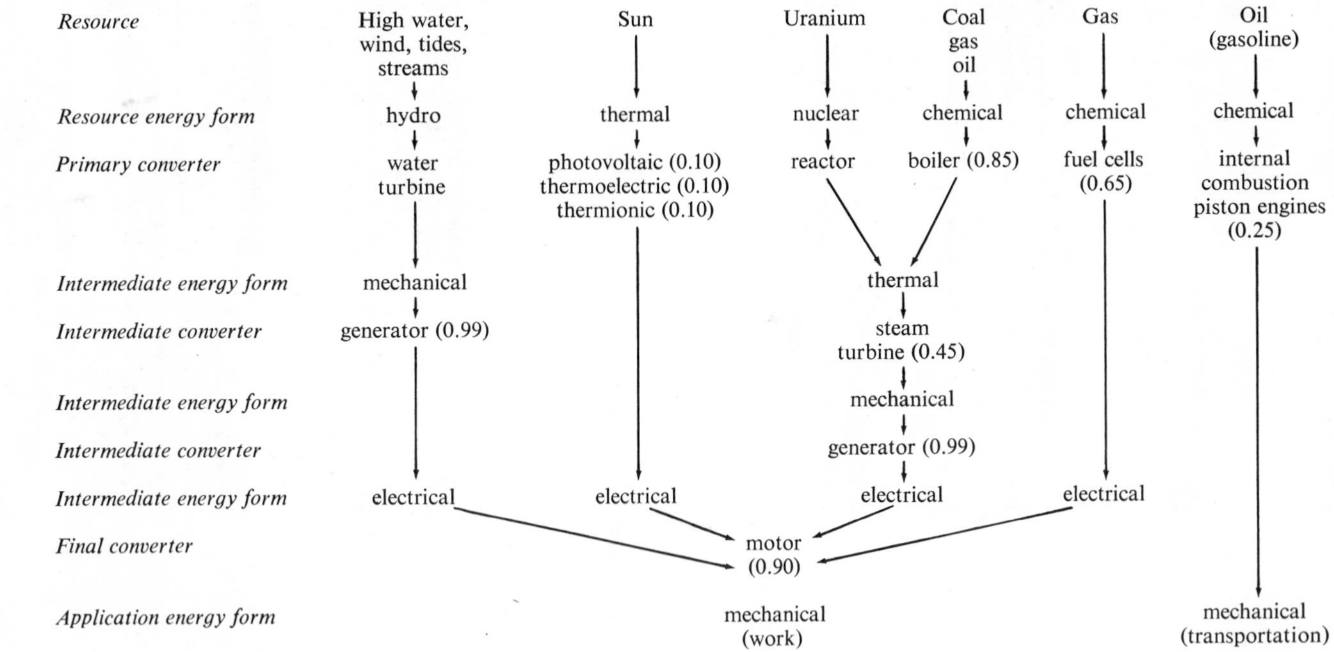

Efficiency here means the *useful* output divided by the *costly* input, both expressed in energy units. Thus, when it is said that a home oil furnace is 70% efficient, it means that 70% of the total energy input of the fuel is actually used to warm the air in the house and the other 30% goes up the chimney as waste.

In Table 1-5, the numbers shown in parentheses after several of the converters give approximate efficiencies for large sizes. (In general, efficiencies improve with size. A small electric motor, for example, can be made only 50 to 70% efficient, whereas large motors are 90% or more efficient.) To find the overall efficiency of a chain, one must multiply the efficiencies along the chain. For example, from chemical energy to electricity by means of a steam turbine, the efficiency is about $0.85 \times 0.45 \times 0.99 = 0.37$. There are, of course, transmission and other losses not explicitly dealt with in the table, which is intended only to give readers a qualitative appreciation for the length and efficiencies of the various chains. Also not shown in the table is the fact that nuclear generated electricity is less efficient than energy from fossil fuels, the reasons for which will be explained later in the book.

Some other approximate efficiencies of converters not shown in the table are

Chemical to electrical	
Dry cell batteries	0.90
Storage batteries	0.70
Chemical to thermal	
Home gas furnace	0.75
Home oil furnace	0.65
Chemical to mechanical	
Gas turbines	0.35
Wankel engines	0.20
Diesel engines	0.35
Steam locomotive	0.10
Electrical to radiant	
Incandescent lamps	0.05
Fluorescent lamps	0.20
Lasers	0.30–0.40
Chemical to kinetic	
Rockets, jet engines	0.45

The efficiency of conversion devices has been increased significantly in the past through scientific and engineering research, development, and design. More of this can be expected but some diminishing returns are beginning to appear in some important areas. In transportation, for example, apparently not much more can be done about the relatively low (25%) efficiency of the

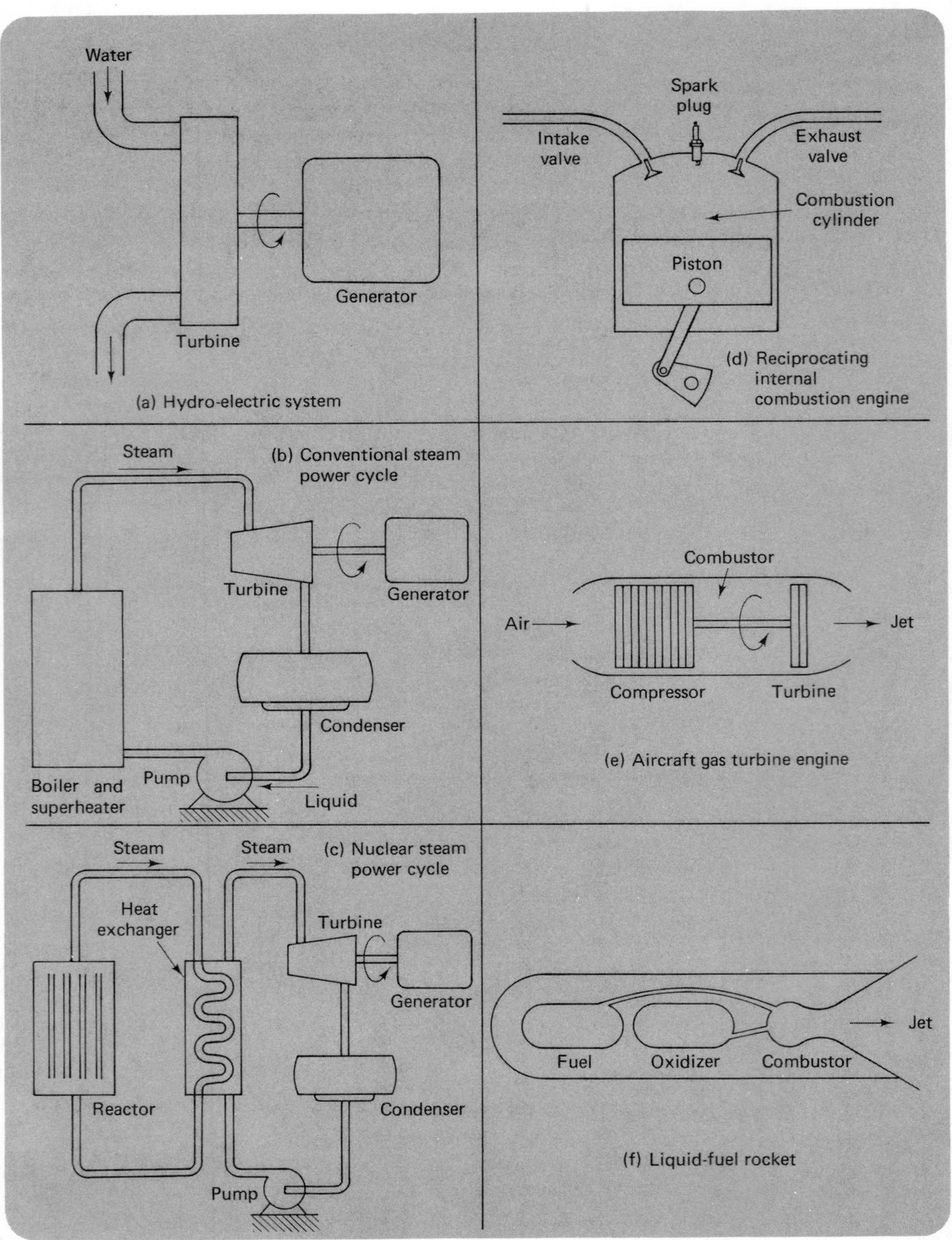

Figure 1.3 Schematic diagrams of some energy conversion systems.

automobile engine. The new Wankel engine is even less efficient, though it may have other benefits. Miles per gallon of gasoline (i.e., per unit of energy consumed) has actually gone *down* since 1940 because of the ever larger cars and their equipment. As noted, nuclear power plants also have a lower efficiency than fossil fired plants. Two bright spots in the past have been the fluorescent lamp and the diesel locomotive, both of which improved the efficiency of their particular conversions.

The next few pages show some schematic diagrams and pictures of some energy conversion systems and devices.

Figure 1.4 A steam boiler-superheater (*courtesy of Combustion Engineering*).

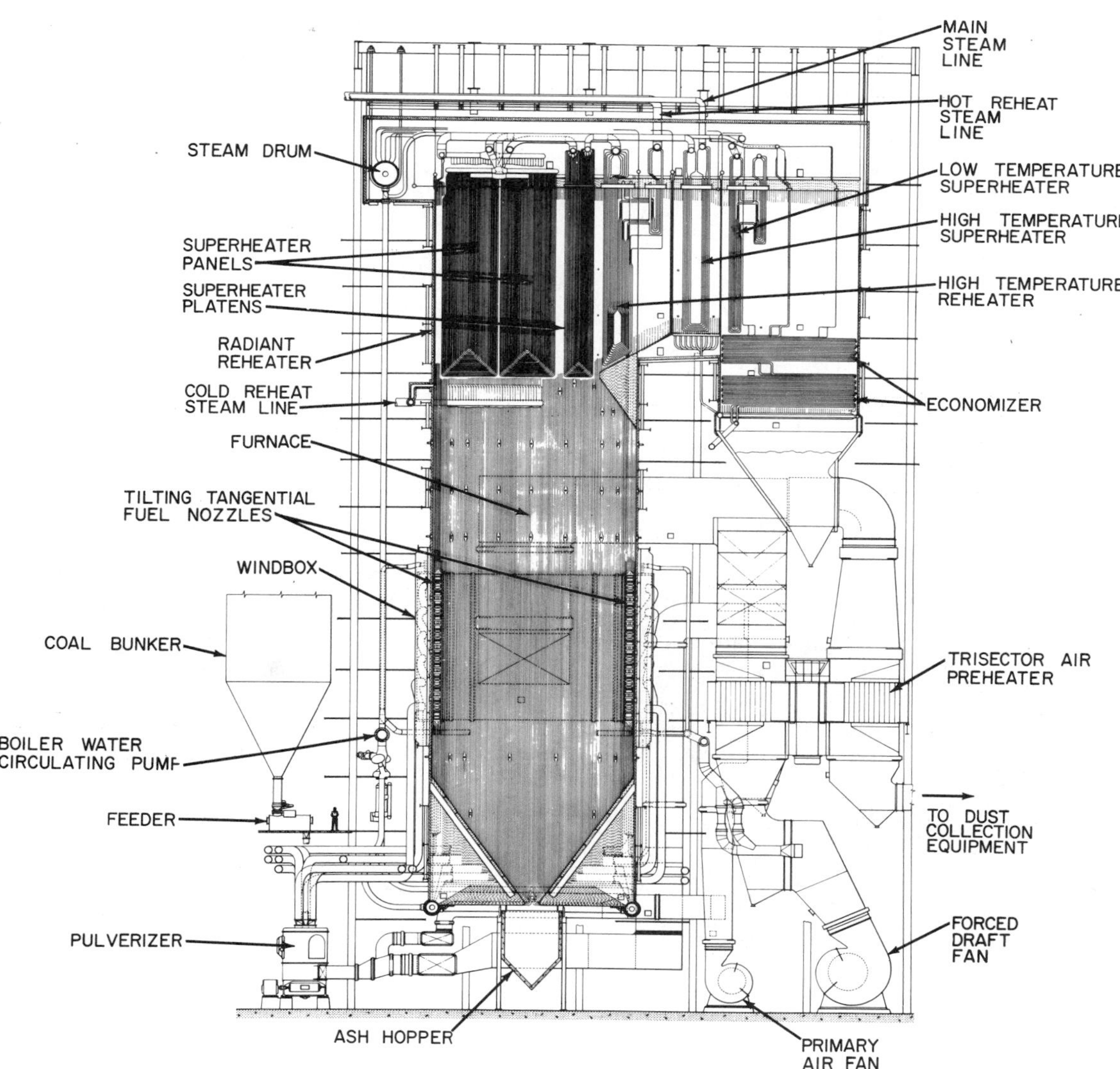

Figure 1.5 (above) A steam generator, 163 ft long, 542,500 kw (*courtesy of Westinghouse Electric*).

Figure 1.6 (below) A nuclear turbine (*courtesy of Westinghouse Electric*).

Figure 1.7 (above) A gas turbine, 68,000 kw (*courtesy of Westinghouse Electric*).

Figure 1.8 (below) A small centrifugal pump (*courtesy of Worthington Corporation*).

Figure 1.9 A turbo jet engine, the JT12(J60) *(courtesy of Pratt and Whitney Aircraft).*

1-6 Energy and the future

There is no question about the need for an energy source to replace fossil fuels. As we have seen, the fossil fuels just aren't there for the long run. Moreover, if nuclear fuel is to become the primary source of energy, the breeder reactor must be developed and problems of safety and cost solved. Breeder reactors use uranium-235 but generate some more fuel in the process. If successful, they would enable us to have nuclear power for several hundreds of years. Breeder reactors are under development in several countries and the United States, though lagging, has made it a "first priority" research and development project.

There is serious opposition to the widespread use of nuclear power and some, such as Ralph Nader, oppose it altogether. The opposition view is that it is too costly and too dangerous (they fear not only accidental catastrophe but also sabotage and very long-term waste disposal problems). Their proposal is that we increase emphasis on research and development of various solar energy sources, utilizing and conserving fossil fuels in the meantime. Of course, this plan also has its dangers.

After the breeder reactor on the nuclear front, the next nuclear frontier is the fusion process. This has not been proved technically (as the breeder has) or economically but if it could be made to work, energy resources will not be a problem for thousands of years. No one is yet betting large sums on it but research is continuing.

Meanwhile, back at the conventional fossil fuel plant, the possibility of improving efficiencies by "topping" them with a magnetohydrodynamic (MHD) cycle is being explored. MHD generates electricity by moving a plasma through a magnetic field and works best at very high temperatures. The discharge from the MHD plant would be used as the hot gas input to the conventional cycle.

Of course even with electricity generated by nuclear fuel, there is still a problem with oil and gas supplies for other uses. The more we convert to nuclear generated electricity, the more of these fuel resources we shall have for other uses, such as home heating, industry, and transportation. Transportation is the biggest worry because of its tremendous growth rate (6 to 7%), its relatively low efficiency (25%), and its huge pollution effects. Alternatives to the internal combustion engine are not yet visible. We may be forced to electric cars simply because we may have the electricity available from nuclear generators. (But if so, look for the thermal pollution problem to be "solved" *first*!)

If we could switch from capital-type sources, like fossil and nuclear fuels, to income-type sources, like hydropower, that would relieve the problem of where the primary energy is coming from. Hydropower, of course, is not available in sufficient quantities and though it is nonpolluting, developing it causes many environmental and social problems. Hydropower in the form of tidal power has been developed successfully in France on the Rance River and there is discussion of reactivating a tidal power project on the Bay of Fundy in Canada where tides average 18 feet. Tidal power, however, is likely to be significant only locally, and it certainly will be significant if other power becomes expensive enough, but it is not likely to contribute much to the solution of the general problem.

The wind, too, may become very important in helping to alleviate some local problems and it might even provide a large-scale solution. Propeller driven turbines on towers could drive electric generators. In fact, such a scheme was tried some years ago in Vermont but a blade broke and the resulting vibrations shook the tower apart, thus discouraging the promoters. One trouble with wind is its variable nature but this might be solved by using the electrical energy generated to decompose hydrogen and oxygen. The elements can then be stored, transported, and used in fuel cells to reconvert the energy to electricity. There is ample wind power and indications are that environmental effects (e.g., on local weather) are not significant. The towers might be unsightly in some locations but offshore locations have appeal because of the strong, steady winds. Cost is still a very big obstacle but as other sources of energy become more expensive, the relative cost should improve.

Fuel cells may turn out to have another important role. The construction of huge—HUGE—power generating parks has been suggested. They would be efficient and make possible central handling of pollution problems; however, distribution would become more difficult. Thus, the electrolysis scheme described in the paragraph above may be a solution. The collection of windmills offshore is very much like a large power park.

As we talk of going back to wind and water, we seem to be going back thousands of years. Let's keep going. The sun is where all our energy has and will come from. The sun can also be and is used directly of course by almost every architect designing homes with lots of south side glass, by every greenhouse owner, and by sunbathers. But it also has large-scale, long-range potential as a part of a solution to the energy problem. There is enough solar energy reaching the earth's surface to supply our energy needs for a very, very long time but the cost of the equipment needed is still prohibitive. Land would also be costly. Research continues, as well it should, but it must be increased greatly because sooner or later we have to develop solar energy as our primary energy source. If it could be sooner, we could avoid the dangers and costs of nuclear development—and conserve our precious fossil fuels (especially oil) for more valuable uses.

Research is also continuing on converting the thermal pollution from power plants to constructive use. The 15°F temperature rise is not enough to make use of in any conventional way, for reasons you will learn about later in this book, but it is significant biologically. Most biological growth rates increase *exponentially* with temperature so small temperature changes can be important. Thus, various schemes for using the warmed water and/or its thermal energy to increase the rate of growth of forests, food plants, or even protein-rich algae are being investigated.

In another aspect of possible future development, it isn't likely that the rationing that kept Gerry Smith from his cross-country trip will become a reality. The political and administrative problems are too great. The crunch, if it occurs, will come as oil and gas become scarce, their price goes up, and nuclear power is not yet able to carry the full load. The price rise itself will help reduce demand for energy and hence help avert a no-energy "crisis."

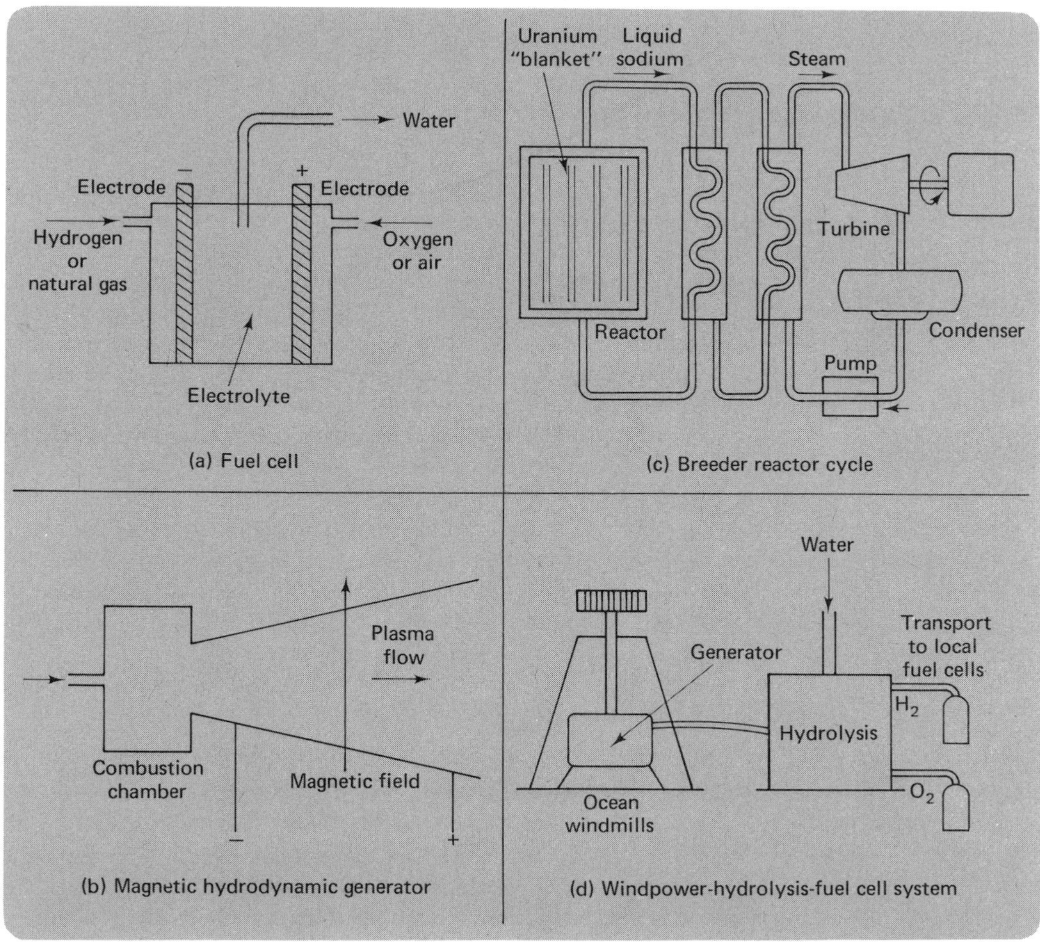

Figure 1.10 Schematic diagrams of some nonconventional systems.

Then Gerry might not have the *dollars* for the expensive gasoline and oil! Along these lines, a new inverted rate schedule is being discussed. An inverted schedule charges more per unit used by large users instead of less as is customary now. Such a schedule would tend to dampen large energy uses without penalizing the little guy through high prices for his necessities. No one yet knows the effect higher prices will have on energy demand.

In summary, the breeder nuclear reactor is currently our best hope for power, with its timing and the transportation/fossil fuel/pollution triad still posing serious open questions. Reducing demand growth may be done by one of several possibilities. Longer range we might hope for fusion power, solar power, and controlled biological uses for waste heat. Somewhere in between we can also hope for more or less help from such income energy sources as tides, winds, ocean currents, or thermal gradients.

1-7 What does it mean to you?

This chapter has tried to give you a look at the overall picture regarding energy. All citizens have a responsibility to understand the nature of our national and global energy problem, or crisis. Engineers share this general responsibility with their neighbors but also have at least three added concerns as professionals. First, you must know your energy engineering well. If you work in the energy field, you are responsible for becoming and remaining competent as an engineer and decision-maker. Even if you do not work directly in the energy field, you cannot help using energy in your work, both as an industrial consumer and as an engineer, so you must understand its fundamental laws and concepts. Second, you must know how your energy engineering knowledge relates to the social and environmental problems surrounding energy. Third, you must help your neighbors, community, and government make wise political and economic decisions involving energy by providing understandable, accurate information and advice. There is no longer any place for engineers who want only to "do their job" with blinders on the social implications, especially regarding energy.

For students who wish to learn more about energy and man, the following reading is recommended:

"Energy and Power," *Scientific American*, September, 1971.

The whole magazine is concerned with energy, is easy and informative reading, and will really give you a strong background on which to become an energy expert. Read it! It'll help you explain to your nonengineering friends, male and female, not only about energy but also how engineering is important to society's needs now and in the future.

Problems

1-1(a) Suppose the rate of growth of usage of a given natural resource is 7% per year, thus doubling every decade. Assume that $\frac{1}{100}$ of the resource has already been used and that the usage during the last decade was $\frac{1}{1000}$ of the resource. How long will the resource last?

(b) Repeat assuming that $\frac{1}{1000}$ of the resource has already been used and that the usage during the last decade was $\frac{1}{10,000}$ of the resource.

Solutions: (a) A little less than 100 years.
(b) A little more than 120 years.

Note: See why estimates of resource depletion are not very sensitive to the amount of resource but are *very* sensitive to the rate of growth of usage.

1-2 Construct the energy conversion chain for an electric home heating application if the power plant uses nuclear fuel.

1-3 Develop an estimate of how much energy you personally use each year and compare it with the national average. Be careful to include *everything*, remembering, for example, that even products, clothes, etc., require energy for materials, manufacture, transportation, etc.

1-4 Go out and find an example of unnecessary energy waste in your community or school. Estimate how much is being wasted and try to find a way to stop the waste. If you can, talk to the person or persons doing the wasting.

1-5 Do you think energy is a "problem" or a "crisis" for the United States and/or the world? Explain.

1-6 Assume you are asked to recommend allocation of 10 billion dollars over the next 10 years for energy research. How would you divide the funds? Explain.

1-7 Where does your community get its energy? How rapidly is energy usage growing in your community? How will your community get its energy 10 years from now? 50 years from now?

1-8 Here is the monthly electric rate schedule for one locality:

	For nonelectric heat customers	*For electric heat customers*
First 12 kw-hr	$1.27	$1.27
Next 88 kw-hr	0.0447/kw-hr	0.0447/kw-hr
Next 250 kw-hr	0.0307/kw-hr	0.0307/kw-hr
Next 650 kw-hr	0.0282/kw-hr	0.0152/kw-hr
Next 1000 kw-hr	0.0181/kw-hr	0.0181/kw-hr

How might an inverted schedule look? Compare bills for customers who use 500, 1000, 1500, and 2000 kw-hr per month for this schedule and for your inverted schedule.

1-9 Here is the estimated monthly kw-hr usage of one new house under construction in the area discussed in Prob. 1-8:

January	5000	July	1000
February	5500	August	900
March	4200	September	1000
April	3300	October	1500
May	2000	November	2400
June	1100	December	3700

The family doing the building is trying to decide between oil and electric heat. The initial cost of the oil system has been estimated at $3000 and the electric system at $1500. Number 2 fuel oil with a heating value of 140,000 Btu/gal costs $0.40/gal. Based on dollars, what decision do you recommend? Why? What other factors should this family consider in making a decision? Discuss.

1-10 In the Spring of 1970, there was considerable discussion in the newspapers about anticipated power shortages in Northeastern United States for the summer ahead.

(a) In planning for the immediate summer, how would you arrive at decisions on whether to cut power at critical times to homes, offices, stores, or factories?

(b) On the longer view, comment on the interaction among population growth, pollution of various kinds, and the availability of energy.

(c) What role do you think engineers should play in helping to make the necessary decisions and to solve the problems associated with the energy and society? (Consider the engineer employed by power companies, auto manufacturers, government, and universities and also the role of the professional society.)

2

Introduction to the laws of energy

The study of energy is often called thermodynamics. To be precise, thermodynamics is the science of the relationship between energy and the properties of substances. It deals with stored energy (chemical, potential, etc.), energy exchanges (heat and work), and the associated changes in properties (pressure, temperature, etc.). This is what we are going to study but there is something not quite right about calling this study "thermo*dynamics*." It does deal with systems in motion but it does *not* deal at all with the *rates* at which processes, energy exchanges, etc., occur. Therefore we have chosen to call our study simply a study of *energy* because energy is at the core of everything we shall do. It is our concern for energy resources, energy conversion, and energy management that motivates us in this study. And it is energy that unites and integrates the study. So we shall call it a study of energy, though you (and others) may call it "thermodynamics" if you (or they) wish.

How much do you already know about energy principles from experience? How much do you know about the zeroth, first, and second laws? What is an operational definition? What is the difference between induction and deduction? By which are scientific laws formulated? What do pressure gauges actually measure? What is meant by the degradation of energy?

This chapter is intended to help you answer these and other questions. It should (1) provide you with a broad, general, overall understanding of energy principles, (2) relate your past experience with energy—what you already know—to the material to come later in this book, and (3) provide you with an understanding of the logical or scientific basis and structure of the study of energy.

2-1 The law of conservation of energy

There are two fundamental laws of energy of practical importance to engineers. Sometimes they are called the *first* and *second laws of thermodynamics.* We shall call them

1. the law of conservation of energy
and
2. the law of degradation of energy

or just the first and second laws. They rank in scientific and engineering importance with Newton's second law ($F = ma$).

Barring nuclear effects, converting mass m to energy E according to $E = mc^2$ (c = speed of light), energy is conserved. We shall consider only such nonnuclear processes in this book because they are the ones of practical importance to most engineers.

Energy is converted from one form to another and transferred from one system to another but its total is conserved. Along any and all the energy conversion chains shown in the preceding chapter, the total energy is conserved. Of the total energy input in the fuel to a furnace, 100% is accounted for as output. Some of that output is useful to the owner to warm the house and some is wasted out the chimney but all the energy that goes into the system comes out somewhere. In a steam turbine, energy enters with the hot, high pressure steam and the same total amount of energy leaves, some as mechanical energy in the rotating shaft, some as heat transfer to the surroundings, and some with the cooler, lower pressure steam. The same is true of all energy conversion systems or devices and of everything that occurs in nature: Energy is conserved. But you know this....

2-2 The law of degradation of energy

What you may not know, or yet understand, is that though the *quantity* of energy is conserved, its *quality* is not. We say it is *degraded.* By *quality,* here we mean the *potential* of the energy to *produce useful work* for us. The term *work potential* is used to refer to the maximum possible useful work that can be obtained from a given quantity of energy in a given environment. Usually the environment is taken to be the earth's, including the air, ocean, and the earth itself. Whenever the work potential of a quantity of energy is reduced, we say that the energy is degraded.

Now the law of conservation of energy is based on the observation that every time energy changes form or is transferred from one system to another, the total amount of energy is constant. The law of degradation of energy is based on the observation that every time energy changes form or is transferred from one system to another, its potential for producing useful work for man is reduced—irretrievably—forever.

Energy conservation and energy degradation work simultaneously. That is, in a change of form of energy, energy is both conserved and degraded. It's a bit like water flowing from a high lake over a dam to a lower lake. When in the high lake that water has work potential because it might be used in a hydroturbine to do work or to make electricity that could do work. After it passes over the dam and falls into the lower lake, we still have the water. Its gravity potential energy will have been converted to an increase in temperature of the water from internal friction as it flowed and was slowed again. The mass and energy are conserved. But the *work potential* is gone. If we took no work at all from the water as it fell, we lost all its original work potential. If we put it through a turbine, we wouldn't obtain quite *all* its *potential* actually converted into some other *useful* energy form because of friction. Hence we shall have had a net loss of work potential even in this case. That's the law of energy degradation at work.

Work potential is a bit like the buying power of money. If a loaf of bread costs 20¢ in 1960 and 40¢ in 1970, you may have conserved your total dollars but you still have lost the power to buy something useful for yourself with them. Inflation degrades money. Conversion degrades energy.

In the energy conversion chains shown in Chap. 1, degradation of the energy is taking place at every stage, even though the total energy is conserved by each device if one counts all its inputs and outputs, both useful and nonuseful. Thus, in the steam turbine the work potential of the incoming steam is greater than the work potential of all the various outgoing energy, including the heat, mechanical work, and the outgoing steam.

It should be noted that it is the *potential* to do *useful* work for man that determines the grade of a quantity of energy. There is emphasis on the words *potential*, *useful*, and *work*.

The law of energy degradation is thus of immense practical importance. It is the reason we face an energy problem or crisis. It probably had occurred to you to ask why there is a possible energy shortage if energy is *conserved.* The reason is that, though conserved, energy is degraded. In fact, all the energy that we "use" ultimately ends up as waste heat transferred to the earth's environment and then to space, degraded to uselessness for our purposes even though it still exists quantitatively.

For many types of energy conversion, the law of degradation says only that work potential output will be less than the work potential input. For one kind of conversion, however, it sets a more restrictive limit. That conversion is the continuous (or cyclic) conversion of thermal energy to any form of work, usually mechanical or electrical. We shall show in a later chapter that the work potential of thermal energy is given by

$$\text{Work potential of an amount of heat } Q \text{ at temperature } T_H = Q\left(1 - \frac{T_0}{T_H}\right)$$

where T_0 is the available environmental temperature and both T's are expressed in absolute terms (i.e., Kelvin or Rankine). Let's see then how much useful work can be obtained from a unit of thermal energy ($Q = 1$) from a fuel. Maximum combustion temperatures are about 3000°R. The atmosphere at about 500°R is the usual available environment. Letting A denote the work potential,

$$A = 1(1 - \tfrac{500}{3000}) = \tfrac{5}{6}$$

This isn't so bad! So far, however, the materials people haven't given us anything that will contain such temperatures in any practical energy conversion machine. About the best we can do in a steam power plant is 2000°R. In this case

$$A = 1 - \tfrac{500}{2000} = \tfrac{3}{4}$$

Worse yet, for a variety of reasons, nuclear plants cannot operate even this high. They are down around 1700°R with an A of about $\frac{2}{3}$.

Remember these A's are the *maximum possible useful* work that can be obtained from a unit of thermal energy at the various temperatures in a continuous or cyclic conversion system. They are thus simply maximum possible efficiencies. That is, the maximum *possible* efficiency of a steam power plant is not 1.0 but actually about 0.75 because the work potential of heat is not equal to the energy of heat. No matter how good a system we built, the law of degradation of energy says the best efficiency we can obtain is $\frac{3}{4}$. For this reason, thermal energy, or heat, is sometimes referred to as low grade energy, though as we have seen, its "grade" is a function of its temperature. Work on the other hand is high grade energy, and conversions from work to work are relatively efficient. Pumps and motors, for example, operate at efficiencies up to 90% or more in their larger sizes. The law of energy degradation says only that *their* efficiency must be less than 1.0.

It's as though the law of degradation of energy discriminates against continuous or cyclic conversion of thermal energy into work. That's of course unfortunate for mankind because we have thermal energy available, but we want work.

In summary, energy is always conserved but it is also always degraded.

2-3 Experience and the laws of energy

The preceding discussion of the laws of energy was very general and mostly qualitative but actually most persons understand a great deal more about the laws of energy just from the experience of living with them. Such understanding may be unconscious and disorganized but it is there. The intent of this section is to make *you more aware* of what *you already know* that relates to the laws of energy.

2-3(a) Experience and temperature

Suppose that you have a simple mercury-in-glass thermometer but *without* the usual scale markings. You place it in a glass of cold water today and mark the level at which the mercury comes to rest. Tomorrow you place it in another glass of water and the mercury comes to rest at the same level. What would you conclude about the temperatures of the water in the two glasses? Write your answer in the space below:

____

____

____

Now imagine that you have three copper blocks labeled A, B, and C. Using an unmarked thermometer as in the previous question, you find that blocks A and B are at the same temperature and the blocks A and C are at the same temperature. What would you conclude about the relative temperature of blocks B and C?

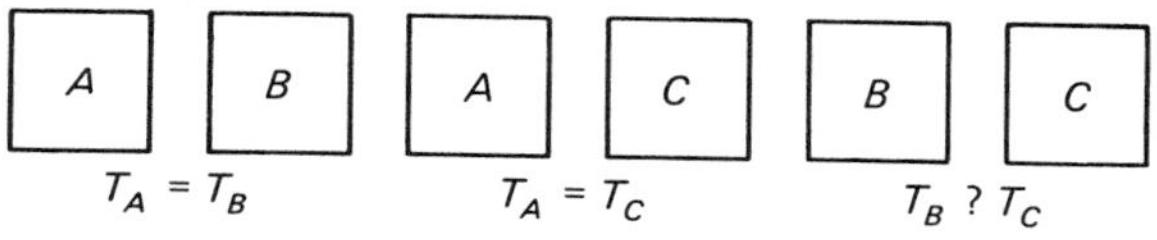

A	B		A	C		B	C
$T_A = T_B$			$T_A = T_C$			T_B ? T_C	

Write your answer here:

____

____

____

The answers to the questions above are obvious to most people. In the first question, the water in the glasses on the 2 days is at the same temperature. In the second, blocks B and C are at the same temperature. If you got these answers, then you have properly used a law called the *zeroth law of energy* even if you didn't know you knew it and even if you don't know how to verbalize it formally.

Roughly, the zeroth law may be stated as follows: If two systems are each equal in temperature to a third system, then the two are equal in temperature to each other. We shall make this definition a bit more rigorous in the next chapter but for now it is enough to note that your understanding of the concept, if not conscious or verbal, is at least sufficient from experience for you to use it properly in a simple real-world situation.

Incidentally, the reason this law is called the *zeroth* law is that the first and second laws had already been named before it was realized that the concept of equal-in-temperature is prerequisite to a logical development of those laws. So, to be logical if nothing else, it was named the *zeroth law*.

Before you think the zeroth law is trivial because it seems to say the obvious—that if *A* relates to *C* in the same way that *B* relates to *C*, then *A* and *B* must be related in the same way—consider the following from *Order and Chaos* by Angrist and Hepler (Basic Books, 1967): If John and Mary love each other, and Bill and Mary love each other, does that mean John and Bill love each other?

2-3(b) Experience and the first law

Now let us see what you know from experience about the first law of energy. It will be assumed here that readers are generally familiar with such terms as *kinetic energy* ($\frac{1}{2}m\mathscr{V}^2$), *gravity potential energy* (mgh), and *specific heat* (the heat required to change a unit weight of a substance by one degree in temperature; the specific heat of water is 1 cal/gm-°C or 1 Btu/lbm-°F). Here, then, are four problems:

1. Two billiard balls labeled *A* and *B* approach each other as shown in the sketch and collide obliquely with each assuming a new, different, unknown direction after the collision.

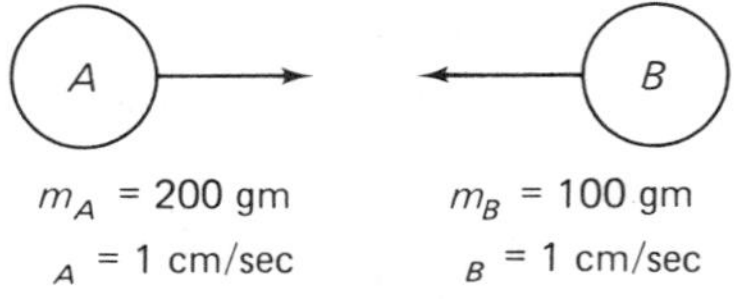

After the balls collide, it is found that ball *A* has a speed of ½ cm/sec in its new direction. What will be the speed of ball *B* in its new direction? Write your answer below:

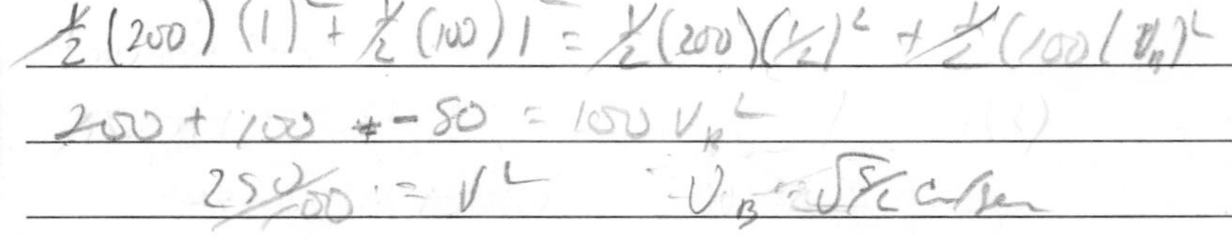

2. A block with a mass m is held at rest above a table at a height h. If the block is released, what will be its velocity when it strikes the table? Write your answer here:

3. Suppose that you have a chamber that is perfectly insulated, that is, one in which the walls neither absorb heat nor allow heat to pass through. You have two copper blocks of equal mass. One block is placed in the chamber at 500°F and the other is placed in the chamber at 700°F. After considerable time has passed, what would you expect to be the temperature of the first block? ______ Of the second block? ______

4. Now imagine that you have two copper blocks A and B such that block A has a mass of 400 gm and block B a mass of 100 gm. The two blocks with different initial temperatures are placed in a perfectly insulated chamber. Initially $t_A = 200°C$ and $t_B = 300°C$. After some time has passed, what will be the temperature of the two blocks in the chamber?

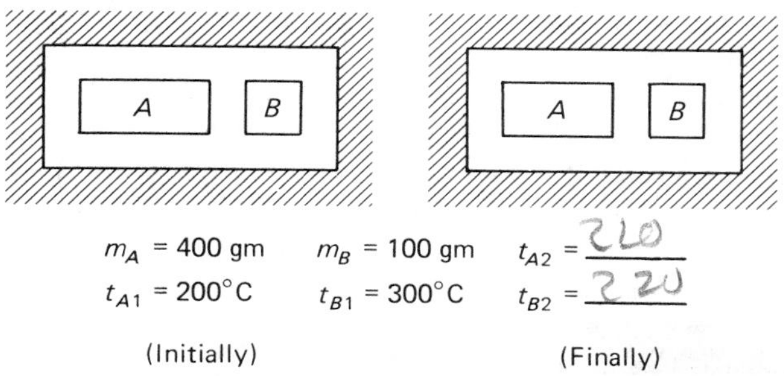

The answers to the questions above can be obtained by application of the first law. If you could work them, then you know how to use at least parts of the first law even if you can't verbalize the law itself. The correct answers are (1) $\mathscr{V}_B = \sqrt{\frac{5}{2}}$ cm/sec; (2) $\sqrt{2gh}$; (3) both will be at 600°F; and (4) $t_{A2} = t_{B2} = 220°C$.

The first two questions are solved by using the principle of conservation of mechanical energy. The third and fourth are solved using the principles of calorimetry, or conservation of thermal energy. Both of these conservation principles are special cases of the first law, which, as we have said is the principle of conservation of "energy," where "energy" is taken in its most general sense. A more rigorous and complete development of the first law will be carried out in a later chapter but it is clear that you have *some* working knowledge of the law from your experience and education to date.

2-3(c) Experience and the second law

Imagine that the following processes all have velocities, temperatures, etc., such that they are *consistent with the principle of conservation of energy:*

1. A block slides *down* a rough plane and becomes *warmer.*
2. A block slides *up* a rough plane and becomes *cooler.*
3. A warm block and a cool block are placed in a perfectly insulated chamber. The warm block becomes cooler and the cool block becomes warmer until they are at the same temperature.
4. A warm block and a cool block are placed in a perfectly insulated chamber. The warm block becomes warmer and the cool block becomes cooler.

Remember that all the processes above are assumed to satisfy conservation of energy. Despite this, will they all occur? If not, which ones will *not* occur? Write your answers here:

____

____

____

You know from your experience with the physical world that processes 2 and 4 will not occur, at least not by themselves. If they did occur, they could easily be imagined to occur in a manner consistent with conservation of energy; still we know they will not occur. Another law is thus needed to describe what we know about the *direction* in which things happen. We call it the *second law.* If you answered the question above about the processes correctly, you know something about this second law even if you aren't aware of it and can't state the law in words. What you know from experience says something about the direction in which the processes occur. Roughly, we might say at this time that the law states that without outside influence processes occur only in one direction. A more rigorous statement will be developed in Chapter 8 and we shall also show how this one-wayness relates to the concept of degradation.

In summary then, you know at least the following about the laws of energy stated roughly from experience with the physical world.

Zeroth law: Two systems each equal in temperature to a third system are equal in temperature to each other.

First law: Energy, generalized to include kinetic, potential, thermal, etc., is conserved.

Second law: Without outside influence, processes occur in only one direction.

2–3(d) Why everyday experience is not enough

In the preceding three sections, we have partly organized and verbalized your natural experience with the laws of energy. But natural experience alone, even when organized and written down, is not sufficient to build the science. Your experience provides a good start but you must add some additional critical thinking, some experimentation that everyday life doesn't encounter, and a little mathematics. To develop a science and to use it in practical engineering ways, you must go beyond the layman's experience with thermometers, pool balls, and blocks sliding down planes. You must go to the laboratory; you must learn to think about and organize what you know; and you must learn there in systematic and useful ways. Otherwise, you will be left with only a little common sense, which, though useful in its place, also has told men at times that the earth is flat, that some other people are witches and should be burned, etc.

2-4 Logic of the science of energy

Before we can proceed to develop and organize the laws of energy more formally so that they may be applied more systematically, there is some background information that you should have in order to put your new knowledge into perspective. This section is devoted to giving you some of this necessary background.

2-4(a) Different approaches to the laws

Broadly speaking, the study of energy can be approached either *macro*scopically or *micro*scopically.

The *macro*scopic approach is empirical (that is, based on experiment) and man-sized. It deals with observable properties and either generalizes the laws from its experiments or states them as axioms.

The *micro*scopic approach is theoretical and atom-sized. It assumes a molecular model, applies statistical analyses, and computes results that have interesting and useful interpretations in terms of the macroscopic properties and laws. The success of theory is justified only by experiment, of course, but the theory makes it possible to suggest useful new experiments and applications that would be extremely difficult or unlikely to come upon empirically. (This is especially true in the case of some of the new direct energy conversion devices.) Thus, though the study of energy may be approached initially either macroscopically or microscopically, a complete understanding of the subject must include both regimes as well as a knowledge of the relationship between them.

Even after one chooses to begin with either a macroscopic or microscopic approach to the logical development of the laws, there remain several ways of carrying out the development. In a macroscopic development, one technique is to define some needed terms and to state the several laws as axioms or first principles. The axioms are justified by the inability of anybody to disprove them and the remaining structure of the subject is *de*duced from the axioms.

Another macroscopic approach, and the one to be used in this book, is best described as operational-inductive. The operational-inductive approach, as the term is used here, has much in common with the axiomatic approach.

The operational-inductive approach, however, insists on the use of a particular kind of definition called an *operational* definition. Also, the operational-inductive approach develops the laws as *in*ductive inferences (or generalizations) from observation and experiments, whereas the axiomatic approach takes the laws as given axioms or first principles. Thus, the two macroscopic approaches are distinguished by (1) whether the laws are stated at the outset as axioms or whether the laws are arrived at inductively as generalizations from observations and by (2) the kind of definitions used for terms.

There are also several microscopic approaches. Just as the two macroscopic alternatives have much in common, so do the various microscopic approaches. There remain certain important distinctions, however.

The microscopic approaches may be identified as (1) kinetic theory, (2) statistical mechanics, or (3) information theory. All assume that matter is made up of discrete particles. Kinetic theory is the oldest microscopic approach. It does not take into account the quantized nature of position and energy of particles and hence has limited usefulness. Statistical mechanics uses the findings of quantum theory and probability theory to determine the most probable state of the various particles subject to the overall constraints on the system. This most probable state is then concluded to be *the* one observed macroscopically. Information theory essentially does the same thing but argues that the observed *macro*scopic state of a system will be the one that yields the *least* information to an observer about the *micro*scopic state of the system. The most probable state and the one yielding the least information are the same by the definition of *information*.

Some scientists and engineers restrict the use of the word "thermodynamics" to the macroscopic laws and properties that are developed from the macroscopic approaches but it is more common to include both microscopic and macroscopic studies under that general heading.

In summary, the various approaches to or ways of thinking about the science of energy may be organized as follows:

Macroscopic	*Microscopic*
Empirical	Theoretical
Man-sized	Atom-sized
Approaches:	Approaches:
Axiomatic-deductive	Kinetic theory
Operational-inductive	Statistical thermodynamics
	Information theory

In this book, energy will be introduced *macroscopically* using the *operational-inductive* point of view. The operational-inductive method has been selected because it begins at a point close to the experiences you have already had with the physical environment and also because it properly emphasizes the inductive nature of scientific laws.

2-4(b) The operational-inductive approach

The operational-inductive approach to the development of thermodynamics begins by defining basic quantities in terms of *how they are measured.* It then considers the results of experience and conducts experiments among these quantities and generalizes the laws from the results. The laws are then used deductively to arrive at new conclusions. Schematically, the process is this:

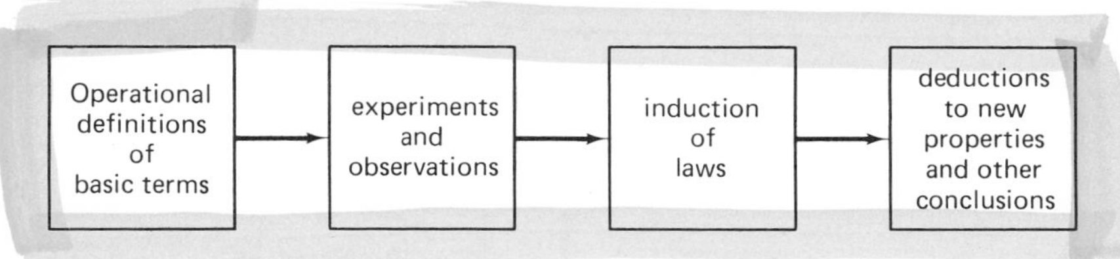

There are two topics in the process outlined above that require attention for those unfamiliar with them. These are (1) the process of induction and (2) the operational definition. The next two sections will treat these topics. Though these sections do not seem to deal with energy or properties, you will find later that understanding induction and operational definitions will help you to understand not only energy but other applied sciences as well.

2-4(c) Induction

Induction is the logical process of drawing *general* conclusions from a collection of specific cases. Schematically, the induction process can be illustrated as follows:

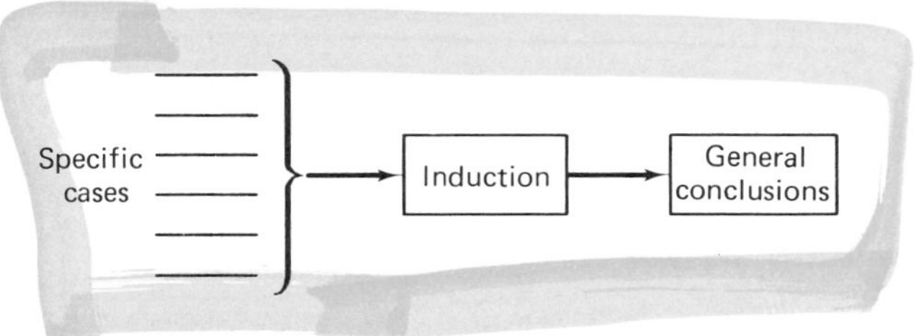

Following are examples of induction and deduction.

Induction

Specific case: Mary, who started to work 4 weeks ago, was late to work on Monday of her first week.
Specific case: Mary was late to work on Monday 3 weeks ago.
Specific case: Mary was late to work on Monday 2 weeks ago.
Specific case: Mary was late to work on Monday 1 week ago.

Inductive Conclusion: Mary is *always* late to work on Monday.

Deduction

General Premise: Mary is always late to work on Monday.
Deductive Conclusion: Mary will be late to work next Monday.

It should be noted that an *in*ductively reached conclusion has some uncertainty associated with it even when the specific cases are all true. (Mary just *might* be on time *next* Monday.) A deductive conclusion, however, is completely certain *if* the given general statement or premise is true.

In the chapters that follow, the words *inductive* and *deductive* will be used to describe certain phases of the development of the laws of energy. It is important, therefore, that students understand the elementary ideas of the two processes as outlined above.

The laws of science are inductive conclusions made by people. Thus, such laws should not be thought of as something *real* or *true*. An inductive conclusion can never be completely certain (unless every possible specific case has already been examined) and a scientific law can never be anything more than an inductive conclusion concocted by scientists themselves.

Induction usually results in abstraction. In science, repeated induction can and does lead to extremely abstract concepts. In a very simple way, the concept of abstraction can be illustrated in the following diagram:

Most abstract: Life
Animals
Vertebrates
Dogs
Newfoundlands

↑ Increasing degree of abstraction

Least abstract: My black Newfoundland called Guinevere

In the example, "my black Newfoundland..." is something that I experience with my senses. "Newfoundland" is more abstract. I can still experience "Newfoundlands" collectively and totally but it is a less detailed experience. "Newfoundlands" is an abstract concept that I have learned by generalizing all the common features of all the individual Newfoundlands I have encountered. "Dogs" is even more abstract, and so on.

Much the same sort of abstraction process can be illustrated in simple scientific terms. Both *height* and *mass* are low abstraction terms. They are close to experience. Multiplying them together gives a quantity called *potential energy*, a concept that is considerably more abstract than either height or mass.

It is typical of science to look for more and more abstract relationships ("laws") because they are more general. As the laws of energy are developed in the next chapters, you will see how this process of cascading abstractions leads to highly useful and to such highly abstract concepts called *entropy*, *degradation*, etc., and to such very general laws as the first and second laws.

2-4(d) Operational definitions

The operational-inductive approach uses a type of definition called an *operational* definition. The use of the word *operational* in this sense goes back to P. W. Bridgman, a well-known scientist and philosopher of science. An operational definition is one that describes in detail how to measure or observe the defined term. Because an operational definition describes a measurement method, the observer or scientist and his instruments and procedures are included as a part of the definition. Using operational definitions, we are never allowed to forget that science is something *people do* and that our laws and theories are man-made. New observations can then easily lead to new laws or theories because the old laws have never been accepted as anything more than the *best* inductive conclusions we can make with the data at hand at any given time.

Operational definitions eliminate the circularities that accompany other types of definitions. For example, most people when asked to define *length* will say something about *distance* or *how far* one point is from another. Dictionaries are frequently also circular. Length is distance and distance is length. These purely verbal replies do nothing to tell one about the concept of length (or distance, or how far. . .). An operational definition of length, on the other hand, describes *how to measure* the length of something. The concept is thus precisely defined by how to measure it rather than in terms of verbal synonyms.

For example, to measure a length or distance, we select a standard (such as a former English king's foot!) and by placing the standard end to end along the unknown length and counting we can arrive at the length of an object in terms of the standard unit. This works fine as long as we are concerned with man-sized objects (in an order-of-magnitude sense) and as long as we do not need to worry about the velocity of the object. For larger lengths we may use indirect methods that make use of trigonometry or radar (or both) or color (as with stars). It should be noted that these different methods of measurement really give us a different operational definition for length since an operational definition is a method of measurement.

For very small distances, the use of a standard ruler and counting is also not practical, though this method can be extended indirectly with a microscope. For much smaller distances, the electron microscope is used together with some assumptions about the structure of matter.

Time is another extremely fundamental term that is difficult to define except with an operational definition. Operationally, we do not ask "What *is* time?" but rather "How do we measure time?" For man-sized time the rotation of the earth provides an adequate standard and we do much the same things with it that we do with the standard distance. Using a pendulum, we can divide the standard into equal small parts. For very short times, we use electric circuit oscillators; certain nuclear processes are used for times of the order of 10^{-20} sec. For longer times, radioactive dating is used. Again, the student should note that each measurement method is really a different operational definition of time. Scientists take many pains to assure themselves, however, that the various methods are consistent.

For the purposes of this study, the student should note that the definitions given above are operational because they describe *how to measure* the term being defined. This is a key concept in physical science.

It is left as an exercise to develop an operational definition for mass. Very small and very large masses should be considered as well as man-sized masses.

Students should also note that once distance and time are defined operationally, then velocity and acceleration can be defined as *paper and pencil* operations, perfectly valid in an operational definition. Similarly, momentum, potential energy, and kinetic energy are all defined in terms of paper and pencil operations (multiplication and division) using the basic quantities already defined. These paper and pencil operations provide more abstract concepts but each is nevertheless soundly based on measurements and clearly defined operations.

The operational definitions described above are somewhat sketchy and abbreviated. Complete operational definitions are long and detailed. Also, of course, a skeptic can point out that any definition written in words is ultimately circular. If we are to use words at all, however, some things must be accepted on an experimental, nonverbal level. In building a science, these undefined words should be simple everday words that cause no confusion and thus are the lowest of all on the abstraction levels. New or abstract concepts should not merely be defined in terms of synonyms or in terms of equal or higher abstraction but rather operationally using lower abstractions and terms closer to experience.

2-5 Pressure

2-5(a) Operation definition

As an example of how operational thinking helps us to keep things straight and of something we need to know anyway, let us consider the concept of *pressure*. As a general concept, it is defined operationally as the paper and pencil operation of dividing the force by the area over which the force acts. We assume now that force and length have been previously defined in terms of how they are measured and area has been defined as a paper and pencil operation with length. Thus, we know operationally what we mean by the concept *pressure*.

But now let us consider several specific pressures. What is atmospheric pressure?

We hope you did not say something circular like "the pressure of the atmosphere" but rather gave an operational definition. One way would be to talk about the weight of a column of air from the earth's surface to infinity, the area of the column at the surface, etc. But what do you do if you want actually to *measure* the pressure of the atmosphere? You use an instrument called a *barometer* that literally does the weighing and dividing for you. So a fair operational definition of atmospheric, or barometric, pressure is *the reading on a barometer.*

Now consider pressure gauges used to measure the pressure of gases, liquids, or other systems in tanks, pipes, etc. If you think operationally, then you will want to know how these things work and what they actually measure so that you will know really (operationally) what is meant by the term *gauge pressure.* If you investigate, you will find that gauges measure the *difference* between the pressure in the system and atmospheric pressure. In symbols, that is

$$p_g = p - p_0$$

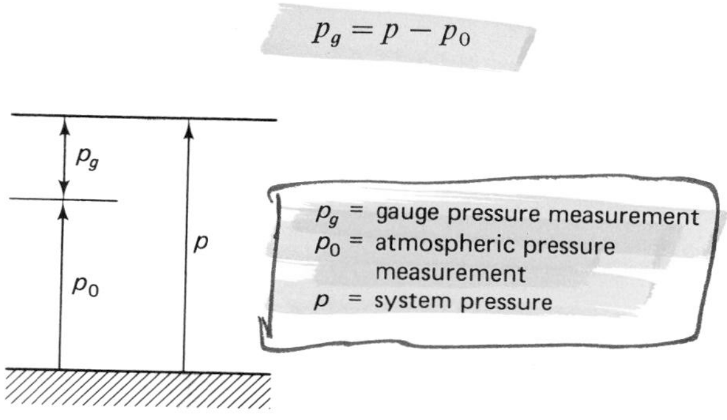

If we rewrite the equation above as

$$p = p_g + p_0$$

then we see clearly that the system pressure is the sum of the gauge pressure measurement plus the atmospheric pressure measurement. This sum is often given the name *absolute* pressure to distinguish it from gauge or atmospheric pressure.

Students should note how the habit of operational thinking leads to clarification of meanings. By asking "What do gauges measure?" instead of "What is gauge pressure?", the whole matter is clear.

What do you suppose the term *vacuum* means? See Prob. 2-8.

2-5(b) Constant pressure processes

A process that takes place at constant pressure is called *isobaric.* Constant pressure, or nearly constant pressure, processes occur in a variety of situations. For example, many processes take place under only atmospheric pressure, which can usually be considered constant during a given process. Others take place in devices designed to produce *almost* constant pressure processes, at least part of the time. For example, part of the cycle in a diesel engine cylinder takes place at almost constant pressure as the piston moves to change the volume. It is important for students to recognize constant pressure and other types of processes when they occur because knowing the process is almost always essential to the solution of thermodynamics problems.

EXAMPLE 2.1. Which of the following processes would you consider to be isobaric for analysis purposes?

1. A valve is opened in a tank of compressed air.
2. Heat is added to boiling water on the stove.
3. Air is compressed in an air compressor.
4. A tank of compressed air leaks air through a tiny pinhole leak.
5. The air in the cylinder with a frictionless piston held by a constant weight on it is heated.

SOLUTION: Processes 1 and 3 are certainly *not* isobaric. Process 2 is isobaric. Process 4 is not isobaric but in *some* circumstances (perhaps when short time periods are involved or something more is known about the temperature) it *might* be practical to assume it is. Process 5 *is* isobaric.

2-6 Solving problems

This book will stress not only the logical scientific development of the energy laws and their consequences but also the application of these laws to engineering problems. The object is to relate the theory to its applications. To accomplish this, a particular problem-solving procedure is advocated and illustrated in the example problems. Nothing sounds sadder to an engineering professor than a student who pleads "I really understand the material but I just can't work the problems!" Such students *know* the equations but either they do not *understand* them or else they do not know how to *use* what they know in problem situations. The problem-solving procedure presented in this book is an attempt to help students bridge this gap between the theory and application of the subject. Obviously, no one procedure can be perfect or all-inclusive. Problem solving often requires human ingenuity. The procedure used here, however, does help by focusing the problem solver's attention on the proper questions and the steps to be taken in many typical problems. The procedure is outlined briefly below and students should note how it is applied in the examples done in the remainder of the text.

Step 1: Translation of the given information and problem statement *into pictures, diagrams, and symbols*, noting any assumptions made. This is a tremendously important step and one which students try to bypass in their haste to write some equation or another which they hope will solve the problem instantly. Good problem solvers, on the other hand, take great pains to understand and set up the problem carefully and completely, to understand the problem, before they begin to write equations. And professors take great delight in making up exam problems that are solved easily by pictures and diagrams but are almost impossible using equations.

Step 2: Selection and definition, in words and pictures, of the *system* to be analyzed. This step is analogous to the selection and sketching of a free body in statics. It is absolutely essential to the solution of most energy problems and many student errors in problems can be traced to a confused or nonexistent definition of the system being studied. More appears about systems in Chap. 5.

Step 3: Selection of the *equation of state* to be used for the substance or substances involved in the problem, noting any assumptions made. Equations of state are discussed in detail in Chap. 4.

Step 4: Description of the *process equations* and/or *process assumptions* that are to be used. This is a statement, using words or the diagrams, of *how* things happen. Often processes are described by some parameter that doesn't change (e.g., a constant pressure process) but we shall see that there are also other factors that are used to describe processes.

Step 5: Apply the fundamentals. This is, finally, the step in which the basic laws are brought to bear on the problem. Notice that it is step 5, not step 1, as so many would-be problem solvers try to make it.

Step 6: Application of general *relations among properties.* In addition to the laws, theory provides some very general relationships among properties that are sometimes very helpful in solving problems. These are discussed in Chap. 10 and 16.

Step 7: Checking the results. There are a lot of quick and practical ways to check for possible errors in your work. Here are some: (a) Do the dimensions and units check? (See Appendix A for units in this book.) (b) Do the equations and results make sense if you imagine the possible limiting cases? (c) Do the equations and results make sense if you examine effects of changes of one variable on the others? (d) Do the results make reasonable engineering sense?

Every problem, of course, does not require each and every step outlined above. We shall use the procedure in the examples in this book as often as practical. Thus, *the examples are a part of the text and should be studied as a part of the reading material.* They contain information, especially about how to make assumptions and apply the new material, that is vital to students who are interested in being able to *use* what they know about energy and properties.

EXAMPLE 2.2. This example assumes students are familiar with the ideal gas equation of state, $pV = MRT$. For those who are not, equations of state are discussed in more detail in Chap. 3.

A tank in a factory is filled with air from a high pressure line and at the conclusion of the filling process the pressure and temperature in the tank are 100 psia and 140°F. If the tank is allowed to stand for several days, what will you expect the pressure to become?

SOLUTION: *Translation to sketches and symbols:*

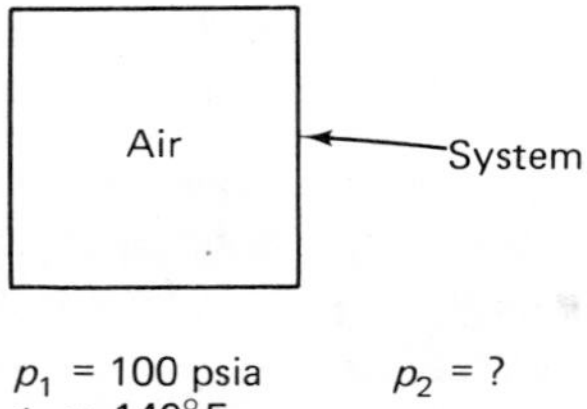

p_1 = 100 psia $\quad$ p_2 = ?
t_1 = 140°F

System: The air in the tank.
Equation of State: Assume the air is an ideal gas: $pV = MRT$. At state 1, $p_1 V_1 = M_1 R T_1$; at state 2, $p_2 V_2 = M_2 R T_2$.
Process: Constant volume: $V_1 = V_2$
Fundamentals:
Conservation of Mass: $M_1 = M_2 =$ constant
Zeroth Law: Final temperature of air will be the same as room temperature (T_r). (Add to sketch: $T_2 = T_r$)
Using the above, we find that

$$\frac{p_1 V_1}{RT_1} = \frac{p_2 V_2}{RT_2}$$

or

$$p_2 = p_1\left(\frac{T_2}{T_1}\right) = p_1\left(\frac{T_r}{T_1}\right)$$

Since we do not know the room temperature, we must make a reasonable assumption. We assume the factory is kept at about 70°F. Thus,

$$100(\tfrac{70}{140}) = 100(\tfrac{1}{2}) = 50 \text{ psia}$$

First Law: Not needed.
Second Law: Not needed.
Checking:
Units: Only ratios were used.
Limiting Cases: If $T_r \to \infty$, $p_2 \to \infty$.
Trends: 1. If T_1 had been larger, p_2 would be smaller.
2. If p_1 had been larger, p_2 would be larger.
3. If T_r were lower, p_2 would be lower.

Engineering Sense: Does a drop in pressure from 100 to 50 psia, a factor of 2, seem reasonable? It seems a lot, too much for such a small temperature change. Where does this result come from? It comes from where we use 70 and 140°F in the ratio of temperatures. But, we must use *absolute* temperatures in $pV = MRT$ and so it should be

$$p_2 = p_1\left(\frac{T_r}{T_1}\right) = 100\left(\frac{70+460}{140+460}\right)$$

$$= 100\left(\frac{530}{600}\right) \text{ psia}$$

$$p_2 = 88.3 \text{ psia}$$

This makes better engineering sense. And we should remember always to use *absolute* temperatures with T from $pV = MRT$.

EXAMPLE 2.3. A 5-m³ tank of compressed air at 300,000 N/m² absolute (3 bars) and 25°C develops a small leak such that the pressure falls to 280,000 N/m² (2.8 bars) absolute in 24 hr. How much air has leaked of the tank? For air, $R = 287$ j/kg-°K.

Note: Students not familiar with SI metric units should see Appendix A.

SOLUTION: Translation to sketches and symbols.

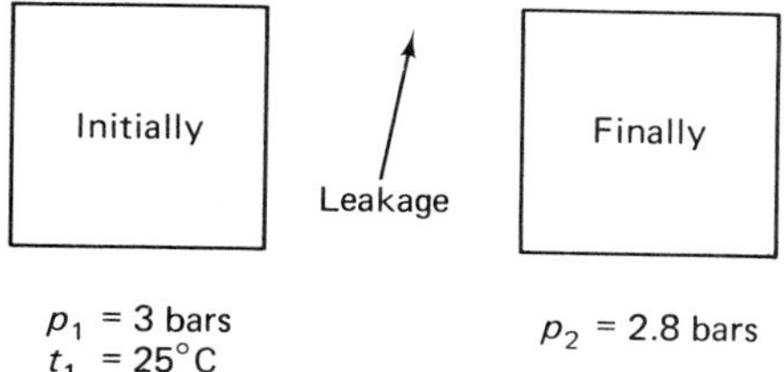

Control Volume: The tank.
Equation of State: Assume the air is an ideal gas.
Process: Since the leak is a very slow one, it is reasonable to assume that there is time for sufficient heat transfer to keep the air that is in the tank (control volume) essentially isothermal. Thus, $t_1 = t_2 = 25°\text{C}$.
Application of Fundamentals:
Conservation of Mass: $M_1 = M_2 + M_{\text{leak}}$

Applying the equation of state, we can find both M_1 and M_2 and thus compute M_{leak}.

$$M_1 = \frac{p_1 V_1}{RT_1} = \frac{(300{,}000)(5)}{(287)(298)} = 17.5 \text{ kg}$$

$$\frac{(\text{N/m}^2)(\text{m}^3)}{(\text{j/kg-°K})(\text{°K})} \rightarrow \frac{\text{N}\cdot\text{m}\cdot\text{kg}}{\text{j}} \rightarrow \text{kg}$$

$$M_2 = \frac{p_2 V_2}{RT_2} = \frac{(280{,}000)(5)}{(287)(298)} = 16.4 \text{ kg}$$

$$M_{\text{leak}} = M_1 - M_2 = 17.5 - 16.4 = 1.1 \text{ kg}$$

2-7 Summary

The two most fundamental and important laws of energy are (1) the law of energy *conservation* and (2) the law of energy *degradation.* They work simultaneously. Energy is always conserved but its potential to do useful work for man is degraded in any energy conversion or transfer.

From experience most persons know more than they realize about the laws of energy. They know, for example, that if two systems are each equal in temperature to a third system, then they are equal in temperature to each other. That is called the *zeroth law of energy.* They also know that mechanical energy is conserved in certain kinds of situations (called *conservative*, of all things!) and that thermal energy is conserved in certain other situations. The law of conservation of energy is more general and includes these and other processes as well, including cases where mechanical, thermal, and other forms of energy are all involved.

Most people also know that some things just do not occur in nature, at least by themselves, even though their occurrence would not violate conservation of energy. In a sense, this is the second law. It turns out that the directions in which things do happen always result in a loss of work potential so we call this *the law of energy degradation.*

The science of energy may be approached in two basic ways: macroscopically or microscopically. In this book, we shall take a macroscopic view. We shall also adopt an operational-inductive approach, which means definitions will be in terms of how measurements or numbers are obtained, and laws will be taken as inductive conclusions from observations rather than axioms or first principles.

Pressure (force per unit area) is an important energy-related property. Atmospheric pressure is the force per unit of a column of the earth's atmosphere and is measured by a barometer. Gauges measure pressure differences.

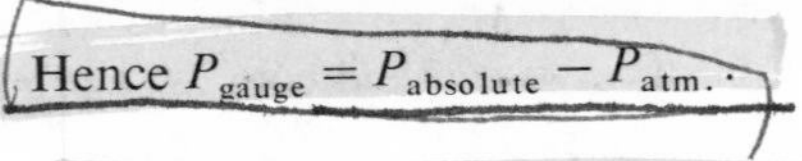

Hence $P_{gauge} = P_{absolute} - P_{atm.}$.

Understanding means more than knowing. Understanding also includes the ability to *use* what is known. In this book, the problem examples are therefore an important part of the text because they often show you how to use the subject material. A problem-solving procedure is recommended, not to be used religiously, but to be used as a guide to help apply what you know. The steps in the procedure are (1) translation into pictures, diagrams, and symbols; (2) definition of the system; (3) selection of the proper equation of state; (4) description of processes and process equations; (5) application of fundamental laws; (6) applications of relations among properties; and (7) checking the results.

The following books are strongly suggested reading for interested students who wish to extend their education and experience beyond textbooks:

1. *Language in Thought and Action* by S. I. Hayakawa. New York; Harcourt, Brace, and Co., Inc., 1949. A *must* for anyone who reads, speaks, hears, or writes. It shows how the words we use affects what we think and do and vice versa—a very practical book.
2. "Standards of Measurements," by Allen V. Astin. *Scientific American*, **218**, No. 6, June, 1968, p. 50. An up-to-data summary of the measurement standards—and hence operational definitions—for length, mass, time, and temperature.

Problems

2-1(a) Give examples from you *personal* experience of the zeroth law, the first law, and the second law.

(b) Would you say that your observations are the result of the laws or are the laws the result of your (and others) observations?

2-2 See if your girl friend or boy friend who is a nonscience or nonengineering major also "knows" the laws from experience by asking them some questions such as those in Sec. 2-3.

2-3(a) Compare the main characteristics of macroscopic and microscopic approaches to energy theory.

(b) Compare the main characteristics of the axiomatic-deductive and the operational-deductive approaches to macroscopic thermodynamics.

2-4 Give an operational definition for mass. Consider also very small and very large masses.

2-5 Given that mass, time, and distance have been operationally defined, what paper and pencil operation can be used to define force?

2-6 Give an operational definition for

(a) "success in college"

(b) the "health of the economy"

(c) "success in life"

2-7(a) Define the term *vacuum* operationally by finding out what a vacuum gauge measures. Write an equation relating vacuum, absolute pressure, and atmospheric pressure.

(b) A vacuum gauge in a tank reads 10 psi vacuum. The barometer is 15 psi. What is the absolute pressure of the system in the tank?

2-8 The following definition for heat is sometimes given: "Heat is that which transfers from one system to another at a lower temperature by virtue of the temperature difference between them." Is this operational? Explain.

2-9 Give examples of inductive and deductive logic.

2-10 Are scientific laws arrived at by induction or deduction?

2-11 Revise the problem-solving methodology given in this chapter so that it applies to statics instead of thermodynamics. Then try it on a tough statics problem (e.g., one you missed on an exam). How does it work?

2-12 A 10-ft^3 tank of compressed hydrogen at 200 psia and 70°F springs a very tiny leak and the pressure falls to 195 psia in 24 hr. What is the average rate of leakage of the hydrogen? For H_2, $R = 773$ ft-lbf/lbm-°R. *State all assumptions.*

2-13 A cylinder fitted with a piston contains air at 30 psia and 80°F. Its initial volume is 1 ft^3. If the process equation is known to be $pV^{1.2} = \text{constant}$, compute the final pressure and temperature as the air expands to twice its initial volume.

2-14 A cylinder fitted with a well-oiled but heavy piston contains an ideal gas at 30°C. The piston area is 0.01 m^2 and the weight of the piston is 50 kgf. Atmospheric pressure outside the cylinder is 0.01 kgf/mm^2.

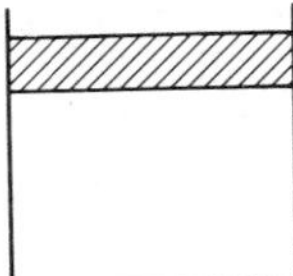

(a) What is the pressure inside the cylinder?
(b) If the gas is heated so that its volume is doubled, what will be its new temperature?

2-15 Would you elect a course in energy (or thermodynamics) even if one were not required? Explain.

2-16(a) Give at least three examples of processes in which energy is conserved.
(b) Is energy degraded in your examples? Explain.

2-17 What is the work potential of 1 million Btu of energy available at 100°F? Atmospheric temperature is 80°F. What does your answer mean regarding the usefulness of the thermal discharge from power plants?

2-18 What is the loss of work potential if a 10-kgf weight is dropped onto a floor from a distance of 2 m?

3

Thermal equilibrium, the zeroth law, and temperature

Thermal equilibrium, despite the use of the word *thermal*, is established operationally without the use of any temperature scale. That is, we establish the concept of thermal equilibrium before we talk about temperature itself. In a sense then, our discussion of *temperature* in Chap. 2 was premature. To be logically rigorous, we should not have discussed even *equal in temperature* until we had established *thermal equilibrium* operationally. But, of course, in that introductory discussion we were simply trying to become aware of our experience, which includes *temperature*, but now we want to place the logic and meanings on a more scientific basis.

As this material is developed and presented, here are some things students should look for: What is an operational definition for temperature? Why is temperature different for different thermometers? Why is an International Temperature Scale needed? How does a constant volume gas thermometer work? How does the constant volume gas thermometer help resolve the theoretical question of temperature? What are isothermal processes? What is absolute temperature? What does thermal equilibrium have to do with temperature measurement?

3-1 Thermal equilibrium

To discover operationally whether two systems are in thermal equilibrium, the two systems are placed together in isolation from their environment and all available instruments except thermometers with scales are used on both systems. If no changes are observed in either system when they are brought together, then the two systems are said to be *in thermal equilibrium*. It should probably be mentioned that the "bringing together" mentioned above is to be done in such a way that such mechanical interactions as might be caused by pressure differences or electrical or magnetic interactions potential differences are not allowed. Sometimes it is said that the systems in this operation are to be brought into *thermal* communication but this becomes rather circular. It is more precise to specify that mechanical, electrical, or magnetic interactions are not to be allowed. Then if no changes are observed when two systems are brought together, they are said to be in *thermal* equilibrium.

3-2 The zeroth law

As almost everyone knows, observations and experiments lead readily to the conclusion that if two systems are each in thermal equilibrium with a third system, then they are in thermal equilibrium with each other. This statement, as noted earlier, is called the zeroth law. Notice that, logically, it is an inductive conclusion based on observations or experiments with systems in thermal equilibrium, where thermal equilibrium has already been defined operationally.

3-3 Temperature

Temperature is an important property in the study of energy. Its definition and measurement must be preceded, in the logical and scientific sense, by the concept of thermal equilibrium and the zeroth law. Since these have been described above, we can now turn our attention specifically to a discussion of temperature.

3-3(a) Operational definition

Since an operational definition is a description of how to measure the item defined, we must describe how to measure temperature. One way to do this is as follows using a volume of mercury in a small, evacuated glass tube (mercury-in-glass thermometer):

Allow the mercury–glass system to come to thermal equilibrium with a mixture of ice and water under atmospheric pressure and mark the position of the mercury on the glass. Now allow the mercury–glass system to come to thermal equilibrium with a mixture of steam and liquid water under atmospheric pressure and mark this position. Divide the distance between the marks into 100 equal parts and call the ice point 0° and the steam point 100°. Intermediate temperatures are denoted between 0 to 100 and temperatures below 0 or above 100 are found to be extending the scale and keeping the degree size the same.

Obviously the description above could be applied as well to electric resistance wires, thermocouples, etc., except that ohms or emf would be substituted for the position or length of the mercury column.

The procedure described above makes it possible for us to establish a temperature scale and thus to measure temperature. *Now* we can call the mercury–glass system a *thermometer*.

Two points should be made by way of elaboration. First, the numbers on the scale are arbitrary. The scale called *Celsius* takes the ice point to be 0 and the steam point to be 100. The Fahrenheit scale takes these as 32 and 212, respectively. There is no special reason why these numbers could not have been anything that seemed convenient. In fact, the steam point need not even have been given the larger number, but it was.

The second point of the elaboration is more fundamental. It was noted that in carrying out the procedure of establishing a scale, many different "thermometers" could be used. That is, we could use mercury in a small glass tube, a thermocouple, an electrical resistor, and so on. The resulting problem is that, except at the calibration points (e.g., ice point and steam point), these thermometers will not generally agree. The reason of course is that each of these thermometers depends on how a different physical property—the volume of mercury, the emf of a thermocouple, the electrical resistance of a wire—varies with temperature. There is no reason why all these real physical properties of various real materials should be the same function of temperature and, in fact, they are not.

This situation presents a fairly serious problem because it means that any temperature measurement is a function of the thermometer used to measure it. As a practical matter in order to enable scientists to know exactly what others mean when a temperature is reported, an International Temperature Scale (ITS) has been established fixing not only the ice and steam points but also the melting point of sulfur (444.6°C), the boiling point of oxygen (−182.97°C), the melting point of silver (960.8°C), and the melting point of gold (1063.0°C). The ITS also establishes a series of operational definitions for temperature in certain ranges. The ITS definition calls for use of a platinum resistance thermometer with a specified relationship between resistance and temperature in the range between the oxygen point and the ice point. Other ranges have different thermometers and/or equations. The establishment of this scale thus makes it possible for scientists to agree on *temperature* measurements. In most practical engineering situations, of course, the variations among thermometers are not large enough to be a serious problem, though calibration of a given thermometer with the ITS is occasionally needed. More information on the ITS can be found in standard references such as the *Handbook of Physics and Chemistry*.

3-3(b) Absolute temperature

It was pointed out in the preceding section that temperature is dependent on the choice of the thermometer and the materials used in the measurement. The practical problems created by this situation are solved by the agreement on the International Temperature Scale. The theoretical aspects remain somewhat distressing, however. Temperature still seems so *arbitrary*. In this section, we describe a thermometer that helps relieve some of our distress and in a later chapter we can put the theoretical problem completely to rest with the help of the second law.

A partial solution to our theoretical worries about temperature lies in the fact that there is a type of thermometer that is independent of the physical properties of the substances or materials used in constructing the thermometer. It is called the *zero pressure–constant volume gas thermometer*. We shall henceforth call it simply the *gas thermometer*. It works on the well-known principle that the pressure of a gas held at constant volume will vary with temperature and thus the pressure can be used to indicate the temperature. The effect of the difference in properties among the various gases is eliminated by taking a series of readings with decreasing amounts of gas, keeping constant total volume, and extrapolating the results to zero pressure (hence the name *zero pressure–constant volume* gas thermometer). See Fig. 3.1.

The extrapolation to zero pressure is possible only in terms of a ratio; otherwise the result is just zero. A temperature scale is defined by using the steam point and ice point as before and specifying that

$$\frac{T_s}{T_i} = \lim_{p_i \to 0} \frac{p_s}{p_i} \tag{3-1}$$

where T_s and T_i refer to the steam and ice points, respectively. Using such a thermometer, *regardless of the gas used*, it is found experimentally that

$$\frac{T_s}{T_i} = 1.3661 \tag{3-2}$$

EXAMPLE 3.1. Compute the unknown temperature from the data recorded in the graph in Fig. 3.1.

SOLUTION: From the graph,

$$\lim_{p_i \to 0} \frac{p}{p_i} = 1.4$$

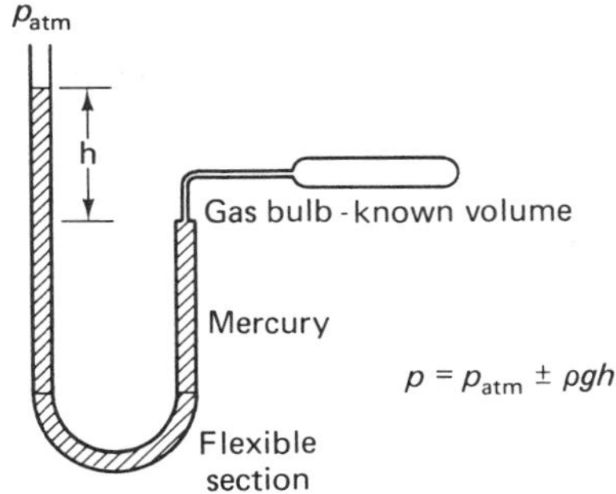

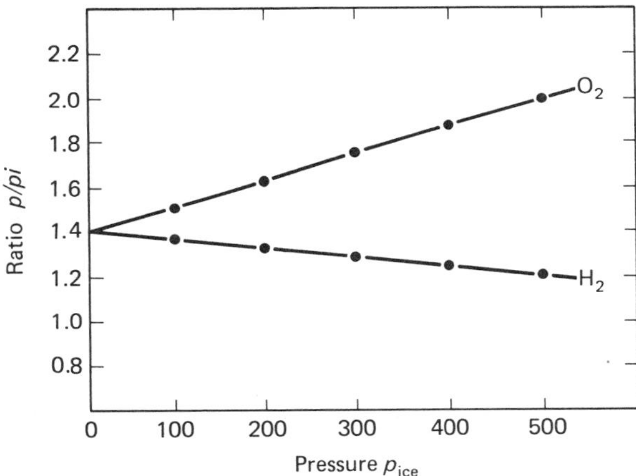

Figure 3.1 Constant volume gas thermometer.

From definition of the Kelvin scale,

$$\frac{T}{T_i} = \lim_{p_i \to 0} \frac{p}{p_i} = 1.4$$

$$T = (1.4)(273) = 382.2°\text{K}$$

From definition of the Rankine scale,

$$T = (1.4)(460) = 644°\text{R}$$

Combining this with the definition of the size of the Celsius degree ($T_s - T_i = 100°\text{C}$) allows one to solve for T_s and T_i. The result is $T_s = 373$ and $T_i = 273$. This conflicts with the previously but arbitrarily assigned values of $T_s = 100°\text{C}$ and $T_i = 0°\text{C}$ but only by a scale shift of 273. The new scale is sometimes called Celsius Absolute but is more commonly called the Kelvin scale.

A Fahrenheit Absolute scale is established in exactly the same way except that the size of the Fahrenheit degree ($T_s - T_i = 180°F$) is used. This scale is usually called the Rankine scale and on this scale $T_s = 672°R$ and $T_i = 492°R$.

To measure an unknown temperature T with a gas thermometer, one finds the ratio T/T_i experimentally by measuring $\lim_{p_i \to 0} p/p_i$. That is,

$$\frac{T}{T_i} = \lim_{p_i \to 0} \frac{p}{p_i} \tag{3-3}$$

$$= \text{experimentally determined value}$$

Combining this with $T_i = 273°K$ (or $T_i = 492°R$) allows easy calculation of the unknown temperature. Though the calculation is easy, however, the experimental work is hard, long, and tedious. Hence, gas thermometers are seldom used in either engineering practice or research. Their importance is primarily theoretical: We do have, if we want it, a temperature scale independent of the substance used in the thermometer (at least as long as we use a gas).

In addition to providing us with a more satisfying concept of temperature, the Kelvin and Rankine scales show something else that is extremely important. *Only positive absolute temperatures are defined.* Negative Kelvin or Rankine temperatures do not exist. It is for this reason that these scales are called *absolute*. In fact, the zero point on these scales is always called *absolute zero* to distinguish it from the zero points on the Celsius and Fahrenheit scales. Absolute zero is at approximately −273°C or −460°F.

Our study of the second law will enable us to say more about temperature but in the meantime we can proceed with the knowledge of how to measure it and with some confidence in its integrity in a theoretical sense provided by the absolute scales of the gas thermometer.

3-3(c) A comment on temperature measurement

Engineers often desire to measure the temperature of objects or substances. Though it is perhaps obvious, it should be noted that thermometers really measure only their *own* temperature. It is up to the experimenter to ensure that the thermometer is at the same temperature as the object whose temperature is wanted. The student should note here how the logic goes: The thermometer measures only its own temperature, its scale having been established by calibration; if it can be put into thermal equilibrium with the unknown, then the zeroth law can be used to compare the unknown with the calibration standard. Thus, in a practical temperature measurement situation, the thermometer is the "third object" in the zeroth law statement: If two objects (the unknown and the calibration standard) are equal in temperature to a third object (the thermometer), then they are equal in temperature to each other. Thus, every time you measure a temperature you use the zeroth law and the concept of thermal equilibrium. It is clearly up to the experimenter to make *sure* that the thermometer is in fact in thermal equilibrium with the unknown temperature.

3-4 Isothermal processes

A process that takes place at constant temperature is called *isothermal*. Isothermal, or nearly isothermal, processes occur in a variety of situations. Usually if a substance is undergoing some changes of state, heat must be transferred to or from it in order to keep it from changing temperature. Since heat transfer is a relatively slow rate process, processes that occur very rapidly are seldom isothermal. On the other hand, relatively slow processes are often very nearly isothermal because there is ample time for heat transfer from the environment to keep the temperature constant.

We shall see in the next chapter that processes involving a two-phase mixture at constant pressure are also isothermal, and in later chapters we shall discuss other situations where isothermal processes occur.

EXAMPLE 3.2. Which, if any, of the following processes would you assume to be isothermal for analysis purposes?

1. A tank of compressed air at room temperature springs a tiny pinhole leak.
2. A large valve in the tank above is opened allowing the pressure to fall to atmospheric pressure.
3. Heat is added to water boiling on a stove.
4. Air is compressed in an air compressor operating at 300 rpm.

SOLUTION: Processes 1 and 3 are isothermal. In process 1, the rate of gas escape is so slow that time is ample for heat transfer to keep the temperature constant. Students who do not understand process 3 should read the next chapter especially well. Processes 2 and 4 are not isothermal because they are too rapid for the significant heat transfer required to keep the temperature constant.

3-5 Summary

The logical progression is from an operational definition for thermal equilibrium to induction of the zeroth law to temperature as a property. Thermal equilibrium is established when two systems are brought together without mechanical, electrical, or magnetic interactions and no changes are observed in either system. The zeroth law is observed as follows: If two systems are each in thermal equilibrium with a third system, then they are in thermal equilibrium with each other.

Temperature scales are established by assigned numerical values to various instrument readings at arbitrary calibration points. The Celsius and Fahrenheit scales use the ice and steam points. This leaves temperature to be a function of the measuring system at points other than at the calibration points. An International Temperature Scale has been established so that temperatures can be compared on a consistent basis despite this problem. Also, the gas thermometer solves this problem by establishing a scale that is independent of the gas used in the thermometer. It establishes absolute temperature scales, always positive, with zeros at about −273°C and −460°F.

Engineers who measure temperature should be aware that thermometers actually record only their own temperature. Hence it is up to the engineer to ensure that his thermometers are in thermal equilibrium with the system whose temperature he desires to know.

Constant temperature processes are called *isothermal*. They occur in nature and in engineering work under several circumstances. One is when the system executing the process is always in thermal equilibrium with a constant temperature environment. This requires sufficient heat transfer between the system and environment.

Students interested in more reading about temperature should obtain a copy of the following little paperback book:

Temperatures Very Low and Very High by Mark W. Zemansky. A Van Nostrand Momentum Book. Princeton, New Jersey, 1964.

You will find a few concepts and equations that you have not yet had in your study of energy but there is nothing that will prevent your profiting and enjoying the book as a whole. Most of it you will understand and find interesting.

Problems

3-1 Give an operational definition for thermal equilibrium.

3-2 Temperature is sometimes said to be "the reading on a thermometer." Is this an operational definition?

3-3(a) Derive equations for converting degrees Celsius to degrees Fahrenheit and vice versa.

(b) At 0°F, what is the Celsius temperature?

(c) At 70°F, what is the Celsius temperature?

(d) At what point are the two scales numerically equal?

3-4 List as many measurable or observable quantities as you can that are functions of temperature. (A good one is the amount of liquid left in a Coca-Cola bottle after opening when it is well shaken before opening.)

3-5 If $(T_s - T_i)$ is taken as 260, what are the values of T_s and T_i on an absolute scale?

3-6 What is the value of absolute zero on the scale developed in Prob. 3-5?

3-7 According to Deep-River Jim's *Wilderness Trail Book*, "crickets are the woodsman's thermometer. To get the temperature accurately you need a watch with a second hand. Count the number of chirps that the crickets make in fifteen seconds, then add forty and there you have it!" (in Fahrenheit). It doesn't work below about 50°F because the crickets stop chirping. Is Jim's definition operational? Is it accurate?

3-8 Consider and discuss the implications of assigning the value of 100° to the ice point and 0° to the steam point.

3-9 Suppose the emf of a particular thermocouple is 4.2 mv at 100°C and 1.8 mv at 0°C.

(a) If this thermocouple reads 2.2 mv when exposed to an unknown temperature, what is the unknown temperature? State your assumptions.

(b) What is the unknown temperature when the emf reads 1.0 mv?

3-10 If in Prob. 3-9 you probably assumed the thermocouple emf and temperature are related linearly (as is commonly done), you probably made a slight error depending on the type of materials used in the couple. Suppose it is known that the emf and temperature are more accurately related according to emf $= a + bt^2$. What is the temperature at 2.2 volts?

3-11 Discuss the human hand as a thermometer. What does it actually measure when touching an object?

3-12 A mercury-in-glass thermometer is taken from a room at 70°F and put into a refrigerator. How would you establish when the thermometer was in thermal equilibrium with its new environment?

3-13 Why can't a thermometer in the sun be used to measure air temperature? What *does* it measure?

3-14 The following data are obtained from a constant volume gas thermometer for the boiling point of an unknown substance:

Pressure p_i	p/p_i
1000 mm Hg	1.05
800 mm Hg	0.95
600 mm Hg	0.86
400 mm Hg	0.78
200 mm Hg	0.71

Compute the unknown temperature.

3-15 Make a list of several isothermal (or nearly isothermal) processes that you have observed in nature, in the laboratory, in a factory, or at home. Also list some nonisothermal processes.

4

Equations of state

An equation of state is a relationship among the pressure (p), specific volume (v), and temperature (T) for a substance. Occasionally the relationship is expressed as an explicit mathematical *equation* (e.g., $pv = RT$). Often the relationship is too complex to be approximated accurately in analytical form and in such cases graphs, tables, or computers are used. Regardless of how the p-v-T relationship is presented, however, it is called an *equation of state.*

The equation of state for a substance is a fact of nature. What will be discussed in this chapter is how scientists and engineers record and describe the facts—using equations, tables, graphs, or the computer as convenience, accuracy, and the availability of data allow.

The engineer who would understand energy and how to use it must know something about the properties of substances as well as about energy. Often the key step in the solution to an energy problem is the decision about the equation of state to be used to help determine the properties of the substance involved. And how embarrassing it would be to use an ideal *gas* equation for a *liquid* or for a *non*ideal gas, or to have to admit that because we know so little about equations of state, we really don't understand how, say, a pressure cooker works.

As you study this chapter, here are some things to look for: What makes a property a property? What makes a pure substance pure? What do phase diagrams look like for real substances? What is meant by the "quality" of a two-phase mixture? When is $pv = RT$ valid? What are expansivity and compressibility? What is the compressibility factor? What is the law of corresponding states? What is meant by phase equilibrium? Why can't you get water temperature over 212°F or 100°C in an open kettle on the stove?

4-1 Properties

4-1(a) What is a property?

The word *property* has a very special and specific meaning. A property is identified by the fact that its change of value in a process is dependent only on the end states of the process and not on the nature of the process itself. The meaning of this can be illustrated symbolically as follows: If the quantity X is a property of the system, then its change in value when the system changes from state 1 to state 2 is the same whether process A or B is used for any and all processes A and B. That is,

$$\Delta X_{1A2} = \Delta X_{1B2}$$

where

$$\Delta X_{1A2} = (X_2 - X_1) \qquad \text{process } A$$

and

$$\Delta X_{1B2} = (X_2 - X_1) \qquad \text{process } B$$

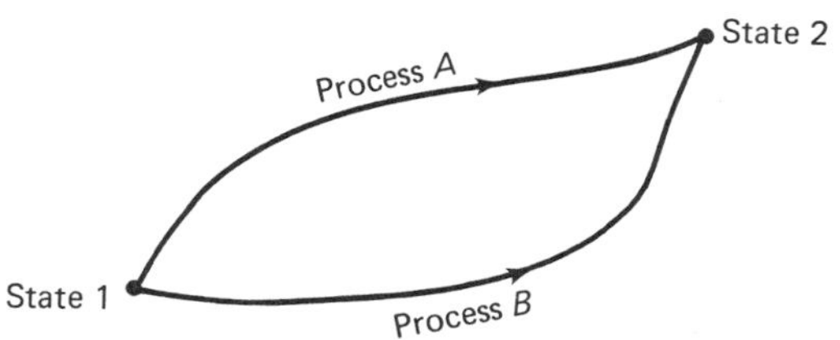

Some examples of properties are volume (V), temperature (T), pressure (p), and mass (M). As we shall see soon, heat and work are examples of *non-properties* because process A may require more or less heat (or work) to accomplish than process B, even though the end states are the same.

4-1(b) Extensive and intensive properties

Properties are classified as *extensive* or *intensive*. Examples of extensive properties are mass (M) and volume (V). Examples of intensive properties are pressure (p) and temperature (T).

The distinction between properties called *extensive* and those called *intensive* is as follows. Imagine a whole system divided into a number of parts. If the value of a property for the whole system is equal to the sum of its values for the various parts of the system, then it is called *extensive*. Notice how this "whole equals the sum of the parts" feature applies to extensive properties like mass and volume but not to intensive properties like pressure and temperature.

Another way to think about the difference between extensive and intensive properties is that intensive properties have meaning at a point or locally. That is, we can talk about the local pressure or temperature but not about the local mass or volume because these latter have no meaning at a *point*.

When an extensive property is divided by the total mass of the system, the result is called an *average specific property*. An example is the average specific volume:

$$\bar{v} = \frac{V}{M} \tag{4-1}$$

If the same process is carried out in limiting fashion in the region of a point, then the resulting specific property has local meaning and is an intensive property. Thus, the local specific volume is defined as

$$v = \lim_{\Delta V \to 0} \frac{\Delta V}{\Delta M} \tag{4-2}$$

and is an intensive property even though the total volume V is extensive. Note that the specific volume is the reciprocal of density ρ and may vary from point to point within a system. If a system has uniform properties throughout, then of course the average specific volume ($\bar{v}$) and the local specific volumes (v) will all be the same.

4-1 (c) Pure substances

A *pure substance* (also sometimes called a *single component system*) is defined as one that is uniform and invariant throughout in chemical composition. Thus, a steady state system of ice and water is called a pure substance because its chemical composition is the same everywhere, in both the solid and liquid portions, at all times. Air and other mixtures may be called pure substances as long as they remain uniform and invariant in chemical composition. An example of a system that may *not* be called a pure substance is an equilibrium mixture of liquid and gaseous air. In such a case, the liquid will be richer in one component and hence the system is not uniform in composition. Also systems undergoing combustion may not be called pure substances because their composition is not invariant with time.

We have introduced the concept of a pure substance for a particular reason. With the help of the second law (and hence we can't prove it yet), it can be shown that in the absence of magnetic, electrical, or surface tension effects *a pure substance has two, but only two,* ***independent intensive*** *properties.* That is, in most circumstances if *two* independent intensive properties are fixed and determined, then *all* other intensive properties are also fixed, and the state of the system is completely determined.

Though not proved (yet), the previous statement *is* consistent with our experience. The intensive properties we have introduced so far are p, v, and T. Consider a gas with its volume and temperature determined and fixed. Then we know that its pressure will also be determined. We can change the pressure, by changing either the temperature or volume, but *only* if we make *some* such change. The same is true of pure substances other than gases.

Mathematically, therefore, for a pure substance, we can write

$$p = p(T, v) \quad \text{or} \quad v = v(p, T) \quad \text{or} \quad T = T(p, v)$$

The above are equivalent to

$$f(p, v, T) = 0 \tag{4-3}$$

This last equation is the most general form of the equation of state. It says only that a p-v-T relation exists, but it says it elegantly.

Please notice that we have said the pure substance has but two *independent* intensive properties. There are very common circumstances (e.g., boiling, melting, or other phase changes) when pressure and temperature are *not independent* and hence may not *both* be counted as independent.

Most of the systems dealt with in this book are composed of pure substances.

4-2 Phase diagrams

4-2(a) Introduction

All of us are familiar with the usual solid, liquid, and gaseous *phases* that substances assume as their pressure, temperature, and volume are varied. Our experience is quite limited, however, compared with all that is known about phases of substances. From experience we know about the phases of water. We experience seeing liquid change to gas in the teakettle, liquid change to solid on the winter lake, gas change to liquid on the cold tumbler in summer, etc. Most of us recognize the "disappearance" of dry ice in the sink as the sublimation of carbon dioxide (CO_2) from solid to gas. We know about metals being melted and resolidifying as castings. And we have heard about liquid oxygen, nitrogen, and helium. This experience is valuable as a place to begin but we must not let it limit us, for there is more to learn about phases. The author knows from the unhappy experience of previous students that some students find certain concepts in the study of phases difficult. Often later troubles with solving problems using the first and second laws turn out to be caused by a lack of understanding of how phases and phase changes behave or of how to describe their behavior with equations or by using tables. These difficulties can be reduced by a careful and complete study of the material in this chapter. In particular, two things must be kept in mind by students who wish to avoid difficulties later with phases and phase changes. One is that the information reported here is in the nature of observed fact. Some of it will seem strange, even illogical, because it is outside normal experience. Don't let limited experience keep you from learning something new. Try not to be one of those who wants to be taught only what he or she already knows! Remember that all we are doing in this chapter is reporting some *facts of nature* to you that you may not have heard about before but which you will need later.

Another thing that will help avoid difficulties is for you to learn quickly and well the language and symbols used to talk about phases and phase diagrams; then you will be able to understand the words being used when something is explained to you. The next two sections introduce many of the new words and concepts you will need to know and understand.

4-2(b) Phase diagrams

Phase diagrams show regions of properties where the different phases of a substance occur and they show the lines where phase changes take place.

For example, if pressure and temperature are taken as the coordinates, most substances have a phase diagram of the general form shown in Fig. 4.1(a). Another phase diagram, this one with pressure and volume as coordinates, is shown in Fig. 4.1(b).

Students should understand that *all* substances have phase diagrams more or less of the general shape of those in Fig. 4.1. Of course, the numerical values of the coordinates will be vastly different for different substances but all substances have these phases in more or less the same arrangement. Thus, the terms we shall define apply very generally.

The lines on the diagrams with arrowheads extend as far as we can measure. In Fig. 4.1(a), line *a-b*, however, terminates at a point *b* called the *critical point*. (To help you become familiar with it and remember it, label it on the diagram.)

Here now is a fact of nature that some people find hard to accept at first. Consider a substance that starts at the letter L in Liquid on the *p-T* diagram, goes up and around point *b* so as not to cross line *a-b* and ends at the letter G in Gas. Note that this process has taken the substance from a liquid to a gas without any *distinct* point of change between the two phases. The process did not cross a phase change line. This is *not* what happens when you boil water on a stove but it still can happen if you control p and T to follow the path from L to G around and above the critical point. Point *b* is called the *critical* point because above that pressure, or temperature, the change from liquid to gas or vice versa is continuous rather than discontinuous as in the usually observed phase change. This seems strange because in our daily lives we don't see this taking place but it can be done in the laboratory and does occur in some engineering processes.

In Fig. 4.1(a), point *a* is called the *triple point*. Please also label it on your diagram. It is called the triple point because it connects regions of the *three* phases: solid, liquid, and gas.

The processes of passing from liquid to gas, solid to liquid, and solid to gas and vice versa all have names that should be known. Liquid to gas: evapora-

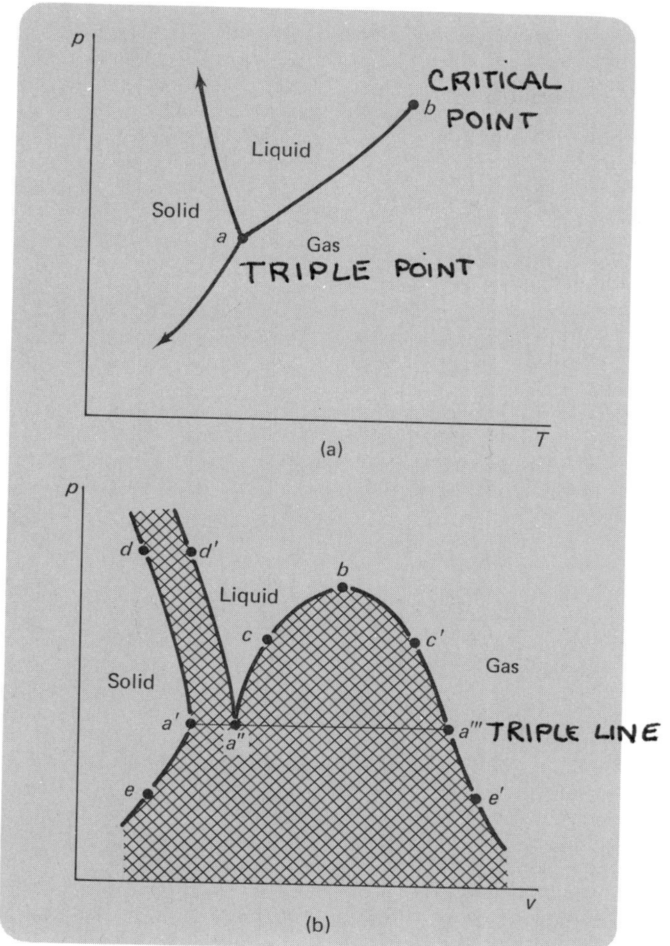

Figure 4.1 Phase diagrams.

tion or boiling; solid to liquid: melting; liquid to solid: freezing; gas to liquid: condensation; solid to gas: sublimation.

In Fig. 4.1(b), the same data are shown but with pressure and volume as independent properties. Note the location of the critical point b. Also note that point a from the p-T plot is a line (a'-a''-a''') on the p-v plot. It is now called the triple *line*. Also, all the *lines* in the p-T plot become *planes* in the p-v diagram. See Fig. 4.2 where the p-v-t surface is shown in three dimensions.

It can be seen that the phase change from a liquid to a gas below the critical point is really a skip from c to c' though this is seen as continuous on the p-T plot where c and c' coincide. Substances do not actually exist on the planes that are crosshatched in the figures but they do exist along the lines bounding

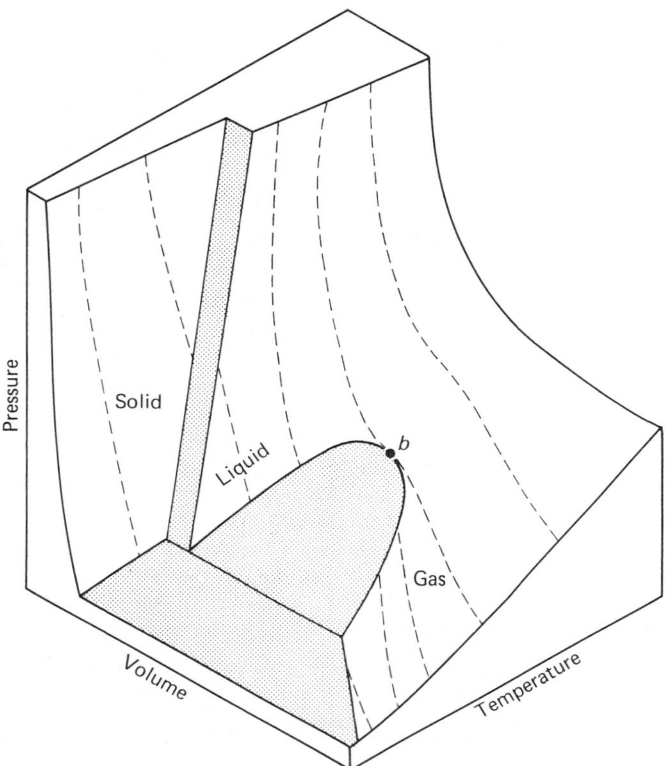

Figure 4.2 A typical phase diagram in three dimensions. The dotted lines are isotherms.

these planes. These lines are called *saturation lines*. Depending on whether the line borders the solid, liquid, or gaseous state, it represents a *saturated* solid, *saturated* liquid, or *saturated* gas (or vapor), respectively. That is, in Fig. 4.1(b) the line a''-b is called the saturated liquid line and line b-a''' is called the saturated vapor line. (Notice that line a''-d' is also a saturated liquid line.)

4-2(c) Phase equilibrium

The saturated states—those that exist along one of the lines on the p-v plot—can coexist with corresponding saturated states *at the same pressure*. That is, saturated liquid at c can coexist with saturated vapor at c'. This coexistence is called *two-phase* liquid-vapor *equilibrium* or sometimes it is said that the vapor is in equilibrium with its liquid or vice versa. Such phase equilibrium can also take place with solid and vapor (e and e') and with solid and liquid (d and d').

Adding or removing heat from an equilibrium mixture of phases at constant pressure causes the relative amounts of mass in the two phases to change. Thus, if heat is added to a constant pressure container that contains a mixture of saturated liquid at c and saturated vapor at c', more of the vapor will be formed. As long as both phases are present and in contact with each other and the process is not too rapid, the phases will remain in equilibrium and the only effect of the heat addition will be the evaporation of some of the liquid to form gas. As long as there is liquid present, the temperature of neither phase can go above, or below, the equilibrium temperature and the pressure of the mixture also remains constant.

The concept of *phase equilibrium* is an especially difficult but important one. Engineers must recognize that when two phases of a substance are present, unless there is some rapid local process upsetting equilibrium, there will be essentially constant temperature and pressure. An example of this occurs when water is boiled in a teakettle on the stove. Once the first gas formed drives the air out of the kettle, the kettle thereafter contains a two-phase *equilibrium* (very nearly) mixture of liquid and gaseous H_2O. Since pressure and temperature are not independent, and pressure in this case is fixed by atmospheric pressure, the temperature also remains constant at about 212°F or 100°C until all the liquid is gone. Then, and only then, do you melt the kettle.

Another example is illustrated by the design of steam generators for power plants. There is a *boiler* in which the liquid is converted to gas but since the gas formed there is saturated vapor and is in the presence of the liquid, it cannot be heated above the two phase equilibrium temperature corresponding to the boiler pressure. Therefore, the vapor generated is taken from the boiler to a unit called a *superheater* where there is no liquid and where it can be heated to a much higher temperature.

Equally practical examples of two-phase equilibrium mixtures in action can be found in refrigeration systems and in many other plant processes. As we said, engineers must recognize two-phase equilibrium situations, know that this means p and T are not independent, and be able to understand what can or cannot happen.

We have said that substances do not exist in the crosshatched regions of p-v-type phase diagrams. This is true but it is nevertheless common practice to call these *two-phase regions* and to represent two-phase mixtures of the saturated liquid and saturated vapor at the same pressure by points in between the saturation lines. Thus, in Fig. 4.3(a) a mixture of saturated liquid at a and saturated vapor at b is shown at some intermediate point c. In Fig. 4.3(b) the process of cooling saturated vapor at constant volume from d to a lower pressure (and temperature) is drawn as a straight downward vertical line to e, though we understand the final state is *really* an equilibrium *mixture* of saturated liquid at f and saturated vapor at g.

We shall show how to handle these two-phase equilibrium mixtures more quantitatively in the next section.

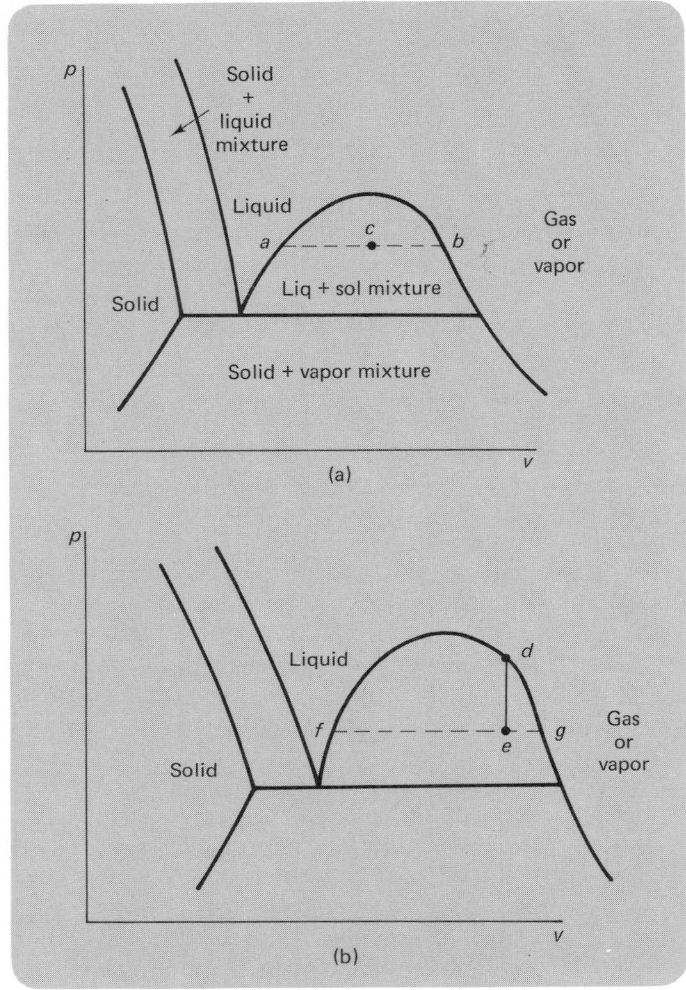

Figure 4.3 Phase diagrams showing use of two-phase region.

4-2(d) Phase symbols and notation

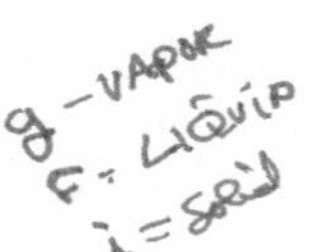

For many substances, tables exist that give the properties along the various saturation lines. In order to use these tables, students must be familiar with the meaning of terms such as *saturated liquid* and *saturated vapor*. Also, it should be noted that the gas or vapor region is often called *superheated* vapor and that the liquid (or solid) regions are called *subcooled* or compressed liquid (or solid). These terms are shown on Fig. 4.4. The subscripts g, f, and i are usually used to denote properties of the saturated vapor, liquid, and solid states, respectively. That is, when referring to the specific volume of the substance that is saturated liquid, v_f is used; for a saturated vapor, v_g is used. The difference between v_g and v_f is denoted by v_{fg}. That is, $v_{fg} = v_g - v_f$.

It is helpful to look at the general shape of lines of constant temperature on the p-v plot. These are shown in Fig. 4.5. Students should note that in a

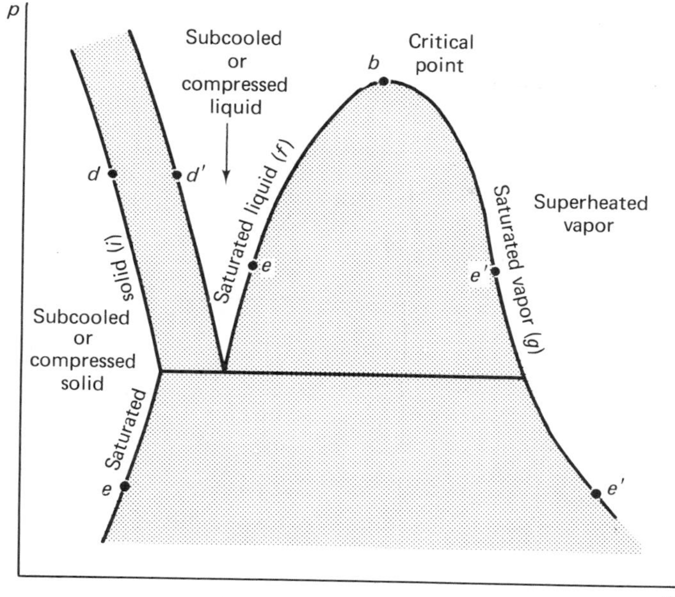

Figure 4.4 Pressure volume-phase diagram showing definitions of terms.

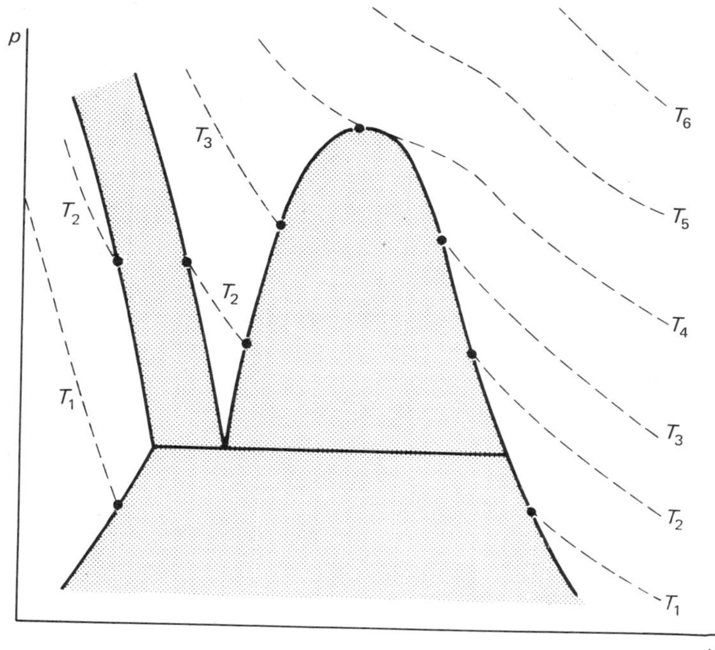

Figure 4.5 Pressure-volume phase diagram showing isotherms.

constant pressure change of phase (e.g., from e to e') there is no change in temperature. This is apparent on the p-T plot and is also known from experience with boiling water—it won't get over 212°F or 100°C until all the liquid is gone. The point to remember here is that *p and T are not independent in a two-phase mixture.*

In an equilibrium two-phase mixture of liquid and vapor, the term *quality* is used to describe the fraction *by mass* of vapor in the mixture. Thus, a mixture of $\frac{1}{2}$ lbm of vapor and $\frac{1}{2}$ lbm of liquid has a quality of 0.5. Saturated vapor has a quality of 1.0; the quality of saturated liquid is 0. In terms of symbols,

$$\text{Quality} = x = \frac{M_g}{M} = \frac{M_g}{M_g + M_f} \tag{4-4}$$

where M_g = mass of vapor
M_f = mass of liquid
M = total mass, liquid plus vapor ($M = M_f + M_g$)

The quality of a two-phase mixture is often an important factor in thermodynamic applications (e.g., the quality in a steam turbine must be kept high to reduce erosion of the turbine blades by the liquid droplets). The total volume V of a two-phase mixture is the sum of the volumes of the liquid portion V_f and the vapor portion V_g.

$$V = V_f + V_g \tag{4-5}$$

Noting that total volume is the specific volume times the mass, we may write

$$V = Mv = M_f v_f + M_g v_g \tag{4-6}$$

where the *un*subscripted symbols stand for the mass and specific volume *of the mixture* and the subscripted symbols stand for the liquid and vapor phases. That is,

$$M = M_f + M_g \qquad v = \frac{V_f + V_g}{M_f + M_g} = \frac{V}{M} \tag{4-7}$$

where V_f = volume of saturated liquid
V_g = volume of saturated vapor

Dividing Eq. (4-6) through by M gives

$$v = \frac{M_f}{M} v_f + \frac{M_g}{M} v_g \tag{4-8}$$

Noting that $M_f/M = (1 - x)$ and $M_g/M = x$, we obtain

$$v = (1 - x)v_f + xv_g \tag{4-9}$$

or, taking into account the notation $v_{fg} = v_g - v_f$, this can be rewritten as

$$v = v_f + xv_{fg} \tag{4-10}$$

These equations are often useful in problems that involve two-phase mixtures.

It should be noted that quality is an intensive property of a two-phase system.

EXAMPLE 4.1. Sketch the following processes on the *p-v* and *p-T* diagrams:

1. The sublimation of dry ice (solid CO_2).
2. A constant pressure cylinder of superheated vapor is cooled until liquid just begins to form.
3. A constant pressure cylinder of superheated vapor is cooled until all the vapor is gone.
4. A liquid–vapor two-phase mixture is heated at constant volume until its quality is 1.0.

SOLUTION:

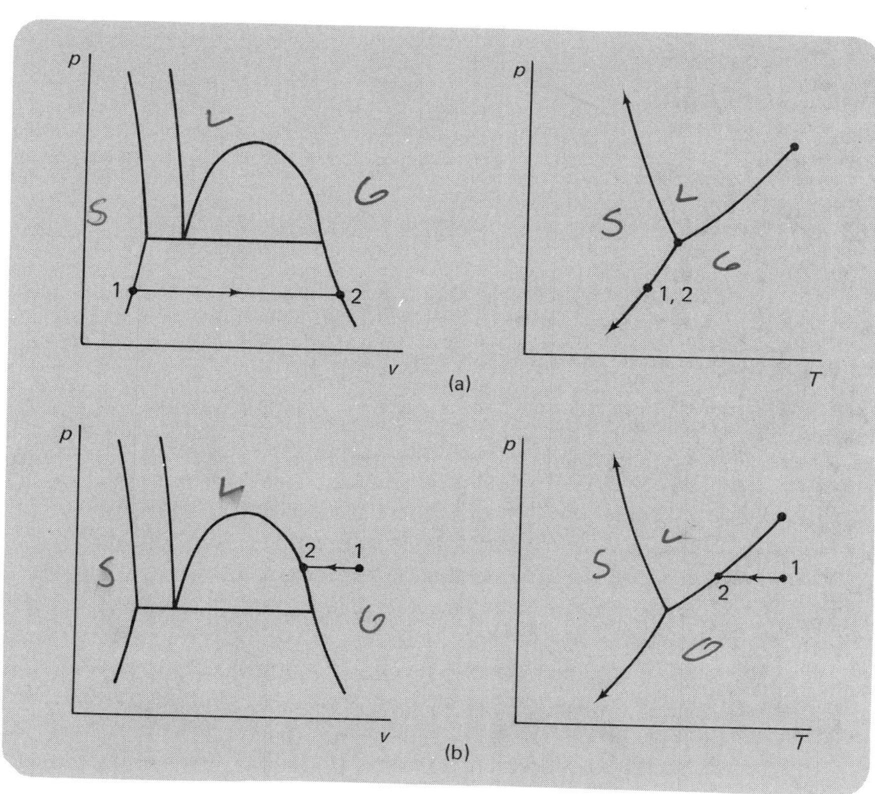

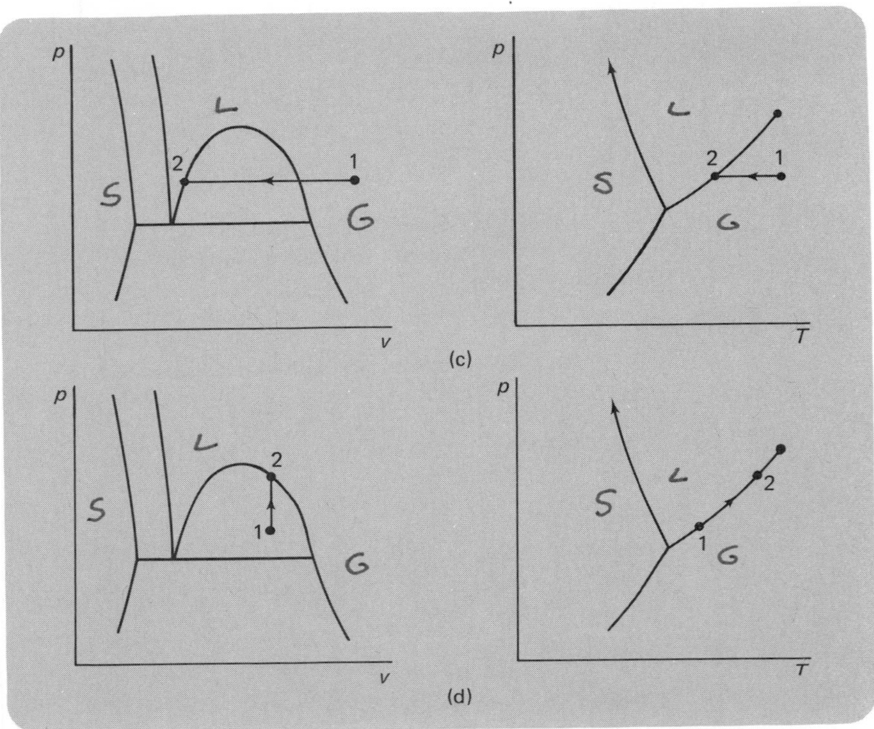

4-3 Tabulated equations of state

The p-v-T relationship for most substances is very complex in the region of the various saturation lines. A mathematical equation, even if one can be found, is usually too complex for practical everyday use. As a result, a great deal of data is presented in tables.

Most tables of properties are entered by knowing either the temperature or the pressure or both. In these cases, we call temperature and/or pressure the *table-entry property*. For superheated substances, the tables are usually as shown in Figs. 4.6 or 4.7. In either case, at a given pressure and temperature one can look up the specific volume (v) and other properties we shall use later such as enthalpy (h), entropy (s), and internal energy (u).

For data along the saturation lines, remember that pressure and temperature are not independent of each other. That is, unlike in the superheated region, if one is fixed, the other is determined. As a result, saturation line data

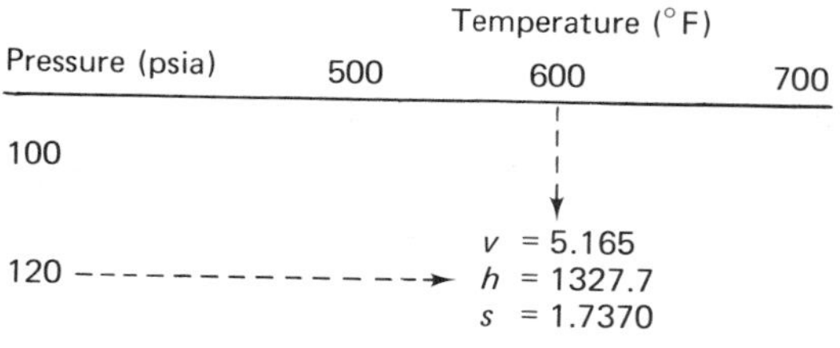

Temperature (°F)
Pressure (psia)
500
600
700
100
120
v = 5.165
h = 1327.7
s = 1.7370
140

Figure 4.6 An example of a typical table of superheated property data.

	p = 28 bars				p = 30 bars			
t (°C)	v	u	h	s	v	u	h	s
240								
250	76.47	2651.6	2865.7	6.3343	70.58	2644.0	2855.8	6.2875
260								

Figure 4.7 Another example of a typical table of superheated property data. Specific volume is in cubic meters per gram.

are usually presented in two tables; one with temperature as the table-entry property and one with pressure as the table-entry property. These are illustrated in Fig. 4.8 and Fig. 4.9.

Only a few values are shown in Figs. 4.6, 4.7, 4.8, and 4.9 in order to make clear the structure of such tables. More complete tables are found in Appendix B.

It is sometimes necessary to use interpolation to obtain the desired data from tables. Unless otherwise stated, data are normally assumed to vary linearly between entry points. This makes the interpolation relatively easy but, of course, introduces some error. Students should especially use care

Temperature (°C)	Pressure (bars)	Specific volume		Internal energy		Enthalpy			Entropy		
		v_f	v_g	u_f	u_g	h_f	h_{fg}	h_g	s_f	s_{fg}	s_g
120											
130	2.701	1.0697	668.5	546.02	2539.9	546.31	2174.2	2720.5	1.6344	5.3925	7.0269
140											

Figure 4.8 An example of another typical table of saturation line property data, with temperature as the table-entry property. Specific volume is in cubic meters per gram.

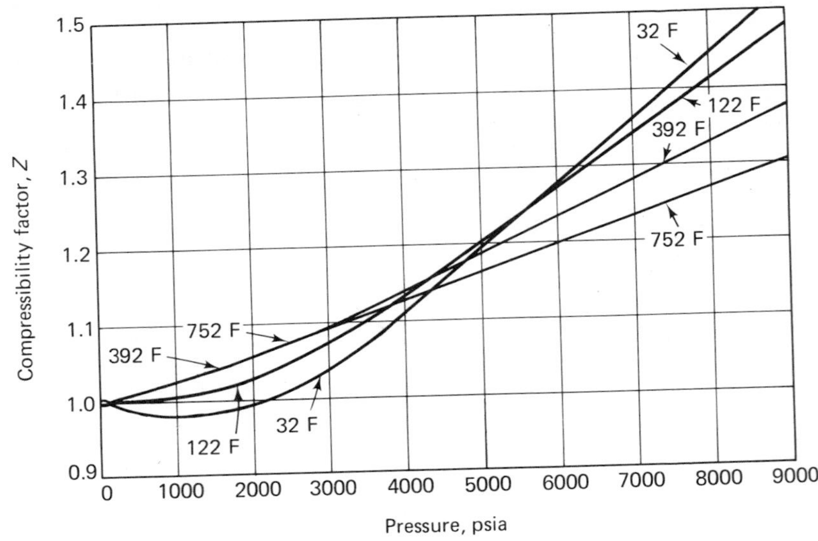

Figure 4.9 Approximate compressibility factors for nitrogen. (Reproduced from *Engineering Thermodynamics* by Jones and Hawkins by permission of John Wiley & Sons, Inc.)

when interpolating for actual professional use from condensed tables (like these in this book!) with widely spaced table-entry values. Always try to use the complete tables from which such condensations are made unless accuracy is not important.

To check their understanding of the table construction, students should find the following data in Figs. 4.6 through 4.9:

1. The specific volume of superheated steam at 120 psia and 600°F.
2. The specific volume of superheated steam at 250°C and 30 bars.
3. The specific volume of saturated vapor at 130°C.
4. The saturation pressure at 130°C.
5. The specific volume of saturated vapor at 120 psia.

The correct answers are (1) 5.165 ft^3/lbm, (2) 70.58 cm^3/gm, (3) 668 cm^3/gm, (4) 2.701 bars, and (5) 3.474 ft^3/lbm. The last value is found by interpolation. The value found from the more complete tables at 130 psia is 3.455 ft^3/lbm. The error introduced by the linear assumption is thus $(3.474 - 3.455)/3.455 = 0.0055$ or about 0.5%.

Several sets of H_2O data (steam tables) now exist. One is an older volume authored by Keenan and Keyes and published by John Wiley & Sons, Inc. Figures 4.6 and 4.8 are from these tables. Another text is published by the American Society of Mechanical Engineers (ASME). Figures 4.7 and 4.9 and the H_2O data in Appendix B are condensed from a more recent version of *Steam Tables* by Keenan, Keyes, Hill, and Moore, also published by John Wiley & Sons, Inc. These tables are available in both English and metric units.

Data for H_2O are available on some computers so that problems can be worked without the numerical "dog-work" and chances for error associated with humans and tables of properties.

Tables also exist for properties in the liquid region (sometimes called *subcooled*) and solid region. Students must be able to learn to use such tables as needed.

EXAMPLE 4.2. At the close of a trial run of a process you have designed, a large tank contains steam at a gauge pressure of 100 psia and a temperature of 400°F. The lines to and from the tank are closed and the tank allowed to cool until it reaches room temperature, 80°F. At this time, the pressure gauge reads 14.0 psi *vacuum* and the thermometer reads 80°F. The barometer reads 14.5 psi. Before starting another run, you want to know how much, if any, liquid (by mass and volume) is in the tank.

SOLUTION: Translation into symbols:

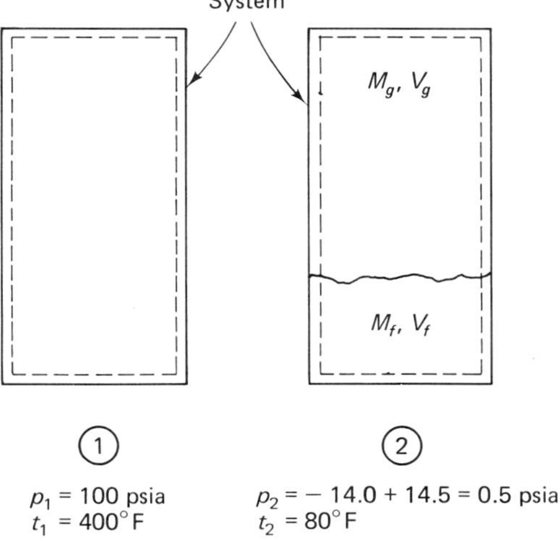

System: The H_2O in the tank.
Equation of State: Use steam tables.
Process Equation: Constant volume cooling, $1 \rightarrow 2$.

We note in the tables that at 80°F the corresponding two-phase equilibrium pressure is 0.5073 psia. Allowing for slight inaccuracies in our instruments, it seems likely that we have a two-phase mixture at state 2. This enables us to draw the process on the p-v plot as follows:

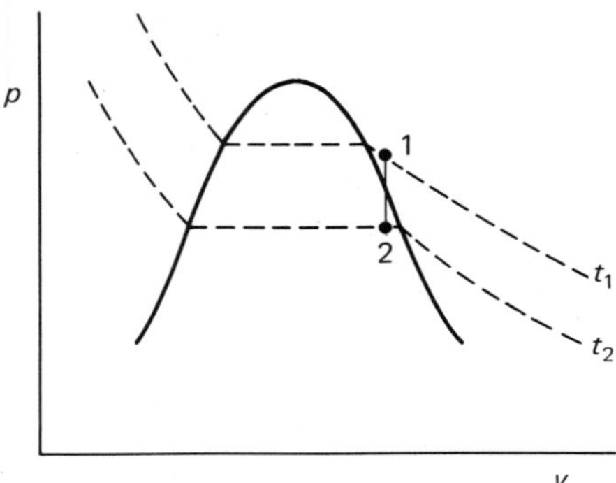

Application of Fundamentals:
Conservation of Mass:

$$M_1 = M_2$$

$$\frac{V}{v_1} = \frac{V}{v_2}$$

or

$$v_1 = v_2$$

where v_2 is the average specific volume of the two-phase mixture. We can find v_1 from the superheated tables; $v_1 = 4.934$ ft³/lbm. From the definition of quality,

$$v_2 = v_{f2} + x_2 v_{fg2} = v_1$$

and at 80°F the tables give $v_{f2} = 0.0161$ ft³/lbm and $v_{fg2} = 632.8$ ft³/lbm. Solving for x_2 gives

$$x_2 = \frac{v_2 - v_{f2}}{v_{fg2}} = \frac{4.934 - 0.016}{632.8} = 0.0078$$

With the quality known we have the percentage by mass by definition since

$$x = \frac{\text{mass of vapor phase}}{\text{total mass of mixture}} = \frac{M_g}{M_f + M_g} = \frac{M_g}{M}$$

We can also find the percentage of liquid by volume as follows:

$$\frac{v_f}{v} = \frac{M_f v_{f2}}{M v_2} = (1 - x)\frac{v_{f2}}{v_2}$$

$$\frac{v_f}{v} = (0.992)\frac{0.016}{4.934} = 0.00322 \text{ ft}^3 \text{ liquid/ft}^3$$

Thus, our answers show that by mass the tank contains 99.2% liquid but by volume it contains 99.7% vapor. Does this result seem reasonable to you?

4-4 Gases

The gaseous phase is an important one because many processes or even entire systems operate with gases alone. Equations of state for gases are typically presented either graphically or by mathematical approximations.

4-4(a) Graphical representation

Phase diagrams include gaseous regions but are usually presented in order to show the relationship among the phases. For gases alone, a common graphical representation of the relationship among p, v, and T is a plot of

Substance	t_c		p_c		v_c	
	°F	°C	psia	bars	ft³/lbm	cm³/gm
H_2O	705	374	3206	221	0.050	3.12
Air	−220	−140	547	38	0.046	2.87
H_2	−400	−240	189	13	0.520	32.5
He	−450	−268	34	2.3	0.230	14.6
SO_2	315	157	1142	79	0.031	1.9
CO_2	88	31	1073	74	0.035	2.18

Figure 4.10 Some critical point data.

pv/RT versus pressure. An example is shown in Fig. 4.10. The ratio pv/RT is given the symbol Z and called the *compressibility factor*.

Approximate equations of state for substances when specific data are unavailable can be found by using what are called *reduced coordinates* and the *law of corresponding states*. The reduced coordinates are defined in ratio to the values of p, v, and T at the critical point. That is, reduced pressure is the actual pressure divided by the critical pressure, etc.

$$P \text{ reduced} = P_r = \frac{p}{p_c} \qquad T_r = \frac{T}{T_c} \qquad v_r = \frac{v}{v_c} \tag{4-11}$$

Some critical point data for substances are shown in Fig. 4.11.

The law of corresponding states says that all substances have the *same* equation of state in terms of reduced properties. This is equivalent to saying that all p-v diagrams are generally alike except for scale factors. Compressibility factor is then plotted against p_r in a generalized plot that applies approximately to all gases at relatively low pressures. See Fig. 4.9. Because actual substances do not have *exactly* similar reduced equations of states, however, the generalized graph can give errors up to about 15%. It should be used with appropriate caution.

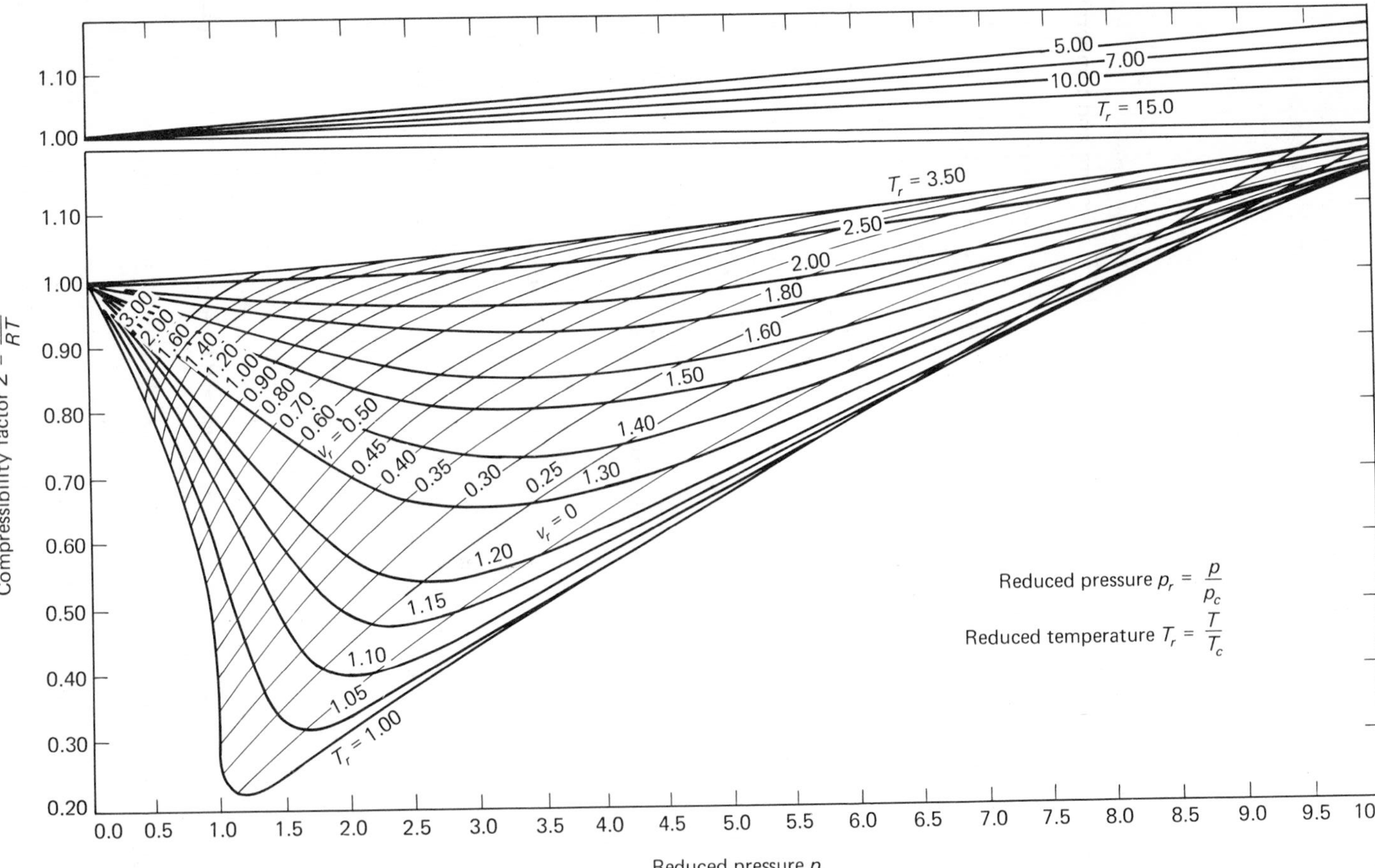

Figure 4.11 Generalized compressibility chart. (From Nelson and Oberts "Generalized *p-v-T* Properties of Gases," *Trans ASME*, 76, 1954, p. 1057.)

4-4(b) Mathematical representation

The most famous, most used, and most misused mathematical equation of state is the ideal gas equation: $pv = RT$. Its origin is suggested by a look at Fig. 4.5. In Fig. 4.11, note that as the pressure becomes very small compared with the critical pressure, the value of pv/RT for all substances approaches unity. Also note in Fig. 4.5 that as higher and higher temperatures are approached, the lines of constant temperature approach hyperbolas (xy = constant), the equation for which on a p-v plot would be $pv \propto T$. Apparently, then at "low" pressures or "high" temperatures, where "low" and "high" mean in comparison with the critical pressure and temperature, respectively, many gases behave approximately according to $pv = RT$. Remember, there are *no real* ideal gases. $pv = RT$ is merely a model that often works well enough for our purposes.

As a brief review, we note that on a per *mole* basis, the gas constant is a universal constant. That is, in

$$pV = n\bar{R}T \tag{4-12}$$

where n is the number of moles, the constant

$$\bar{R} = 1545 \frac{\text{ft-lbf}}{\text{lbmole-°K}} = 8312 \frac{\text{joules}}{\text{kgmole-°K}} \tag{4-13}$$

is the same for *all* gases. Dividing and multiplying by molecular weight, MW = lbm/lbmole or kg/kgmole for a given gas gives

$$pV = (n)(\text{MW})\left(\frac{\bar{R}}{\text{MW}}\right)T$$

$$pV = MRT \tag{4-14}$$

where M is the number of pounds or kilograms and R is the gas constant for the given gas. Dividing by M gives

$$p\frac{V}{M} = RT$$

or

$$pv = RT \tag{4-15}$$

In the analysis of many processes involving ideal gases or nearly ideal gases, it is extremely important to be able quickly to sketch the process on a property diagram. For this purpose one must have readily in mind the general shape of, say, constant temperature lines on a p-v plot, constant volume lines on a p-T plot, etc., for ideal gases. These are shown in Fig. 4.12. See also Prob. 4-9.

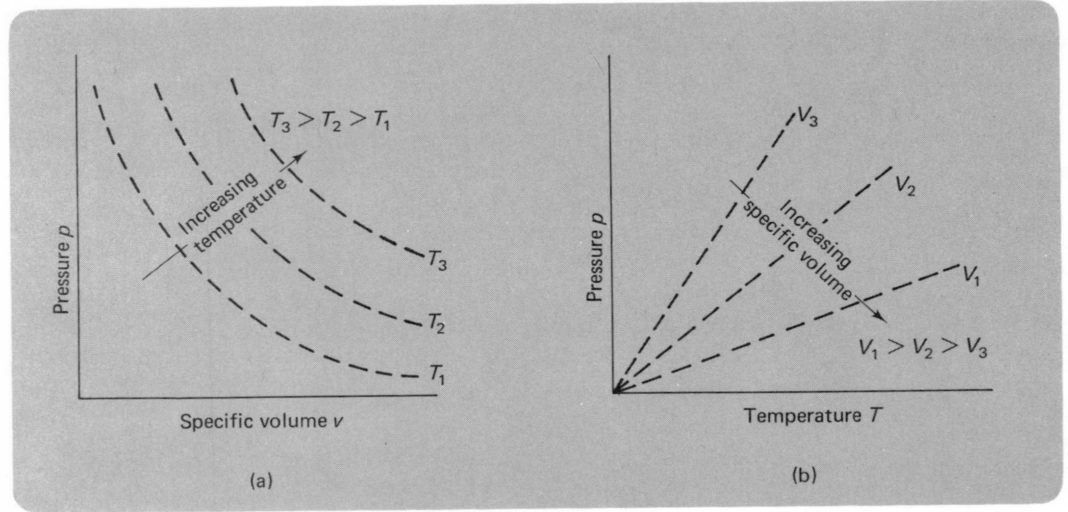

Figure 4.12 A pressure-specific volume plot (a) and a temperature-specific volume plot (b) for an ideal gas.

Essentially, $pv = RT$ is a good approximation to the equation of gases when $p_r < 1$ and/or $T_r > 1$. Often, however, it is necessary to have a mathematical equation of state that has a wider range of applicability. The trouble is that the relationship among p, v, and T is extremely complex for substances near the saturated regions and hence the approximate equations that might do better than $pv = RT$ become quite complex. One that is well-known is called the van der Waals equation:

$$\left(p + \frac{a}{v^2}\right)(v - b) = RT \tag{4-16}$$

where a and b are constants different for each gas. Another equation more accurate over a wider range is the Beattie–Bridgman equation:

$$p = \frac{RT(1 - \varepsilon)}{v^2}(v + B) - \frac{A}{v^2} \tag{4-17}$$

where $A = A_0(1 - a/v)$; $B = B_0(1 - b/v)$; $\varepsilon = C/vT$; and A_0, a, B_0, b, and C are constants different for each gas. A somewhat more general form is the virial equation:

$$pv = A + \frac{B}{v} + \frac{C}{v^2} + \cdots \tag{4-18}$$

where A, B, C, ... are functions of temperature difference for each gas. Engineering students need not make any attempt to memorize these but should remember that they exist and note that the high-speed digital computer makes their use feasible if the need arises despite their complication.

EXAMPLE 4.3. Oxygen is contained in a 0.1-m³ cylinder at 50°C and 23 bars gauge. Atmospheric pressure is 1 bar. Compute the pounds of oxygen in the tank.

SOLUTION: Translation:

p = 23 bars gauge	
t = 50°C	p_0 = 1 bar

System: O_2 in tank—closed.
Equation of State: Assume ideal gas.

$$pV = MRT$$

$$(23 + 1)10^5(0.1) = M\left(\frac{8312}{32}\right)(50 + 273)$$

Units:

$$(\text{bars})\frac{(\text{N/m}^2)}{(\text{bar})}(\text{m}^3) = (\text{kg})\left(\frac{j}{\text{kg-}^\circ\text{K}}\right)(^\circ\text{K})$$

$$\text{Nm} = \text{j}$$

$$M = \frac{(32)(24)(10^4)}{(8312)(323)} = 2.86 \text{ kg}$$

EXAMPLE 4.4. How well does $pv = RT$ for saturated H_2O vapor?

SOLUTION: Since, in general, the ideal gas equation does not work at all well for substances in the vicinity of their two-phase region, we would not expect it to work well for H_2O right on the very edge of the region. We do some calculations, however, to confirm our suspicions. Let us solve for pv/RT at several different pressures, using the corresponding two-phase equilibrium temperature. At $p = 2000$ psia,

$$\frac{pv}{RT} = \frac{(2000)\,(144)\,(0.1878)}{(1545/18)\,(635.8 + 459.6)} = 0.575$$

For other pressures, we find

p (psia)	p (bars)	pv/RT
1000	68.9	0.745
100	6.89	0.944
10	0.689	0.990
1	0.069	0.995
0.5	0.035	0.998

Thus, we see that though the use of $pv = RT$ for saturated H_2O vapor at high pressures is very inaccurate, if the pressure is low enough compared with critical pressure (3206 psia), the assumption is not a bad one.

This result has a very important application. In air conditioning work, the partial pressure of the water vapor in the atmosphere is usually well below 1 psia (0.069 bars) and so the ideal gas equation is used with great accuracy *and* great savings in computational work.

EXAMPLE 4.5. An ideal gas executes the following processes: isothermal expansion, followed by heat removal at constant volume, followed by constant pressure heat removal to the initial volume. Sketch these processes on p-v, p-T, and T-v plots.

SOLUTION:

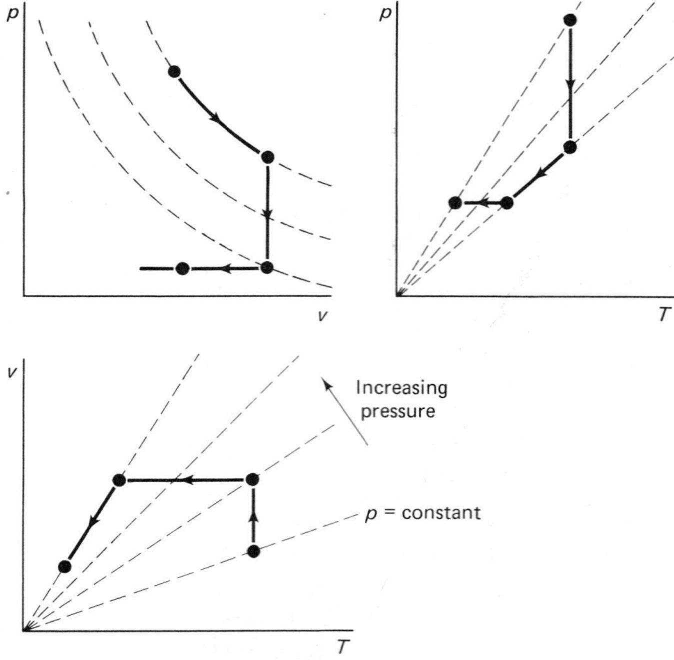

EXAMPLE 4.6. For nitrogen gas at 300°R and $v = 1.1169$ ft³/lbm, what is the pressure as found by

1. Ideal gas equation?
2. Van der Waals equation?
3. Beattie–Bridgeman equation?
4. A table or graph for nitrogen?

SOLUTION:

1.

$$pv = RT$$

$$p = \frac{RT}{v} = \frac{(1545/28)(300)}{(1.1169)(144)} = 102.9 \text{ psia}$$

2.

$$\left(p + \frac{a}{v^2}\right)(v - b) = RT$$

For nitrogen,

$$a = 346 \frac{\text{atm-ft}^6}{(\text{lbmole})^2} - 7.315 \times 10^5 \frac{\text{lbf-ft}^4}{(\text{lbmole})^2}$$

$$b = 0.618 \text{ ft}^3/\text{lbmole}$$

Solving for p gives $p = 102.4$ psi.

3.

$$p = \frac{RT}{v^2}(1 - \varepsilon)(v + B) - \frac{A}{v^2}$$

where

$$\varepsilon = \frac{C}{vT^3}$$

$$A = A_0\left(1 - \frac{a}{v}\right)$$

$$B = B_0\left(1 - \frac{b}{v}\right)$$

For nitrogen,

$$C = 391.7 \times 10^4 \frac{\text{ft}^3\text{-R}^3}{\text{lbmole}}$$

$$A_0 = 344.3 \frac{\text{atm-ft}^3}{(\text{lbmole})^2}$$

$$a = 0.419 \text{ ft}^3/\text{lbmole}$$

$$B_0 = 0.809 \text{ ft}^3/\text{lbmole}$$

$$b = -0.111 \text{ ft}^3/\text{lbmole}$$

Solving for p gives $p = 99.15$ psia.

4. From "Thermodynamic Properties of Nitrogen" by O. T. Bloomer and K. N. Rao, Institute of Gas Technology Research Bulletin No. 18, Chicago, 1952, we find

$$p = 100 \text{ psia}$$

In summary, at $T = 300°\text{R}$ and $v = 1.1169 \text{ ft}^3/\text{lbm}$, we found the pressure to be as follows:

Ideal gas:	$p = 102.9$ psia
Van der Waals:	$p = 102.5$ psia
Beattie–Bridgeman:	$p = 99.15$ psia
Tabulated:	$p = 100$ psia

4-4(c) Gas tables

In addition to mathematical equations and graphs, some properties of gases are also given in tabular form. Substances in the superheated state are gases so tabulations of their properties are a kind of gas table. When a gaseous substance is in a state near its two-phase region, however, it is common practice to call it a *vapor* rather than a *gas*. The term *gas* then is generally reserved for substances like air, hydrogen, carbon monoxide, etc., under conditions where they do not condense. This distinction between gas and vapor is not rigorous or formal and exceptions exist. For example, many refrigeration engineers use *gas* to describe certain states of a refrigerant in a refrigeration cycle even though the substance is not far from its two-phase region.

In any event, tables such as those discussed in Sec. 4-3 and found in Appendices B-3 and C-3 are not called gas tables even though they present properties of what, strictly speaking, may be called gases. Tables of gas (air, hydrogen, etc.) properties do exist, however. The most well-known is the *Gas Tables* by Keenan and Kaye published by John Wiley & Sons, Inc., New York. These tables assume that the gases included are ideal gases obeying $pv = RT$ and so, of course, they cover a range where that assumption is a good one. We shall have more to say about the use of these gas tables later.

4-5 Introduction to kinetic theory of ideal gases

Students who wish to learn about the kinetic theory model of ideal gases should now study Sec. 17-1 and 17-2.

4-6 Expansivity and compressibility

4-6(a) The chain rule

Given a relation $f(x, y, z) = 0$, it can be written as

$$x = x(y, z) \quad \text{or} \quad y = y(x, z)$$

Differentiating x and y according to the rules of partial differentiation,

$$dx = \left(\frac{\partial x}{\partial y}\right)_z dy + \left(\frac{\partial x}{\partial z}\right)_y dz \tag{4-19}$$

$$dy = \left(\frac{\partial y}{\partial x}\right)_z dx + \left(\frac{\partial y}{\partial z}\right)_x dz \tag{4-20}$$

Eliminating dy and rearranging gives

$$1 - \left[\left(\frac{\partial x}{\partial y}\right)_z \left(\frac{\partial y}{\partial x}\right)_z\right] dx = \left[\left(\frac{\partial x}{\partial y}\right)_z \left(\frac{\partial y}{\partial z}\right)_x + \left(\frac{\partial x}{\partial z}\right)_y\right] dz \tag{4-21}$$

Now remember that two properties may be selected as independent (say, x and z) and thus the equation above can be true only if the coefficients of dx and dz are zero (since dx and dz may themselves be anything). Setting each equal to zero, we find

$$\left(\frac{\partial x}{\partial y}\right)_z = \frac{1}{(\partial y/\partial x)_z} \tag{4-22}$$

and

$$\left(\frac{\partial x}{\partial y}\right)_z \left(\frac{\partial z}{\partial x}\right)_y \left(\frac{\partial y}{\partial z}\right)_x = -1 \tag{4-23}$$

In terms of p, v, and T, this last result can be written as

$$\left(\frac{\partial p}{\partial v}\right)_T \left(\frac{\partial T}{\partial p}\right)_v \left(\frac{\partial v}{\partial T}\right)_p = -1 \tag{4-24}$$

4-6(b) Expansivity (β) and compressibility (κ)

If an equation of state is the relation among p, v, and T, then the partial derivatives $(\partial p/\partial v)_T$, $(\partial T/\partial p)_v$, $(\partial v/\partial T)_p$ above are certainly a part of or significant to the equation of state because they involve p, v, and T. What is perhaps more important is that they are readily measurable and hence often available for solids, liquids, and gases. The derivatives themselves are not reported, however, but rather the data are given in terms of the coefficient of expansion β and the coefficient of compressibility κ.

The coefficient of expansion β is a measure of the way a substance changes volume as a result of a temperature change at constant pressure. Symbolically it is defined as

$$\beta = \lim_{\Delta T \to 0} \left(\frac{1}{V} \frac{\Delta V}{\Delta T} \right) = \frac{1}{V} \left(\frac{\partial V}{\partial T} \right)_p = \frac{1}{v} \left(\frac{\partial v}{\partial T} \right)_p \tag{4-25}$$

The compressibility is a measure of the way in which a substance changes volume as a result of a pressure change at constant temperature. Symbolically,

$$\kappa = \lim_{\Delta p \to 0} \left(- \frac{1}{V} \frac{\Delta V}{\Delta p} \right) = - \frac{1}{V} \left(\frac{\partial V}{\partial p} \right)_T = - \frac{1}{v} \left(\frac{\partial v}{\partial p} \right)_T \tag{4-26}$$

The minus sign is included here to keep κ positive. (*Almost* nothing that we know of gets larger when squeezed.)

Notice that to measure β and κ requires only measurement of volume, volume changes, pressure changes, and temperature changes so that *any* experiment involving p, v, and T easily yields data on β and κ also. For most substances, β and κ are functions of temperature and pressure. Some data are shown in Fig. 4.13.

Going back to Eq. (4-24) and incorporating the definitions of β and κ, it is easily shown that

$$\left(\frac{\partial p}{\partial T} \right)_v = \frac{\beta}{\kappa} \tag{4-27}$$

EXAMPLE 4.7. As an illustration of the use to which such results may be put, suppose it is required to find the pressure that will result if a piece of copper is heated from 20 to 30°C at constant volume.

SOLUTION: Assuming β and κ to be constant over this range, Equation 4.27 can readily be integrated to give

$$(p_2 - p_1) = \frac{\beta}{\kappa} (T_2 - T_1) \qquad v = \text{constant}$$

For copper,

$$\beta = 5.5 \times 10^{-5}\ (°\text{K})^{-1} \quad \text{and} \quad \kappa = 8.0 \times 10^{-12}\ \text{m}^2/\text{N}$$

$$p_2 - p_1 = \frac{5.5 \times 10^{-5}}{8.0 \times 10^{-12}} (10) = 7 \times 10^8\ \text{N/m}^2.$$

$$p_2 - p_1 = 700\ \text{atm}$$

It appears that anyone desirous of doing this should first build a strong box.

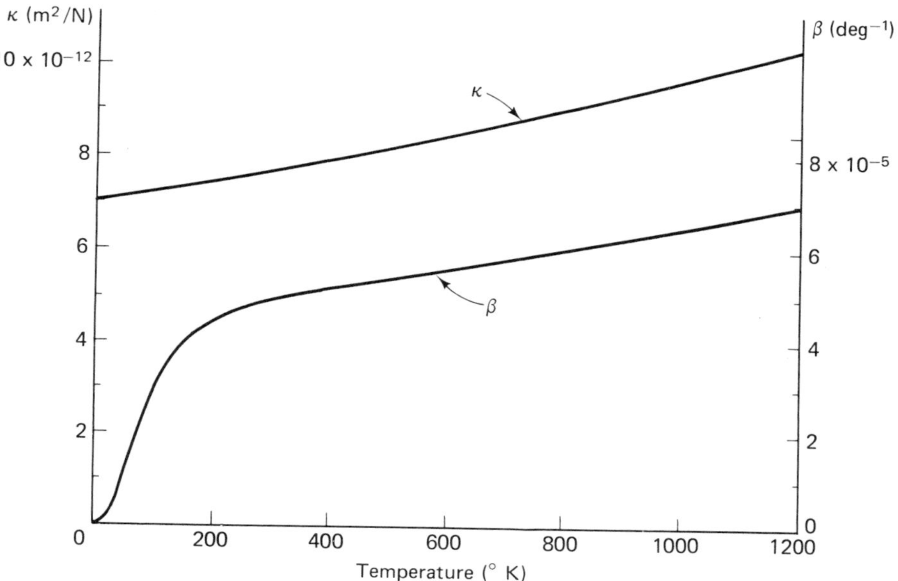

Compressibility κ and coefficient of cubical expansion β of copper, as functions of temperature at a constant pressure of 1 atm.

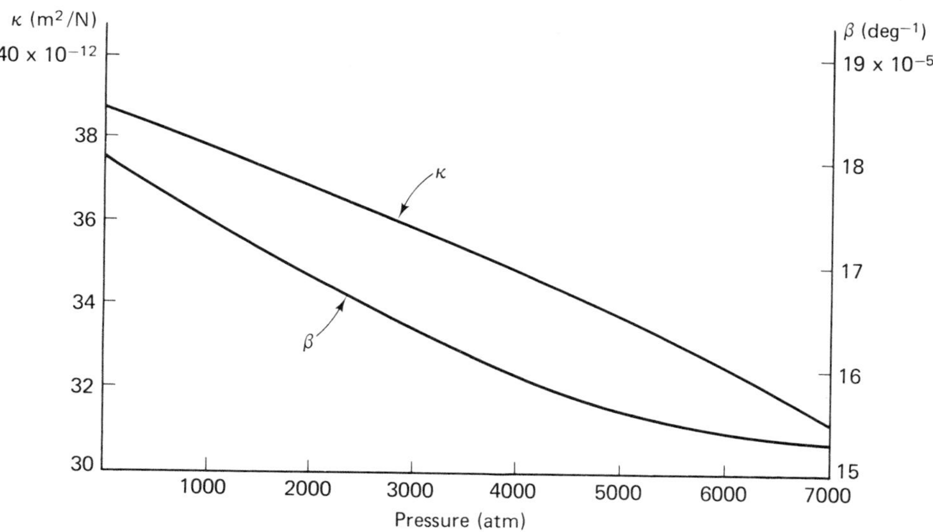

Compressibility κ and coefficient of cubical expansion β of mercury, as functions of pressure, at a constant temperature of 0°C.

Figure 4.13 Compressibility and coefficient of cubical expansion of copper and mercury, as functions of temperature at a constant pressure of 1 atm. and pressure at a constant temperature of 0°C respectively. (Reproduced from *Thermodynamics* by F. W. Sears by permission of Addison-Wesley Co.)

4-7 Summary

The relationship among pressure, volume, and temperature of a substance is called its *equation of state*. Data on *p-v-t* relations are reported in tables, in graphs, and by empirical equations.

A property is an observable characteristic of a substance. The change in the value of a property in a given change of state is independent of the path or process of the change. Furthermore if the change in an observed characteristic is independent of the path of state change, then that characteristic is a property.

Properties whose values for the whole system are equal to the sum of the values for the parts are called *extensive*. Properties whose values have meaning at a point are called *intensive*.

It has been found that pure substances in the absence of magnetic, electrical, and surface tension effects have only two independent intensive properties. A pure substance is one that is uniform and invariant throughout in chemical composition.

Phase diagrams are graphical *p-v-t* relations for substances showing locations of the different phases. When two phases are in equilibrium with each other, then pressure and temperature are *not* independent.

The "quality" x of a liquid–vapor-phase mixture is defined as the ratio of the mass of vapor to the total mass. In terms of quality, the value of an intensive property for a two-phase mixture is as given in the following equation for specific volume:

$$v = v_f + xv_{fg} = (1 - x)v_f + xv_g \tag{4-28}$$

where v_f and v_g are the specific volumes of the liquid and vapor phases, respectively.

The compressibility factor Z for gases is defined as

$$Z = \frac{pv}{RT} \tag{4-29}$$

Reduced coordinates for gases are defined as

$$p_r = \frac{p}{p_c} \qquad T_r = \frac{T}{T_c} \qquad v_r = \frac{v}{v_c} \tag{4-30}$$

where the subscript c refers to critical point values. The law of corresponding states, accurate to perhaps 15% and shown in Table 4-3, says that all gases have the same equation of state in terms of reduced coordinates.

Some gases behave under certain conditions according to the equation of state $pv = n\bar{R}T$ (or $pV = MRT$ or $pv = RT$). The gases under these conditions are called *ideal* gases. Many gases at very low pressure compared with their critical pressure, and/or at very high temperature compared with their critical temperature, obey the gas law very well, or at least well enough for practical use to be made of the ideal gas equation.

For students who wish to extend their study and become energy experts, a really good place to learn more about equations of state is in Chap. 2 and 6 of the following book:

Thermodynamics, The Kinetic Theory of Gases and Statistical Mechanics by F. W. Sears. Addison-Wesley Co., Reading, Mass. 1955.

Problems

4-1 Using the data in Chap. 4 and in Appendix B, fill in the numbers for the question marks in the following table:

Substance	*Phase*	p (psia)	t (°F)	v (ft³/lbm)	v_f	v_g	x
H_2O	?	10	500	?			
H_2O	Saturated liquid	40	?	?			
H_2O	Saturated vapor	300	?	?			
H_2O	Saturated liquid	?	100	?			
H_2O	Saturated vapor	?	600	?			
H_2O	?	100	650	?			
H_2O	?	20	?	25.43			
H_2O	?	20	?	24.0			
H_2O	Liquid vapor equilibrium	20	?	15.0	?	?	?
H_2O	?	?	200	?	?	?	0.20
H_2O	Liquid vapor equilibrium	50	?	?	?	?	0.80
Freon-12	?	?	0	?			0.50
Freon-12	Saturated vapor	100	?	?			
Freon-12	?	100	200	?			

4-2 Using the data in Chap. 4 and in Appendix B, fill in the number for the question marks in the following table:

Substance	*Phase*	p (bars)	t (°C)	v (cm³/gm)	v_f	v_g	x
H_2O	?	2.5	150	?			
H_2O	?	20	550	?			
H_2O	Saturated liquid	16	?	?			
H_2O	Saturated liquid	?	200	?			
H_2O	Saturated vapor	100	?	?			
H_2O	Saturated vapor	?	10	?			
H_2O	?	5	?	424.9			
H_2O	?	5	?	450.0			
H_2O	?	?	250	232.7			
H_2O	?	?	250	200.0			
H_2O	Liquid vapor equilibrium	2.5	?	500.0	?	?	?
H_2O	Liquid vapor equilibrium	2.5	?	?			0.3
H_2O	Liquid vapor equilibrium	?	100	?			0.9

4-3 For each of the following systems, how many independent intensive properties are there?
(a) An equilibrium mixture of solid–liquid–vapor H_2O at the triple point.
(b) An H_2O–NaCl solution in equilibrium with its vapor.
(c) A mixture of oxygen and helium gas at room temperature and atmospheric pressure.

4-4 Which of the following gaseous systems can be approximated by the ideal gas equation? Justify your answers quantitatively.
(a) Air at 400 psia and 80°F.
(b) Helium at 20 psia and −300°F.
(c) Saturated H_2O vapor at 15 bars.
(d) Discuss the use of $pv = RT$ for air generally.
(e) Discuss the use of $pv = RT$ for steam generally.
Support your view with numbers.

4-5(a) Derive an expression for β and κ for an ideal gas.
(b) Find β and κ for air at 550°F and 1 atm pressure.
(c) Find β and κ for air at 500°C and 2 bars pressure.

4-6 Use compressibility charts to obtain the value of pv/RT for steam at 600°F and 1000 psia. Compare with values from steam tables.

4-7 The following data for specific volume in cubic feet per pound mass are taken from *Steam Tables*. Use them to compute β and κ for steam at 420°F and 91 psia.

	400°F	420°F	440°F
90 psia	5.505	5.651	5.796
92 psia	5.381	5.524	5.666
94 psia	5.262	5.403	5.542

4-8(a) Heat is removed from a superheated vapor at constant volume until liquid just begins to form. Show this process on a *p-v* diagram and on a *p-T* diagram.

(b) Heat is added to a saturated liquid at a constant pressure until it is at the critical temperature. Show this process on a *p-v* diagram and on a *p-T* diagram.

(c) A solid sublimes. Show this process on a *p-v* and on a *p-T* diagram.

(d) Heat is removed from a substance at its critical point until the quality is $x = 0.5$. Show this process on *p-v* and *p-T* plots.

(e) Heat is added to subcooled liquid until it is saturated. Show this on *p-v* and *p-T* plots.

(f) A saturated liquid is compressed isothermally to its critical pressure and then expanded at constant pressure to its initial volume. Show these processes on *p-v* and *p-T* diagrams.

(g) A saturated liquid is heated at constant volume to its critical temperature and then expanded isothermally to its initial pressure. Show these processes on *p-v* and *p-T* diagrams.

(h) A saturated liquid is heated to its critical pressure and then expanded isothermally to its initial pressure. Show these processes on *p-v* and *p-T* diagrams.

4-9 For an ideal gas, show the following processes on *p-v*, *p-T*, and *T-v* diagrams.

(a) Isothermal compression.

(b) Constant volume addition of heat.

(c) Constant pressure removal of heat.

4-10 You have just filled a large tank in your factory with compressed air and find that the pressure of the air in the tank is 15 bars absolute and that its temperature is 50°C. What will you expect its pressure to be tomorrow?

4-11 Explain a pressure cooker. Use diagrams to support your discussion.

4-12 Most substances contract when freezing. Water is different. It expands when freezing. Sketch *p-v* and *p-T* phase diagrams for both kinds of substances. How are they different?

4-13 As a new engineer for the Badday Tire Company, you now have a chance to earn your pay. Nalph Radar's army has launched an attack on your company's latest tires, claiming that they are unsafe and will blow out when they get hot. Specifically, he refers to some high-speed tests where the temperature of air inside the tire increased from 25 to 250°C. His engineers claim that this factor of 10 times increase in temperature will cause a 10 times increase in pressure inside the tire and blow it out.

(a) Is Radar correct about the pressure rise?

(b) Another bright young engineer in the company has suggested a valve that lets the air out of the tire when it gets hot. Assuming that the temperature increase you have is correct, calculate the amount of air that must be expelled in order to maintain an inside tire pressure of 2.0 bars gauge. Assume that the inside volume of the tire is constant at $0.02\ m^3$.

4-14 Water is brought to a boil in a pressure cooker at atmospheric pressure. The steam drives out all air and the valve is then closed. The cooker is then left on the stove until the pressure reaches 180 psia, at which time the steam is superheated.

(a) Show this entire process on *p-v* and *p-T* diagrams, identifying (i) the point where boiling just begins; (ii) the point where the valve is closed; (iii) the point where the last of the liquid is vaporized; (iv) the final state.

(b) If the volume of the pressure cooker is 0.5 ft^3 and there is 0.113 lb of H_2O in the cooker when the valve is closed, what is the final pressure when the liquid phase disappears?

(c) What is the final temperature?

4-15 A block of copper at 1000°K is constrained to prevent expansion. To raise the pressure from 1 to 100 atm, what temperature increase is required?

4-16(a) Use the compressibility charts and critical point data on pp. 75 and 76 to estimate the values of the question marks:

Substance	*p* (psia)	*T* (°F)	*v*	*pv/RT*
Air	1100	700	?	?
He	100	80	?	?
H_2O	2000	800	?	?

(b) For the H_2O in part a, what is *v* from the tables in Appendix B?

5

Systems, conservation of mass, and continuity

With the operational definition of work and heat in Chap. 6, we begin the operational-inductive development of the first law. But a clear understanding of the first law, especially in applications, requires first that the concept of the energy *system* be established.

What is the same and what is different about closed and open energy systems? Why is it so important to define a system? How do you decide whether to select an open or closed system? What is the continuity equation? What exactly is meant by steady flow? How do you recognize steady flow situations?

5-1 Systems

The word *system* is used to denote either (1) a particular collection of matter or (2) a particular region of space.

When definition 1 is elected, the collection of matter specified is called a *closed system* or *control mass*.

When definition 2 is elected, the region of space specified is called an *open system* or *control volume*.

A system, whether open or closed, is always defined by careful description of its boundaries. Sometimes the boundary is called a *control surface*. The boundaries may be real or imaginary. When a closed system or control mass is defined, material does not cross the boundary or control surface, though the boundaries themselves may move and distort. When an open system or control volume is defined, material may cross the boundary or control surface. The open system or control volume may move but it is nearly always defined as rigid.

Whenever a system is defined, the universe outside the prescribed boundary is called the *environment*.

As a practical matter in solving energy problems, the careful and explicit definition of the system by its boundaries is of utmost importance. It is equivalent in importance to isolating a free body in statics for a force balance. For heaven's sake, when working problems, *define the system*! Otherwise you literally will not know what you are talking about.

Energy analyses always involve definition of a system, open or closed, followed by application of the laws and the equations of state to the defined system. In most problems, systems interact with their environment in such a way that heat, work and (in open systems) mass cross the boundary. In the remainder of this chapter we shall discuss one of these three kinds of interactions: mass. In Chap. 6, we shall discuss heat and work.

EXAMPLE 5.1. A mass of gas is contained in a cylinder at a given pressure and temperature (p_1 and T_1). Heat is added and the gas is allowed to expand at constant pressure to a new volume (v_2). If you were going to try to solve for the final state and the amount of heat, would you choose an open or closed system? Define your system in words and with a schematic sketch showing states 1 and 2. Also, assume the gas is ideal and show the process on a *p-v* plot.

SOLUTION: A closed system definition is usually best for nonflow confined systems. In this case, the system selected is the gas in the cylinder.

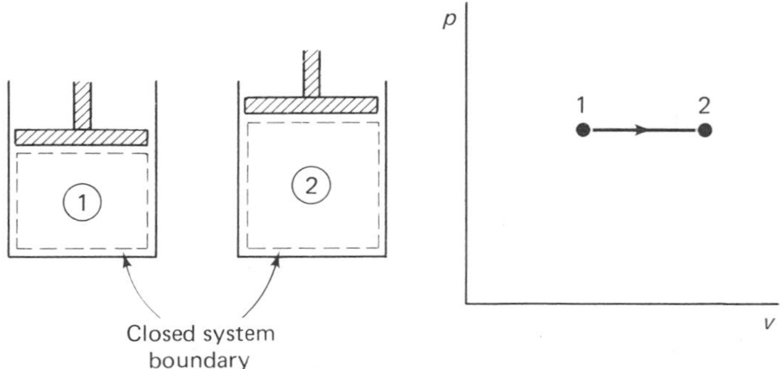

EXAMPLE 5.2. A stream of gas at 80°F in steady flow is compressed from 15 to 60 psia in an isothermal compressor. If you were going to try to solve for the power required, would you choose an open or closed system? Define your system with a schematic sketch. Also show the compression process on a *p-v* plot.

SOLUTION:

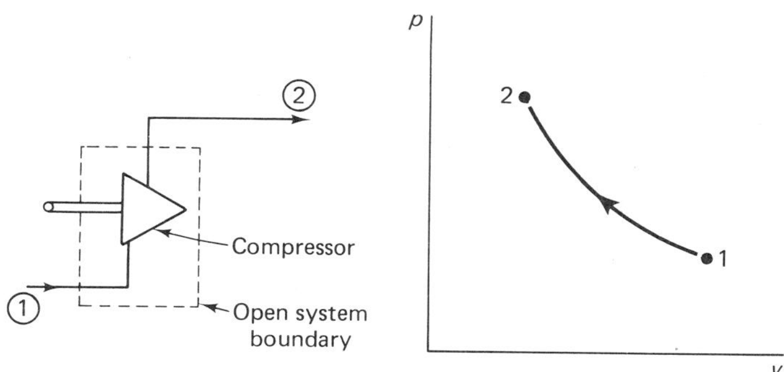

For steady flow processes, an open system is usually the most practical.

EXAMPLE 5.3. An electronic component for space use has a power input of 5 watts. To keep the contents at constant temperature, the container is pressurized and a valve installed that operates automatically to allow gas to escape at a rate such that the temperature inside the container remains constant. If you were going to try to solve for the required rate of flow of gas from the container, would you define a closed or open system? Define your system with a schematic sketch.

SOLUTION: This is an unsteady problem. Sometimes these are best solved using a closed system and sometimes an open system is best. In this case, since it is the flow rate that concerns us, we should use an open system because that term will appear in the equation. Usually, when you want to find a rate, steady or unsteady, the open system is best. If all you care about are end states, then the closed system is best.

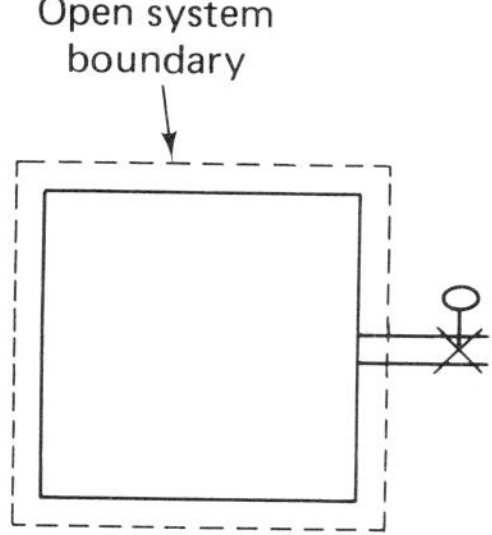

EXAMPLE 5.4. A tank of compressed gas is initially at 27°C and 3 bars absolute. A valve is opened and the pressure in the tank falls rapidly to 1 bar absolute. If you were to try to find the final state of the gas in the tank, what open or closed system would you define? Show your system in a sketch and describe it in words.

SOLUTION: Here is a case where a closed system is best although, as in every case, either will work. We are looking for the state of the gas that *remains* in the tank.

Therefore, we select a closed system consisting of the gas that *remains* in the tank. Note that initially this system occupies only a portion of the tank as shown in the sketch. This is one of the most difficult situations in which to define a closed system and students should think about the rationale. If we chose the system to be the gas *initially* in the tank, part of our system would be outside the tank at the completion of the process. This would be most difficult to analyze since the portion outside would be at a pressure different from the part inside. Again, since we are concerned only with the state of the gas that remains in the tank, we choose that gas for the system.

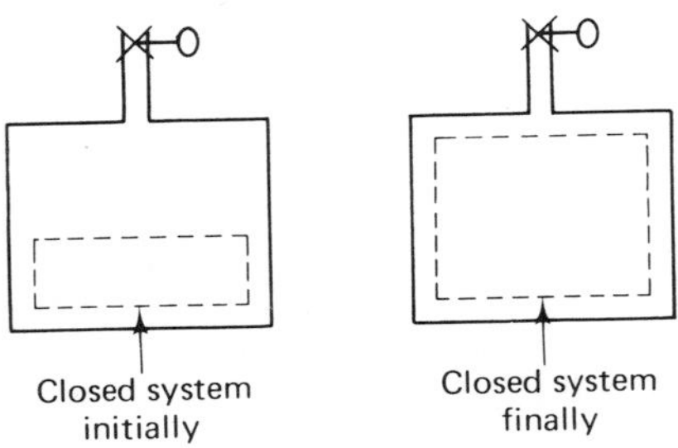

5-2 Conservation of mass in closed systems

Mass is operationally defined by comparison with a standard. The principle of conservation of mass (in the absence of nuclear mass–energy conversions) is an inductively arrived at conclusion based on experience, observation, and experiment. For a closed system or control mass, the principle simply says that the mass is invariant or constant. In symbolic notation;

$$M = \text{constant} \tag{5-1}$$

or

$$dM = 0 \tag{5-2}$$

or

$$\frac{\partial M}{\partial \theta} = 0 \tag{5-3}$$

where M stands for the total mass of the system and θ stands for time. We use partial derivatives here to emphasize that we do not wish to consider variations over space but only with time.

Sometimes, usually in theoretical developments, it is helpful to use a somewhat more general notation. The system is assumed divided into a large number of differential volume elements of volume dV.

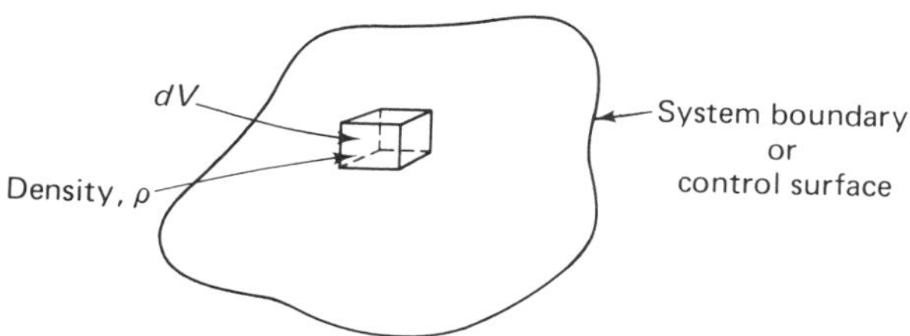

The density of a given element, which may vary throughout the system, is given by ρ. Thus, the mass of an element dV is

$$dM = \rho\, dV \tag{5-4}$$

and the total mass of the system is found by integrating over the entire system volume. Symbolically,

$$M = \int_V \rho\, dV \tag{5-5}$$

and conservation of mass is written as

$$\boxed{\frac{\partial M}{\partial \theta} = \frac{\partial}{\partial \theta} \int_V \rho\, dV = 0} \tag{5-6}$$

5-3 Open systems

5-3(a) The continuity equation

In words, the principle of conservation of mass applied to an open system or control volume is as follows:

$$\begin{bmatrix}\text{The rate of flow}\\ \text{of mass } \textit{into} \text{ the}\\ \text{control volume}\end{bmatrix} - \begin{bmatrix}\text{the rate of flow}\\ \text{of mass } \textit{out of} \text{ the}\\ \text{control volume}\end{bmatrix} = \begin{bmatrix}\text{the rate of change}\\ \text{of mass } \textit{inside} \text{ the}\\ \text{control volume}\end{bmatrix} \tag{5-7}$$

(Or, in different words, what goes in but doesn't come out must accumulate inside!)

There are several ways of writing this equation in symbols. Referring to Fig. 5.1, we can write

$$\boxed{\dot{m}_1 + \dot{m}_2 + \dot{m}_3 - \dot{m}_A + \dot{m}_B = \frac{\partial}{\partial \theta}(M_{\text{c.v.}})} \tag{5-8}$$

where the $\dot{m}$'s stand for various mass flow rates into or out of the region as shown and $M_{\text{c.v.}}$ stands for the mass inside the control volume at any time.

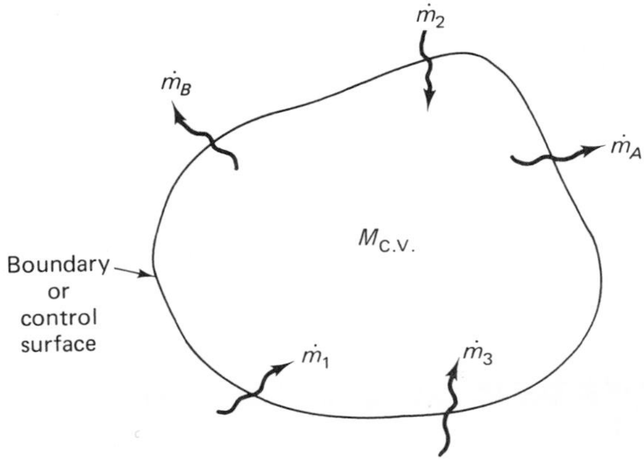

Figure 5.1 Open system for continuity equation.

Generalizing the above to any number of in and out flows gives

$$\sum_{\text{in}} \dot{m}_{\text{in}} - \sum_{\text{out}} \dot{m}_{\text{out}} = \frac{\partial}{\partial \theta}(M_{\text{c.v.}}) \tag{5-9}$$

where $\sum_{\text{in}} \dot{m}_{\text{in}}$ and $\sum_{\text{out}} \dot{m}_{\text{out}}$ stand for the sum of all mass flow rates into and out of the region, respectively.

Often in problems, mass flow rates are not given directly but rather indirectly in terms of velocity, density, and area. In a pipe, for example, if the average velocity is $\mathcal{V}_{\text{av}}$ normal to an area A and the fluid has average density ρ_{av}, then the volume flow rate across the area is given by

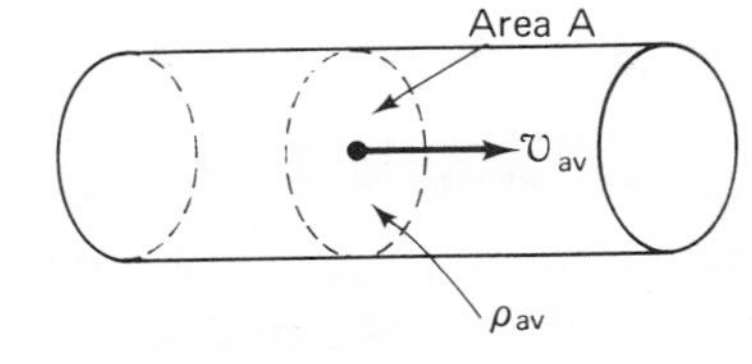

$$\text{Volume flow rate} = A\mathcal{V}_{\text{av}} \tag{5-10}$$

$$\text{Mass flow rate} = \dot{m} = \rho_{\text{av}} A\mathcal{V}_{\text{av}} = \frac{A\mathcal{V}_{\text{av}}}{v_{\text{av}}} \tag{5-11}$$

where $v_{\text{av}} = 1/\rho_{\text{av}}$, the average specific volume. Obviously, the mass flow is in the direction of the velocity.

In cases where the average velocity and density are not established and in theoretical work the concept above is best applied using a differential area together with local values of velocity and density. That is, consider a small area element dA through which a fluid with local density ρ is passing with local velocity $\mathcal{V}$. Then the mass flow is given by

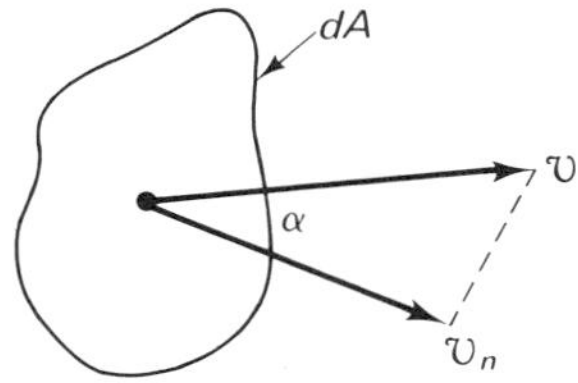

$$d\dot{m} = \rho \mathcal{V}_n \, dA = \rho \mathcal{V} \cos \alpha \, dA \tag{5-12}$$

where $\mathcal{V}_n$ is the component of the velocity normal to the area dA. It is common practice in this kind of situation to use vector notation for convenience. The area element dA is represented by a unit vector $\overrightarrow{dA}$ normal to dA and pointing outward from the control volume. Then

$$\mathcal{V}_n = \overrightarrow{\mathcal{V}} \cdot \overrightarrow{dA} = \mathcal{V} \cos \alpha \, dA \tag{5-13}$$

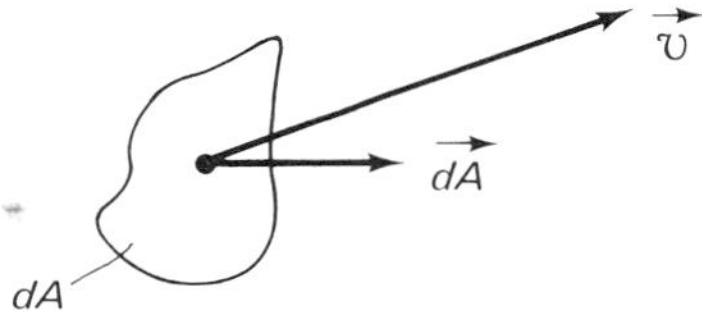

The total flow over an area A is found by integrating:

$$\dot{m} = \int_A \rho \overrightarrow{\mathcal{V}} \cdot \overrightarrow{dA} \tag{5-14}$$

We might now rewrite Eq. (5-9) using the following new notation:

$$\sum_{\text{in}} \dot{m}_{\text{in}} = \int_{A_{\text{in}}} \rho \mathcal{V}_n \, dA_{\text{in}} \qquad \sum \dot{m}_{\text{out}} = \int_{A_{\text{out}}} \rho \mathcal{V}_n \, dA_{\text{out}} \tag{5-15}$$

where A_{in} and A_{out} denote those portions of the control surface or boundary area where the flow is in and out, respectively. The result is

$$\int_{A_{\text{in}}} \rho \mathcal{V}_n \, dA_{\text{in}} - \int_{A_{\text{out}}} \rho \mathcal{V}_n \, dA_{\text{out}} = \frac{\partial}{\partial \theta} \int_V \rho \, dV \tag{5-16}$$

When the notation above is applied to the control volume or open system, $\overrightarrow{dA}$ is always taken as an *outwardly* drawn normal vector at the boundary. Notice then that the vector dot product $\overrightarrow{\mathcal{V}} \cdot \overrightarrow{dA}$ is positive for flow out of the region and negative for flow into the region. Thus, if we integrate $\vec{\mathcal{V}} \cdot \vec{dA}$ over the entire boundary or control surface, we obtain the *net* mass flow rate *out* of the region. That is,

$$\oint_A \rho \vec{\mathcal{V}} \cdot \vec{dA} = \begin{bmatrix} \text{net mass flow rate} \\ \text{out of control volume} \end{bmatrix} = \int_{A_{\text{out}}} \rho \mathcal{V}_n \, dA_{\text{out}} - \int_{A_{\text{in}}} \rho \mathcal{V}_n \, dA_{\text{in}} \tag{5-17}$$

Giving due attention to the sign, we can thus write the conservation of mass principle for a control volume as

$$-\oint_A \rho \vec{\mathcal{V}} \cdot \vec{dA} = \frac{\partial}{\partial \theta} \int_V \rho \, dV \tag{5-18}$$

or, rearranging, we find

$$0 = \frac{\partial}{\partial \theta} \int_V \rho \, dV + \oint_A \rho \vec{\mathcal{V}} \cdot \vec{dA} \tag{5-19}$$

The circle on the integral sign above is used to signify that the integration is to be done over the entire boundary of the control volume.

EXAMPLE 5.5. Air at 80°F and 20 psia flows through a pipeline with an inside diameter of 3 in. at an average velocity of 100 ft/min. Compute the mass rate of flow.

SOLUTION: Translation into diagrams and symbols:

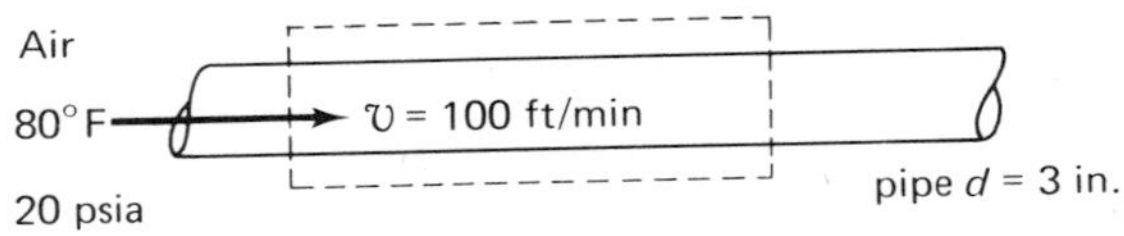

System: Open.
Equation of State: Ideal gas, $pv = RT$.
Process Equation: Not applicable.
Fundamentals:
Conservation of Mass:

$$\dot{m} = \rho A \mathcal{V} = \frac{1}{v} A \mathcal{V}$$

Combined with Equation of State:

$$\dot{m} = \frac{p}{RT} A \mathcal{V}$$

$$= \frac{(20)(144)}{(\frac{1545}{29})(540)} \frac{(\pi)(\frac{1}{4})^2}{(4)} (100)$$

$$\frac{(\text{lbf/in.}^2)(\text{in.}^2/\text{ft}^2)(\text{ft}^2)}{(\text{ft-lbf/lbm-°R})\text{°R}} \left(\frac{\text{ft}}{\text{min}}\right) \rightarrow \frac{\text{lbm}}{\text{min}}$$

$$\dot{m} = 0.491 \text{ lbm/min}$$

EXAMPLE 5.6. An initially evacuated tank develops a small leak. Air leaks in slowly at a mass flow rate proportional to the difference between atmospheric pressure p_0 and the pressure p inside the tank. Derive an expression for the pressure in the tank as a function of time t if the temperature remains constant at T_0.

SOLUTION: Translation into diagrams and symbols:

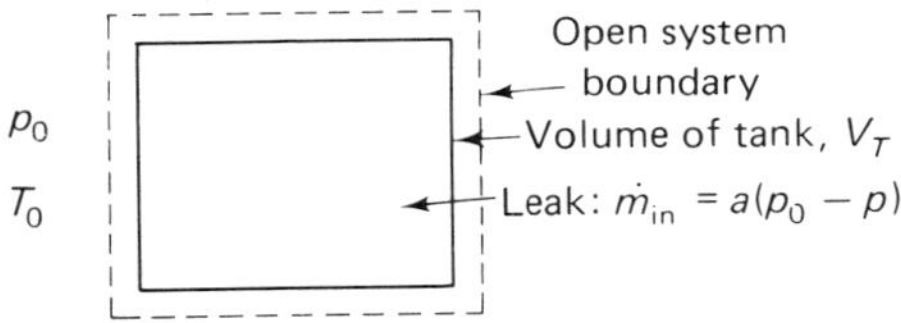

System: Open as shown in sketch.
Equation of State: Ideal gas, $pV = MRT$.
Process Equation: Isothermal, $T = T_0 =$ constant.
Fundamentals:
Continuity:

$$\begin{bmatrix}\text{Rate of mass}\\ \text{flow in}\end{bmatrix} - \begin{bmatrix}\text{rate of mass}\\ \text{flow out}\end{bmatrix} = \begin{bmatrix}\text{rate of change}\\ \text{of mass inside}\end{bmatrix}$$

$$\dot{m}_{\text{in}} - 0 = \frac{d}{d\theta}(M_{\text{c.v.}})$$

Combining with Equation of State and Given Information:

$$a(p_0 - p) = \frac{V_T}{RT_0}\frac{dp}{d\theta}$$

$$p = p_0(1 - e^{-aRT_0/V_T\ \theta})$$

The several equations above that result from application of conservation of mass to open systems are called *continuity equations.*

EXAMPLE 5.7. A liquid storage tank is supplied with water from multiple sources and has water drawn from it from several points as shown in the sketch. Determine the rate of change of the amount of liquid in the tank.

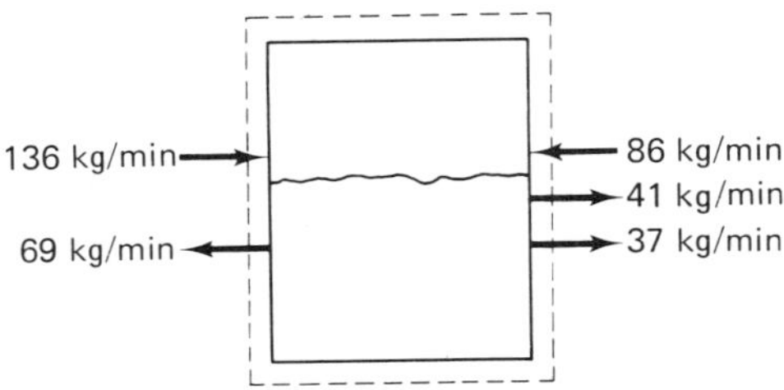

SOLUTION: Translation in diagrams and symbols: Not applicable.
System: Open as shown.
Equation of State: Not applicable.
Process: Steady state.
Fundamentals:
Continuity:

$$\begin{bmatrix}\text{Rate of mass}\\ \text{flow in}\end{bmatrix} - \begin{bmatrix}\text{rate of mass}\\ \text{flow out}\end{bmatrix} = \begin{bmatrix}\text{rate of change}\\ \text{of mass inside}\end{bmatrix}$$

$$\sum_{\text{in}} \dot{m}_{\text{in}} - \sum_{\text{out}} \dot{m}_{\text{out}} = \frac{d}{d\theta}(M_{\text{c.v.}})$$

$$(136 + 86) - (41 + 37 + 69) = \frac{d}{d\theta}(M_{\text{c.v.}})$$

$$+75 \text{ kg/min} = \frac{d}{d\theta}(M_{\text{c.v.}})$$

5-3(b) Steady flow

Steady flow situations are those in which time variations are zero. All manner of things may change with *position* but if one fixes attention on a particular point of matter in space and no changes are observed with time, then the situation is called *steady state.* Many engineering situations are steady state, enough so that we shall write the steady state equations even though they are merely special cases of equations already written.

In steady state, the various symbolic forms of the continuity equations become

$$\sum_{\text{in}} \dot{m}_{\text{in}} = \sum_{\text{out}} \dot{m}_{\text{out}} \tag{5-19}$$

$$\int_{A_{\text{in}}} \rho \mathscr{V}_n \, dA_{\text{in}} = \int_{A_{\text{out}}} \rho \mathscr{V}_n \, dA_{\text{out}} \tag{5-20}$$

$$0 = \oint_A \rho \vec{\mathscr{V}} \cdot \vec{dA} \tag{5-21}$$

5-3(c) Steady flow, single inlet–single outlet systems

Many flow situations reduce to the simple case of a steady flow, single inlet–single outlet system. *Single inlet–single outlet* means that all the flow enters the system at one place and departs at one other place. Examples are nozzles, flow-through pipes or ducts without branches, pumps, and some turbines. In these cases, the continuity equation becomes

$$\dot{m}_{in} = \dot{m}_{out} \tag{5-22}$$

or

$$(\rho A \mathcal{V})_{in} = (\rho A \mathcal{V})_{out} \tag{5-23}$$

EXAMPLE 5.8. An ideal gas enters a constant area duct with a velocity of 100 m/min, at a pressure of 1.2 bars absolute and a temperature of 30°C. At point 2 downstream the pressure is 1 bar absolute and the temperature is 40°C. What is the velocity at point 2?

SOLUTION: Translation to diagrams and symbols:

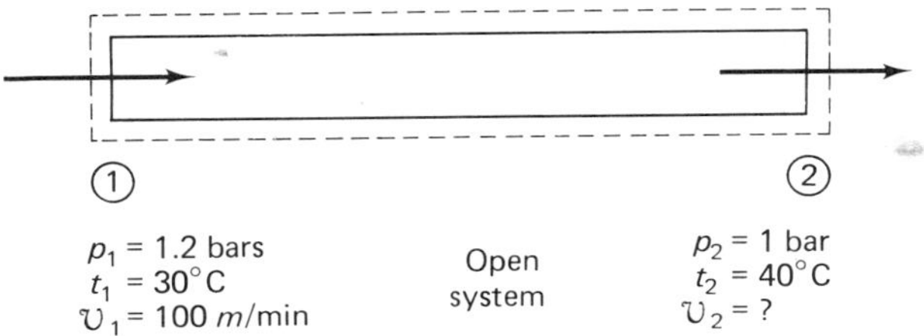

System: Open as shown.
Equation of State: Ideal gas.

$$p_1 v_1 = RT_1 \qquad p_2 v_2 = RT_2$$

Process: Steady state.
Fundamentals:
Continuity:

$$\dot{m}_1 = \dot{m}_2$$

$$\rho_1 A_1 \mathcal{V}_1 = \rho_2 A_2 \mathcal{V}_2 \qquad A_1 = A_2 \quad \rho = \frac{1}{v}$$

$$\frac{\mathcal{V}_1}{v_1} = \frac{\mathcal{V}_2}{v_2}$$

$$\mathcal{V}_2 = \mathcal{V}_1 \frac{v_2}{v_1} = \mathcal{V}_1 \frac{T_2}{p_2} \frac{p_1}{T_1}$$

$$\mathscr{V}_2 = \mathscr{V}_1 \frac{T_2}{T_1}\frac{p_1}{p_2} = 100\left(\frac{40+273}{30+273}\right)\frac{1.2}{1.0}$$

$$\mathscr{V}_2 = 124 \text{ m/min}$$

Checking:
Units:
Trends: If p_2 goes up, $\mathscr{V}_2$ goes down.
If T_2 goes up, $\mathscr{V}_2$ goes up.
Sense: $\mathscr{V}_2$ seems high, but. . . .

5-4 Summary

It is essential in both theoretical and practical work with energy to define the system with which you are working. You may define a closed system, or control mass, which is a specified collection of matter. Or you may define an open system, or control volume, which is a specified region of space. Systems are defined by careful description of the boundary. Everything outside the boundary is called the *environment*.

Mass is conserved. For closed systems, this means no matter crosses the boundary; that is, the mass of the system is constant. Symbolically, this can be expressed in a variety of ways:

$$M = \text{constant} \qquad dM = 0 \qquad \frac{\partial M}{\partial \theta} = 0 \qquad \frac{\partial}{\partial \theta}\int_V \rho\, dV = 0 \tag{5-24}$$

For open systems, the conservation of mass principle is also called *continuity*. In words

$$\begin{bmatrix}\text{The rate of flow}\\ \text{of mass } \textit{into} \text{ the}\\ \text{control volume}\end{bmatrix} - \begin{bmatrix}\text{the rate of flow}\\ \text{of mass } \textit{out of} \text{ the}\\ \text{control volume}\end{bmatrix} = \begin{bmatrix}\text{the rate of change}\\ \text{of mass } \textit{inside} \text{ the}\\ \text{control volume}\end{bmatrix} \tag{5-25}$$

In symbols this equation can be written in several ways as

$$\sum_{\text{in}} \dot{m}_{\text{in}} - \sum_{\text{out}} \dot{m}_{\text{out}} = \frac{\partial}{\partial \theta}(M_{\text{c.v.}}) \tag{5-26}$$

$$\int_{A_{\text{in}}} \rho \mathscr{V}_n\, dA_{\text{in}} - \int_{A_{\text{out}}} \rho \mathscr{V}_n\, dA_{\text{out}} = \frac{\partial}{\partial \theta}\int_V \rho\, dV \tag{5-27}$$

$$-\oint_A \rho \vec{\mathscr{V}} \cdot \vec{dA} = \frac{\partial}{\partial \theta}\int_V \rho\, dV \tag{5-28}$$

For students who wish to see another author's discussion of systems, the following suggested reading is recommended:

Chapter 6, *Introduction to Thermodynamics* by Sonntag and Van Wylen. John Wiley and Sons, Inc., New York, 1971.

Problems

5-1

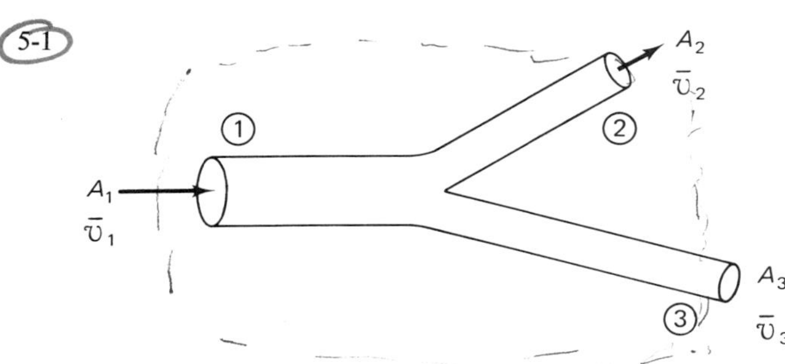

Water enters the Y connection shown at section 1 where the diameter is 3 cm with a velocity of 10 m/sec. The pipe diameter at positions 2 and 3 are 1 and 2 cm, respectively. If $\mathcal{V}_2$ is 20 m/sec, find $\mathcal{V}_3$.

5-2 The sketch shows a normal stationary shock in a gas flowing in a pipe. Find the velocity, $\mathcal{V}_2$.

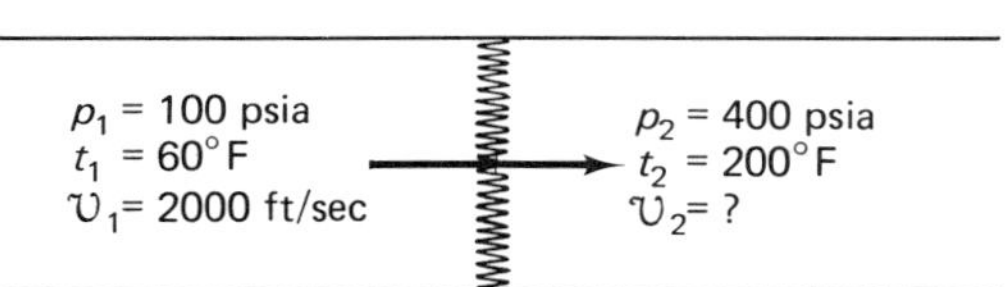

5-3 Suppose the velocity of slow fluid flow in a wide open channel is assumed to be parabolic as shown.

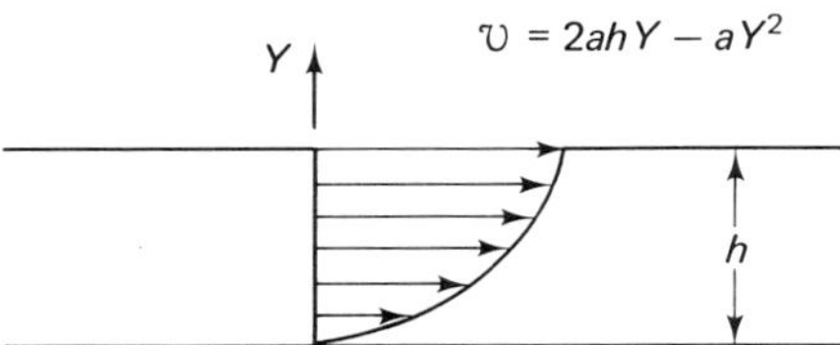

If the total volume flow rate is to be Q, find a in terms of Q and h.

5-4 A tank containing air at 6 bars absolute and 30°C springs a leak such that the mass rate of air leakage is directly proportional to the gage pressure inside the tank.

The leak is slow so the temperature remains constant. After 1 hr the pressure has fallen to 5 bars absolute. $v = 1\ \text{m}^3$ and $p_0 = 1$ bar.

(a) What is the average rate of air leakage during first hour?

(b) What is the current rate of air leakage after 1 hr?

5-5 Air is forced into an isothermal tank at a constant rate of B lb/min. It flows out at a rate proportional to the gage pressure in the tank. Derive an expression for the mass of air in the tank as a function of time.

5-6 Find an expression for the pressure distribution in a tall column of an ideal gas at a uniform temperature. Assume g = constant, column area = A, and pressure at the bottom = p_0.

5-7 A steady flow of air at 20 psia and 70°F enters a nozzle with inlet area of 4 in.^{2} at a velocity of 100 ft/sec. It leaves at 15 psia and 60°F. The outlet nozzle area is 3 in^2. Compute the mass flow rate of air and the average velocity of the air leaving the nozzle.

5-8 A compressor is supplied with nitrogen at 3 bars absolute and 27°C and compresses it in steady flow to 5 bars absolute and 57°C. The inlet pipe diameter is 5 cm. What is the minimum discharge pipe size if the outlet velocity may not exceed twice the inlet velocity?

5-9 A two-phase mixture of liquid and vapor H_2O at 100 psia and a quality of 0.4 are mechanically separated in a separator. Compute the mass flow rates of the separated phases per unit mass entering. Also compute the volume flow rates of the separated phases per unit volume entering.

5-10 Steam leaving a nozzle of area 1 in.^{2} has a pressure of 1 bar absolute, a temperature of 300°C, and a velocity of 30 m/sec. What is the mass flow rate?

5-11 A tank of air contains a valve programmed to allow air to escape in direct proportion to the amount of air in the tank. Derive an expression for the amount of air in the tank as a function of time.

5-12 A tank of air initially at p_i and t_i develops a small leak such that air leaks out *slowly* and at a rate proportional to the pressure difference between the tank and outside (p_0). Derive an expression for the amount of air in the tank as a function of time.

6

Work and heat

Based on the experience we all have prior to a formal study of energy, we know that work is force multiplied by the distance through which the force moves. But work is also transmitted by rotating shafts and by electrical and magnetic means. There is also a kind of work called *flow* work. All of these types of work, important to engineers, are discussed in this chapter. The objective is to extend, formalize, and generalize your understanding of work as a form of energy exchange.

The other way energy is exchanged is by heat transfer. Most of us have some understanding of heat from experience but it tends to be confused with temperature. Also, remnants of an old discredited caloric theory are also often present when the layman talks about heat. This chapter should help you straighten out these matters for yourself and thus set the stage for inducing the first law in Chap. 7.

6-1 Work

6-1 (a) Operational definition

A block weighs 5 lbf. How much work is required to lift it slowly a distance of 7 ft? Write your answer here:

35 Ft/lbs
____

A block weighing 18 newtons sliding along a horizontal plane with friction requires a force of 6 N to keep it in motion. How much work is done as the block slides a distance of 1 m? Write your answer here:

A block weighs 5 kgf. How much work is required to hold it without motion in a position 2 m above the floor? Write your answer here:

Most engineering students, by the time they take an energy course, can easily do the kind of problems just presented. (The correct answers are 35 ft-lbf, 6 N-m, and 0.) That is, in these simple situations, you already know how to compute work. It is the product of the force times distance. More specifically, it is the product of the force multiplied by the distance through which the force moves in its own direction. This can best be expressed symbolically using vector notation as

$$dW = \vec{F} \cdot d\vec{x} \quad \text{or} \quad W = \int \vec{F} \cdot d\vec{x}$$

The definition of work as force times distance is operational since it tells us what to do to obtain a number for the quantity defined—in this case, work. For thermodynamic purposes, however, force times distance as a definition does not go far enough because it does not tell us exactly how to identify the work with a particular system nor does it include work that might be done by electrical, magnetic, or other forces. Force times distance gives a basic starting point on which a more complete operational thermodynamic definition of work is built. Here is the definition:

> Work is done by a system (on another) when the sole effect external to the system could be the rise of a weight. The amount of work done is the product of the weight (force) times the distance lifted. By convention, work done *by* a system (which could lift weights in the environment) is taken as positive for that system; work done *on* a system (the environment lifts weights within the system boundaries) is taken as negative.

This definition does three things: (1) It tells how to identify when an interaction between systems is work (when the *sole* effect *could be* the rise of a weight); (2) it tells how to compute the magnitude of the work (force times distance); and (3) it provides a sign convention for keeping track of work in reference to thermodynamic systems (work done *by* a system is taken as positive).

Some key words in the definition require elaboration. *Sole* effect: *Sole* is included to help distinguish work from heat, which will be defined later. *Could be: Could be* is used to permit us to include forms of work other than linear mechanical (force times distance) work. The *could be* implies the possible use of pulleys, motors, and other devices to convert an interaction into the rise of weight. Thus, the flow of electric current is identified as work because it *could be* used to operate a motor that *could* lift a weight. The amount of the work is always computed as the work that could be done if the imagined motors, pulleys, etc., are idealized without losses such as friction, hysteresis, and electrical resistance etc.

It should be noted and emphasized that work as defined is an *interaction* between a system and its environment or between two systems. Work therefore is identified *at the boundary* of systems. This is another of the reasons it is so important that system boundaries be clearly and explicitly described as discussed in Chap. 5.

Before going on to types of work interactions other than linear mechanical ($\vec{F} \cdot d\vec{x}$), it is helpful to define power because it is so closely related to work. Power is the *rate* at which work is being done (i.e., the rate at which work crosses the boundary of a system). In a linear mechanical system

$$P = \vec{F} \cdot \frac{d\vec{x}}{d\theta} = \vec{F} \cdot \vec{\mathscr{V}} \tag{6-1}$$

where $\vec{\mathscr{V}}$ is the velocity at which the force moves. Power is a scalar (nondirectional) property.

6-1(b) Rotational, electrical, and magnetic work

Now let us consider rational mechanical work. Assume a shaft rotates at ω radians per unit time and transmits a torque τ. If connected to a pulley to lift a weight, this shaft could lift a weight of w pounds with a velocity $\mathscr{V}$ where $\mathscr{V}$ equals the product of the pulley radius times the angular speed ω. See Fig. 6.1.

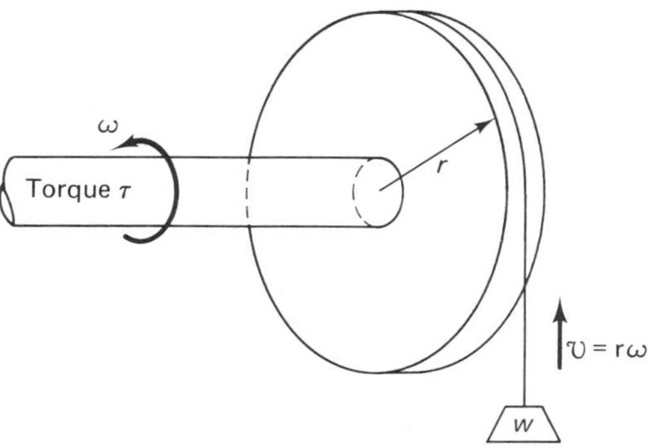

Figure 6.1 Diagram for derivation of rotational work equation.

A force balance on the weight now gives

$$w = \frac{\tau}{r} \tag{6-2}$$

The work W done in time θ in lifting the weight is the distance traveled by the weight times the weight itself:

$$W = w \cdot \mathscr{V} \cdot \theta = \left(\frac{\tau}{r}\right)\mathscr{V}\theta = \left(\frac{\tau}{r}\right)r\theta\omega$$

$$W = \tau\omega\theta = \tau\alpha \tag{6-3}$$

where α is the angle through which the shaft rotates in time θ. Thus, work in rotational mechanical systems is given by the product of torque times angle of rotation. In differential form,

$$dW = \tau\, d\alpha \tag{6-4}$$

A flow of electrical current across a voltage drop can also be converted to work. The amount can be computed from electrical field principles from which it is known that the force on a charge q in an electric field of strength E is equal to E times q (Coulomb's law). Suppose the field is created by a voltage e acting over a length l, then the electric field is $E = e/l$. If the charge is moved the distance l, then the work done is force (Eq) times distance (l). Using the charge as the system, this work is done *on* the system. Hence, the work *by* the system is

$$W = -Eql = -\frac{e}{l}ql = -eq \tag{6-5}$$

Current is charge per unit time so

$$I = \frac{q}{\Delta\theta} \tag{6-6}$$

Hence

$$W = -eI\,\Delta\theta \tag{6-7}$$

Written in differential form, these equations become

$$dW = -e\,dQ = -Ie\,d\theta \tag{6-8}$$

Combining with Ohm's law, we find the electrical work that could be done when a current I passes through a resistor R ($= e/I$) for time $d\theta$:

$$dW = -I^2R\,d\theta \tag{6-9}$$

Note that the resistor is taken as the system, hence the minus sign. In terms of power, which is the rate of doing work, these expressions become

$$P = \frac{dW}{d\theta} = -Ie = -I^2R \tag{6-10}$$

Table 6.1 Electrical terms, symbols, and units

Name	*Symbol*	*Units*	*Definition*
Ampere	I	ampere	coulomb/sec
Charge	q	coulomb	amp × sec
Voltage	e	volt	watt/amp
Resistance	R	ohm (Ω)	volt/amp
Capacitance	C	farad (F)	amp-sec/volt
Inductance	L	henry (H)	volt-sec/amp
Permeability	μ	henry/meter	
Permeability (vacuum)	μ_0	henry/meter	$\mu_0 = 4\pi \times 10^{-7}$ H/m
Magnetic field strength	H	ampere/meter	
Magnetic flux density	B	weber/meter^2	$B = \mu_0 H + M$
Magnetic flux	ϕ	weber (Wb)	volt-sec
Permittivity	ε	farad/meter	
Permittivity (vacuum)	ε_0	farad/meter	$\varepsilon_0 = 8.854 \times 10^{-12}$ F/m
Electric field intensity	E	volt/meter	
Electric flux density	D	coulomb/meter^2	$D = P_e + \varepsilon_0 E$
Polarization (electric)	P_e	coulomb/meter^2	$P_e = D - \varepsilon_0 E$
Polarization (magnetic)	M	weber/meter^2	$M = B = \mu_0 H$

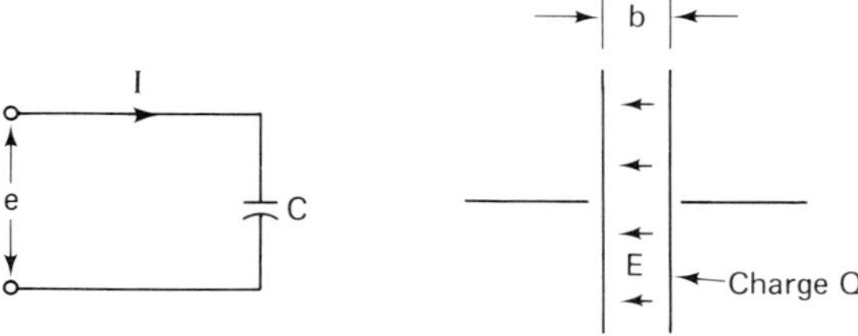

E = Electric field intensity = e/b (volts/m)
C = Capacitance = Q/e (coulombs/volt) (or farads)
D = Electric flux density = Q/A (coulombs/m^2)
I = $dQ/d\theta$ (coulombs/sec)

Figure 6.2 Diagram for derivation of capacitor work.

Once it is established that electrical power is given by Ie, it is not difficult to use electrical fundamentals to find expressions for work in capacitors and inductances.

Refer to Fig. 6.2. The power supplied to the capacitor is Ie as shown. This can be related to capacitor parameters as follows:

$$P_{\substack{\text{on}\\\text{capacitor}}} = Ie = IEb = Eb\frac{dq}{d\theta} = EbA\frac{dD}{d\theta} = VE\frac{dD}{d\theta} \tag{6-11}$$

In terms of work, the work done *on* the capacitor in time $d\theta$ is

$$dW_{\substack{\text{on}\\\text{capacitor}}} = P\,d\theta = VE\,dD \tag{6-12}$$

We usually report work as positive when done by the system so this is conventionally written as

$$dW_{\substack{\text{by}\\\text{capacitor}}} = -VE\,dD \tag{6-13}$$

where V is the volume of the capacitor, Ab. By definition,

$$D = \varepsilon_0 E + P_e \tag{6-14}$$

where ε_0 = permittivity of vacuum
$= 8.85 \times 10^{-12}$ farad/m
P_e = polarization of material, coulomb/m^2
E = electric field intensity, volts/m

With this definition the work becomes

$$dW = -V\varepsilon_0 E\,dE - VE\,dP_e \tag{6-15}$$

The first term on the right is the work required to increase the magnetic field in a vacuum and it is usually omitted so that the work done to increase the polarization of the substance is

$$dW_{\substack{\text{by}\\\text{system}}} = -VE\,dP_e \tag{6-16}$$

For the work done in an inductance, refer to Fig. 6.3.
Ampere's law:

$$NI = Hl \tag{6-17}$$

Faraday's law:

$$e = N\frac{d\phi}{d\theta} = NA\frac{dB}{d\theta} \tag{6-18}$$

$$P_{\substack{\text{by}\\\text{system}}} = -Ie = -NI\frac{d\phi}{d\theta} = -HlA\frac{dB}{d\theta} = -VH\frac{dB}{d\theta} \tag{6-19}$$

$$dW_{\substack{\text{by}\\\text{system}}} = -HV\,dB \tag{6-20}$$

By definition,

$$B = \mu_0 H + M \tag{6-21}$$

where μ_0 = permeability of vacuum
$= 4\pi \times 10^{-7}$ henry/m
H = magnetic field intensity, amp/m
M = magnetization of material, webers/m^2

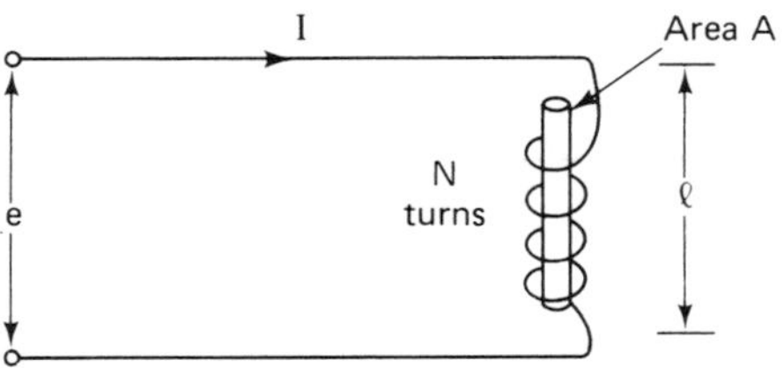

B = Magnetic flux density (webers/m^2)
H = Magnetic field intensity (ampere/meter)
ϕ = BA = Magnetic flux (webers) (or volt-seconds)

Figure 6.3 Diagram for derivation of magnetic work.

With this definition, the work becomes

$$dW = -\mu_0 VH\, dH - VH\, dM \tag{6-22}$$

The first term on the right is the work required to increase the magnetic field in a vacuum and it is usually omitted so that the work done to increase the magnetization of a substance is

$$đW_{\substack{\text{by}\\ \text{system}}} = VH\, dM \tag{6-23}$$

For many substances called *paramagnetic*, a relationship among H, M, and T exists. It is analogous to an equation of state. The relationship, called Curie's law, is

$$CH = MT \tag{6-24}$$

where C is a constant. This expression, like $pv = RT$, has a limited range of applicability, but enough to be important.

A summary of the various work equations is shown in Fig. 6.4.

System	Force	Displacement	Work
Linear mechanical	$\vec{F}$	$d\vec{X}$	$dW = \vec{F} d\vec{X}$
Rotational mechanical	τ	$d\alpha$	$dW = \tau d\alpha$
Electrical charge	$\vec{e}$	$d\vec{Q}$	$dW = -\vec{e} d\vec{Q}$
Electrical field	$\vec{E}$	$d\vec{P}$	$dW = -V\vec{E} d\vec{P}$
Magnetic field	$\vec{H}$	$d\vec{M}$	$dW = -V\vec{H} d\vec{M}$

Figure 6.4 Summary of work equations.

EXAMPLE 6.1

1. Show that for an elastic band the work done in extending its length can be written as $dW = -\sigma V\, d\varepsilon$ where σ is the tensile stress, ε is the strain, and V is the volume.
2. If the equation of state of a rubber band is given by

$$\sigma = \frac{A_0}{A} KT\left(\lambda - \frac{1}{\lambda^2}\right)$$

where K is a constant, A_0 is the initial cross-sectional area, T is temperature, and $\lambda = L/L_0$, show that the work required to stretch the band is given by

$$dW = -V_0 KT\left(\lambda - \frac{1}{\lambda^2}\right) d\lambda$$

SOLUTION:

1.

$$dW_{\text{by}} = F\, dx$$

We are asked to find the work done *on* the system so

$$dW_{\text{on}} = -F\, dx$$

By definition, $F = \sigma A$, so

$$dW_{\text{on}} = -\sigma A\, dx$$

The strain is given by $d\varepsilon = dL/L$, but $dL = dx$, so

$$dW_{\text{on}} = -\sigma AL\, d\varepsilon = \sigma V\, d\varepsilon$$

2. Substituting the given equation of state into the work expression gives

$$dW_{\text{on}} = -V \frac{A_0}{A} KT\left(\lambda - \frac{1}{\lambda^2}\right) d\varepsilon$$

Now we note that $\lambda = L/L_0$ so $d\lambda = dL/L_0$. Since $d\varepsilon = dL/L$ by definition, we have

$$d\varepsilon = \frac{L_0}{L}\, d\lambda$$

Then

$$dW_{\text{on}} = -V \frac{A_0}{A} KT\left(\lambda - \frac{1}{\lambda^2}\right) \frac{L_0}{L}\, d\lambda$$

or

$$dW_{on} = -V_0 KT\left(\lambda - \frac{1}{\lambda^2}\right) d\lambda$$

EXAMPLE 6.2

1. A paramagnetic substance having a volume V is kept at constant temperature while the magnetic field is increased from M_1 to M_2. If the substance follows the Curie equation of state, derive an expression for the work required.
2. Find the work if the following values are used:

$$M_1 = 0$$
$$M_2 = 4\pi \times 10^{-1} \text{ weber/m}^2$$
$$C = 4\pi \times 10^{-9} \text{ weber-°K/amp-m}$$
$$T = 4\text{°K}$$
$$V = 0.1 \text{ m}^3$$

SOLUTION:

Work:

$$dW_{on} = VHdM$$

Equation of state:

$$CH = MT$$

Combining:

$$dW_{on} = \frac{VT}{C} M\, dM$$

1.

$$W_{on} = \frac{VT}{2C}(M_2^2 - M_1^2)$$

2.

$$W_{on} = \frac{(0.1)(4)(4\pi)^2(10^{-2})}{(2)(4\pi)(10^{-9})}$$

$$W_{on} = 8\pi(10^6) \text{ N-m}$$

$$\text{Units} \rightarrow \frac{(\text{m}^3)(\text{°K})(\text{webers/m}^2)^2}{(\text{webers-°K/amp-m})} \rightarrow \text{webers-amp}$$
$$\rightarrow \text{N-m}$$

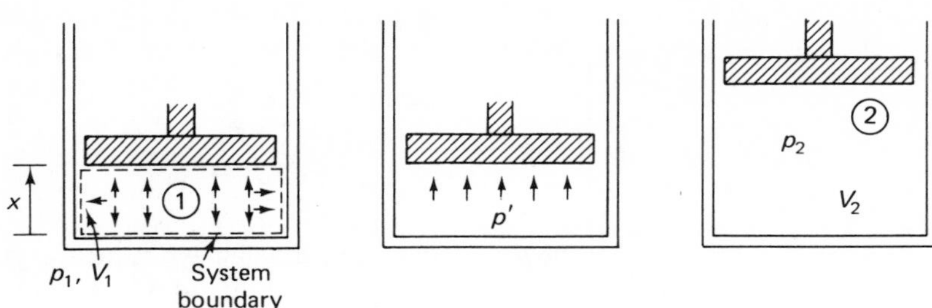

Figure 6.5 Diagram for derivation of $d'W = p\,dV$.

6-1(c) *p dV* work

Consider a closed system contained within a cylinder fitted with a no-leak piston. See Fig. 6.5.

Let the piston area be A and the pressure acting directly on the piston be p'. The force acting on the piston is $F = p'A$. If a differential motion of the piston is allowed, then work is done by the system equal to

$$d'W_{\substack{\text{by}\\\text{system}}} = F \cdot dx = p'A\,dx \tag{6-25}$$

But since the volume of the system is $V = Ax$, $A\,dx = dV$ and

$$d'W_{\substack{\text{by}\\\text{system}}} = p'\,dV \quad \text{or} \quad W_{12\,\substack{\text{by}\\\text{system}}} = \int_1^2 p'\,dV \tag{6-26}$$

Now if the piston moves sufficiently slowly, the pressure of the system will be essentially uniform throughout and can be denoted as p throughout. In this case, p will also be the pressure acting on the piston p' and so $d'W = p'\,dV = p\,dV$. But note that if the piston moves rapidly, say, near the speed of sound in the system, then a uniform system pressure p will *not* exist, the pressure on the piston will be considerably less than in a slow expansion, and the work done by the system will not be so great. In practice this difference does not become significant until the piston speed does, in fact, get close to the sonic velocity. The reason for this is that the velocity of sound is the velocity with which pressure waves move through a substance. Thus, if the piston moves at something less than sonic speeds, pressure changes can continually equalize throughout the cylinder since the piston moves slowly compared with the speed of the pressure waves. But as the piston approaches sonic speed, pressure will build up or fall off (depending on the direction of motion) significantly next to the piston and the pressure throughout the cylinder will be far from "uniform."

An expansion in which the system is essentially uniform at all times, and hence for which $p = p'$, will be called a *restrained* (or "slow") expansion.

We wrote

$$dW_{\substack{\text{by}\\\text{system}}} = p\,dV$$

before but it may be noted that if dV is negative (compression), dW will be negative and work is done *on* the system. Hence the sign convention works automatically in this equation and the subscript on W may be dropped. We normally simply write

$$dW = p\,dV = pm\,dv \tag{6-27}$$

Notice that the work per unit mass is given simply by $p\,dV$ and that this is represented by the area under the process line on a p-v plot.

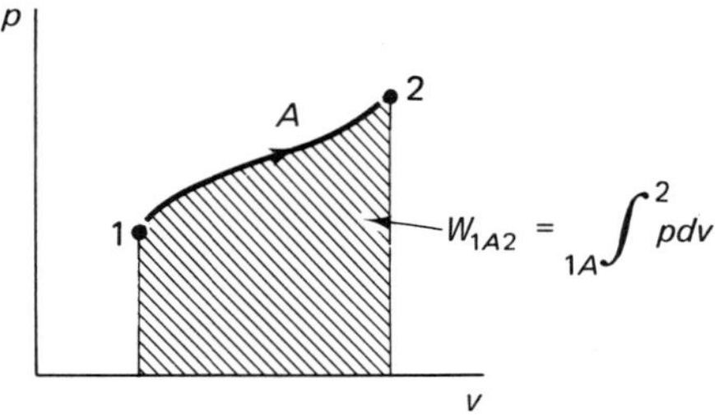

EXAMPLE 6.3. Air initially at 20 psia and 100°F is compressed in a closed isothermal cylinder to 40 psia. Compute the total work required per pound of air and show the area representing the work as a p-v diagram. Also compute the net work required to be done by mass and show this area. Atmospheric pressure is 15 psia.

SOLUTION: We select the air in the cylinder as the closed system. Under these conditions, air behaves more like an ideal gas so we shall use $pv = RT$ for the equation of state. We assume that the compression is slow so that $p = p'$.

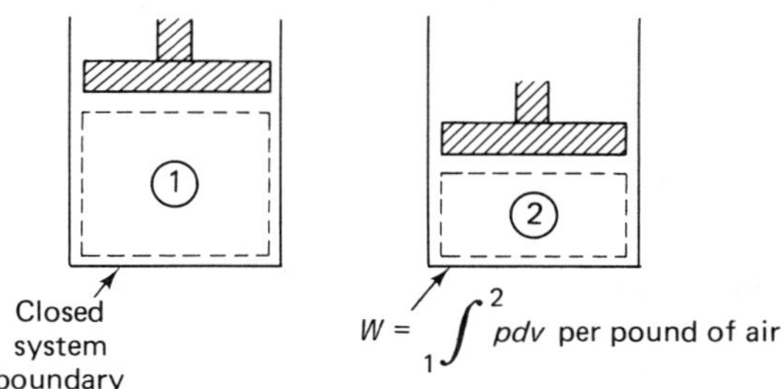

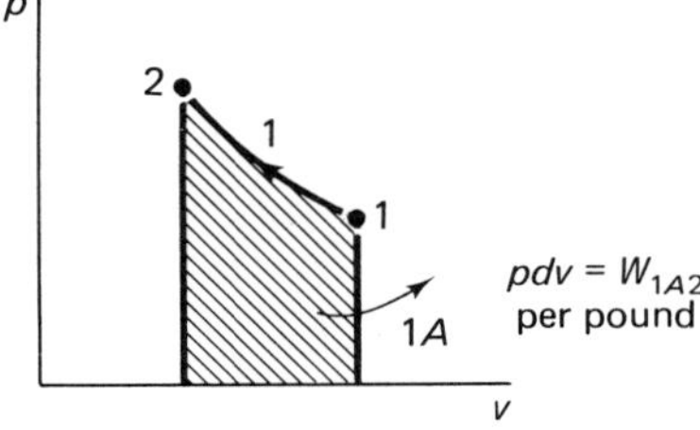

Substituting RT/v for p, noting that T is constant and that $p_1 v_1 = p_2 v_2$, gives

$$W = RT \int_1^2 \frac{dv}{v} = RT \ln \frac{v_2}{v_1} = RT \ln \frac{p_1}{p_2}$$

$$W = \left(\frac{1545}{29}\right)(560) \ln \frac{20}{40} = -20{,}700 \text{ ft-lbf/lbm}$$

Note that the minus sign for W means work done *on* the system, as expected. We can write

$$W_{\substack{\text{on}\\ \text{system}}} = 20{,}700 \text{ ft-lbf/lbm}$$

A portion of the total work above is done by the atmosphere as the piston moves. Another way to look at this is to notice that, to compress the gas, man need only provide a force sufficient to overcome the gas pressure *minus* atmospheric pressure. That is,

$$W_{\substack{\text{by man}\\ \text{on system}}} = \int_1^2 F\,dx = -\int (p - p_0)A\,dx$$

$$= -\int_1^2 (p - p_0)\,dV$$

On a per pound basis,

$$W_{\substack{\text{by man}\\ \text{on system}}} = -\int_1^2 p\,dv + \int_1^2 p_0\,dv$$

We have already found

$$\int_1^2 p\,dv = 20{,}700 \text{ ft-lbf/lbm}$$

To find the work done by the atmosphere,

$$W_{\substack{\text{by}\\ \text{atmosphere}}} = -\int_1^2 p_0\,dv = -p_0(v_2 - v_1) = -p_0 RT\left(\frac{1}{p_2} - \frac{1}{p_1}\right)$$

$$= -p_0 RT\left(\frac{p_1 - p_2}{p_1 p_2}\right)$$

$$= (15)(144)\left(\frac{1545}{29}\right)(560)\left[\frac{(20)(144)}{(40)(20)(144)^2}\right]$$

$$= 11{,}200 \text{ ft-lbf/lbm}$$

Thus, man must provide 20,700 − 11,200 = 9500 ft-lbf/lbm as shown in the following sketch:

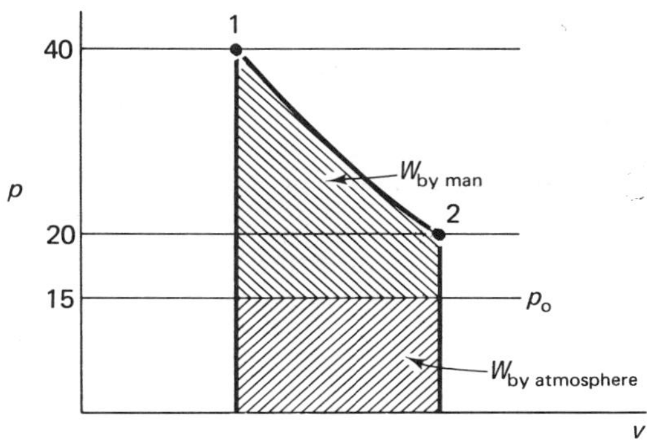

6-1(d) Flow work

Work can flow across the boundaries of an open system or control volume in many forms: rotating shafts, electrical wires carrying a current, etc. $p\,dV$ work does not exist for the open system, of course, because the boundary of the open system does not move. There is, however, a work flow associated with the flow of material across the boundary into or out of an open system. This work is called *flow work*. Let us see how it arises and how it is computed.

Consider a boundary where material flows into a control volume. There will exist a pressure within the fluid.

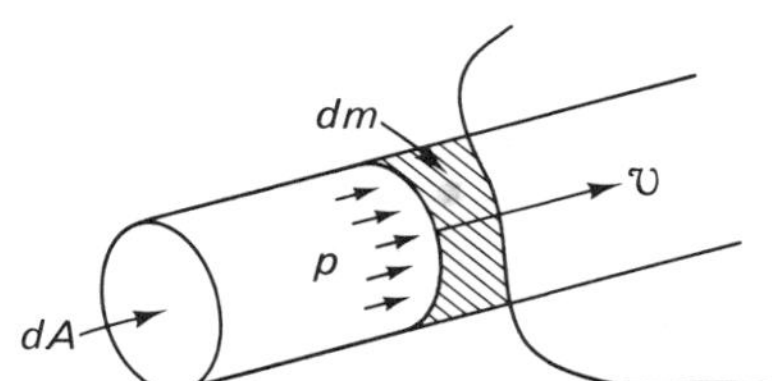

Now the fluid entering must continually push its way into the region against this pressure. Consider the differential mass element dm, for example. In moving across the boundary into the system, it must do work:

$$dW = F\,dx = pA\,dx = p\,dV \tag{6-28}$$

If we use $dm = \rho\,dV$, this can be written as

$$dW = \frac{p\,dm}{\rho} = pv\,dm \tag{6-29}$$

Per unit mass, this work, called *flow work*, is

$$\text{Flow work per unit mass} = \frac{dW}{dm} = pv \tag{6-30}$$

A similar expression is obtained for the work done by a mass element from inside the system moving to the outside. Flow work terms are work done on or by the open system just as surely as if there were little pistons pushing each little mass in or out against the pressure at each point. There is no way to get mass in or out of a control volume without a flow work term appearing.

If the flow work term per unit mass (pv) is multiplied by the mass rate of flow ($\dot{m}$), then the result is a *power* input or output to or from the open system depending on the flow direction.

EXAMPLE 6.4

1. Air at 20°C and 1.5 bars absolute flows steadily into a tank. Compute the flow work per pound.
2. Water at 20°C and 1.5 bars absolute flows into a tank. Compute the flow work per pound.

SOLUTION: Flow work per pound $= pv$.
For the air:

$$W_{\text{flow}} = pv = RT = \left(\frac{8.32}{29}\right)(293)$$
$$= 84.1 \text{ kj/kg}$$

For the water,

$$W_{\text{flow}} = pv$$

For the steam tables, we find the specific volume of saturated liquid at 20°C to be 1.0018 cm³/gm. This is a good approximation to the actual specific volume because liquid H_2O is highly incompressible.

$$W_{\text{flow}} = (1.5 \times 10^5)(1.0018 \times 10^{-3})$$
$$= 150.3 \text{ j/kg}$$

6-2 Heat

6-2(a) Calorimetry and specific heat

Interactions between both closed and open systems and their environments include not only work but also heat. Heat interactions are important in most energy problems and especially in energy conversion systems where man is interested in converting heat into work or in using work to remove heat from his food storage case, experiments, or office.

The empirical science of calorimetry—or, in other words, the measurement of heat—is well over 100 years old. Historically it preceded the concept of operational definitions and so, though the early experimenters were able to measure heat, the words and theory they used to explain the phenomenon were faulty. The theory in vogue at the time postulated the existence of a mysterious fluid substance called *caloric* such that the amount of caloric in a substance determined its temperature. The more caloric, the higher the temperature. Somehow the nature of caloric always made it flow from a higher to a lower temperature.

The caloric theory could not interpret such a common phenomenon as the rise in temperature associated with friction and so it was eventually discarded. A surprising amount of the old caloric theory remains in the language, however. We still talk about the *flow* of heat, for example, and it is usual to talk about the heat *contained in* a substance. No careful engineering student today uses the expression "heat contained in," however, because of its obvious connection with the discredited caloric theory.

The early experimenters did not have a good theory but they were able in practice to measure what we now call heat. The science of calorimetry became, and remains, very sophisticated. The major theoretical deficiency at the time was the lack of operational definitions that we can now set straight.

Calorimetry, or the measurement of heat, makes use of a standard quantity just as there is a standard for length and mass. One standard might be established as follows: If, in the absence of work, the temperature of 1 lb of water is raised from 59.5 to 60.5°F at atmospheric pressure, then the heat transferred to the water is called one British thermal unit (Btu). Students should note that this definition tells *how to measure* a standard unit of heat and hence defines this standard unit operationally.

In order to measure heat flow—at least conceptually—we can now theoretically use a collection of little systems of 1 lbm of H_2O at 59.5 or 60.5°F. (It can be established experimentally that the heat transferred *from* 1 lb of H_2O in changing from 60.5 to 59.5°F is also 1 Btu.) The logically wary student should note, however, that we have further assumed that an experiment can be performed in which the heat transferred *from* one system equals the heat transferred *to* another system. This is done theoretically by performing the experiment in one of those "perfectly" insulated (adiabatic) containers mentioned in Chap. 2. In practice, this is most difficult but the art of calorimetry has developed some ingeniously clever and accurate techniques. It is to be noted that in defining work we required that *all* the work done be converted into lifting weights external to the systems with no losses in friction or other ways. With heat, too, we require that *all* the heat be transferred to a standard so that it can be measured.

Fortunately it is not necessary for heat measurement always to be referred to H_2O between 59.5 and 60.5°F. The concept of specific heat enables us to establish more convenient but equivalent measurement standards. The specific heat (c) of a substance is the amount of heat (in the absence of work or a phase change) required to raise a unit mass of the substance 1° in temperature. From the definition of the standard unit of heat above, it should be obvious that the specific heat of H_2O at 60°F is 1 Btu/lbm-F°.

To compute the heat transferred to a substance (in the absence of work or phase changes) in a given temperature change from t_i to t_f, the specific heat is used as follows:

$$\boxed{\text{Heat} = Q = mc(t_f - t_i)} \tag{6-31}$$

where Q is the heat transferred to the substance and m is the mass of the substance. Because specific heat is often a function of temperature, this equation is best written as

$$dQ = mc\,dt \quad \text{and} \quad Q = m\int_{t_i}^{t_f} c\,dt \tag{6-32}$$

Of course, in order to use this equation, the specific heat must be known. To determine the specific heat of substances other than H_2O at 69°F, the principles of calorimetry can be used.

For example, suppose that 1 lb of water at 59.5°F is placed in a perfectly insulated container together with 1 lb of copper at 80°F. When the water has reached a temperature of 60.5°F, the copper is found to be at 70°F. To compute the specific heat of the copper,

$$\text{Heat transferred to water} = m_w c_w(t_f - t_i)_w = (1)(1)(60.5 - 59.5) = 1 \text{ Btu}$$

$$\text{Heat transferred from copper} = m_c c_c(t_i - t_f)_c = (1)c_c(80 - 70) = 10c_c \text{ Btu}$$

$$\text{Heat transferred } \textit{to} \text{ water} = \text{heat transferred } \textit{from} \text{ copper}$$

$$c_c = 0.1 \text{ Btu/lbm-°F}$$

Specific heats of wide varieties of substances over wide ranges of substances have been determined and recorded. The results are presented in graphical, tabular, or equation form usually with temperature as a parameter. (Specific heats vary rather strongly with temperature but only slightly with pressure for most substances.)

Because the specific heats of substances are different for different processes, it is necessary to specify the processes involved. This is usually done with a subscript indicating what has been held constant in the process. Thus, c_p indicates specific heat in a constant pressure process and c_v indicates the specific heat in a constant volume process. (We shall show in a later chapter that c_p and c_v cannot be the same for a given substance and that $c_p > c_v$ for all substances.)

6-2(b) Operational definition

Calorimetry makes it possible for us to measure heat transfers and hence gives us a useful operational definition. This is what we need to proceed to the first law but let us not leave the subject until we have written a clear verbal statement of our operational definition.

Heat is transferred from one system to another when there is a temperature difference between them, the heat flowing from the higher to the lower temperature region. In the absence of work, the heat transferred is measured using calorimetry by comparison with a standard heat unit (Btu or calorie). Heat transferred *to* a system is taken as positive for that system; heat transferred *from* a system is taken as negative for that system.

In elaboration on the statement above, students should note that heat is an *interaction* between systems and not a property of a system or something a system *has*. We speak only of heat *transferred*, not heat contained. Watch out for the old caloric theory hang-ups!

Another point that sometimes causes trouble is the assumption that the only way a system can change its temperature is by having heat transferred to or from it. This is simply not true. Students should be careful to distinguish between a *difference* in temperature between two systems and a *change* in temperature of one system. It is quite possible for a system to have heat transferred to or from it, while it remains at constant temperature. Consider water boiling on the stove as an example.

6-2(c) Heat transfer: A "slow" rate process

Heat is transferred from one system to another by radiation and/or conduction, with or without the presence of convection flows. Many engineering students will take a course in heat transfer so it is not our purpose here to discuss these mechanisms in detail. All that matters here is how *much* heat is transferred, not the mechanism of the transfer. Sometimes, however, the *rate* at which heat is transferred is important.

Heat transfer, regardless of the mechanism, is a relatively *slow* rate process compared with most mechanical and electrical processes. For this reason, many processes that occur rapidly (in, say, seconds or less) may be solved using the assumption that the process is adiabatic. Even some steady flow processes are substantially adiabatic when looked at on a *per unit mass* of substance flowing. (Careful! Neglecting heat on a per mass basis may in some cases be fine for the energy laws but if you have *many* pounds flowing and you are *paying* for the heat, the *economics* may not be so fine!)

The point here is that for many processes that occur rapidly, heat transfer is often negligible because it requires a rather long time and/or great pains to transfer heat very rapidly.

EXAMPLE 6.5. Assume that the specific heats for air are constant at $c_p = 0.24$ Btu/lbm-°F and $c_v = 0.17$ Btu/lbm-°F. Find the heat required to raise the temperature of 1 lb of air from 60 to 160°F (1) at constant pressure and (2) at constant volume.

SOLUTION: At constant pressure,

$$Q = \int mc_p\, dt = (1)(0.24)(100) = 24 \text{ Btu}$$

At constant volume,

$$Q = \int mc_v\, dt = (1)(0.17)(100) = 17 \text{ Btu}$$

Heat is identified as the interaction between systems *only* when, but also *always* when, there is a temperature difference between the systems. If there is not a temperature difference, then no heat transfer takes place. This fact helps us to identify heat interactions between systems and to distinguish them from work interactions.

EXAMPLE 6.6. An ideal gas is allowed to expand in a closed cylinder without heat transfer to a lower pressure and larger volume according to the process equation $pv^{1.4} = \text{constant}$. Show this process on a *p-v* plot and compare it with an isothermal process to the same final volume. Is the final temperature in the $pv^{1.4}$ process above or below the starting temperature?

SOLUTION: The isothermal process for an ideal gas is $T = \text{constant}$. A process $pv^{1.4} = \text{constant}$ will have a steeper slope downward as shown.

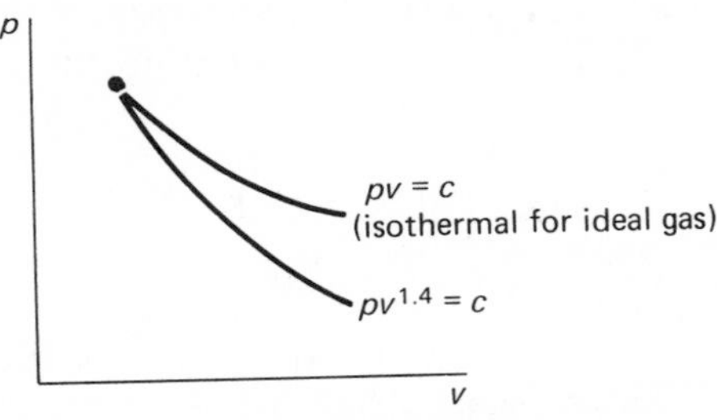

Since for ideal gas the isotherms decrease in value toward the origin, the final temperature is lower in the $pv^{1.4}$ process. (*Note:* The value of 1.4 is fairly typical, as we shall see later, for a process without heat exchange.)

EXAMPLE 6.7

1. A pressurized tank of gas has a large valve opened and the pressure falls to atmospheric pressure in a matter of seconds. If a system is defined as the gas remaining in the tank, what assumption would you make regarding heat transfer to or from the system during the blowdown?
2. Gas at 100°F flows steadily through a valve in a room at 60°F. In applying open system equations to the valve, what assumption would you make regarding heat transfer to or from the gas in the valve?
3. A tank of compressed gas springs a pinhole leak such that the pressure falls from 100 to 99 psia in 24 hr. Would you describe the process of the air in the tank as adiabatic or isothermal?

SOLUTIONS:

1. Because of speed of the blowdown, the adiabatic ($Q = 0$) assumption is justified.
2. Per pound of gas, the heat transferred to the gas in the valve will be very small and negligible.
3. Isothermal because this is a slow process allowing time for the heat transfer necessary to keep the tank isothermal.

6-3 Summary

Work and heat are energy in transit. They are *not* properties.

Work is done by a system (on another) when the sole effect external to the system could be the rise of a weight. The amount of work done is the product of the weight (force) times the distance lifted. By convention, work done *by* a system (which could lift weights in the environment) is taken as positive for that system; work done *on* a system (the environment lifts weights within the system boundaries) is taken as negative.

A summary of work terms for various systems is shown in Fig. 6.4.

Heat is transferred from a system (to another) when there is a temperature difference between them, the heat flowing from the higher to the lower temperature region. In the absence of work, the heat transferred is measured using calorimetry by comparison with a standard heat unit (Btu or calorie). Heat transferred *to* a system is taken as positive for that system; heat transferred *from* a system is taken as negative for that system.

The specific heat of a substance is the amount of heat (in the absence of work or a phase change) required to raise a unit mass of the substance 1° in temperature.

Heat is a slow rate process and hence some rapidly occurring processes may be assumed to be adiabatic, though engineering judgment is required.

For students interested in an enlightening discussion of thermodynamics and its implication, the following reading is recommended:

Order and Chaos by Angrist and Hepler, Basic Books, New York, 1967.

Problems

6-1 In a textbook, heat is defined as "that which transfers from one system, and to another at a lower temperature, solely by virtue of the temperature difference between them." Is this operational? Explain.

6-2 The specific heat for a substance is different for different processes. Thus, we define c_p and c_v. Are c_p and c_v properties? Explain. (Be careful!)

6-3 Air at 80°F is compressed isothermally in a closed cylinder from 20 psia to 40 psia.

(a) How much of man's work is required?

(b) How much total work is done on the air?

(c) Sketch the process on p-v, p-T, and T-v diagrams.

6-4 Nitrogen initially at 100°F is compressed from 20 to 40 psia according to a process equation of $pv^{1.4} = \text{constant}$. Compute the final temperature and the total work done on the nitrogen.

6-5 Saturated H_2O vapor at 200 psia is allowed to expand isothermally to 100 psia. Show this process on a p-v diagram. How much work does the steam do?

6-6(a) How much heat is required to raise the temperature of 1 lbm of air from 80 to 150°F at constant pressure?

(b) At constant volume? Assume $c_p = 0.24$ Btu/lbm-°F, constant; $c_v = 0.17$ Btu/lbm-°F, constant.

(c) Show the processes above on p-v, v-T, and T-p diagrams.

6-7 A hot copper block is suspended in a constant temperature room for cooling. If heat is transferred from the block at a rate proportional to the temperature difference between the block and the room, derive an expression for the temperature of the block as a function of time.

6-8 The specific heat of a substance is given $c_p = 2 + 0.02t$ where t is °F and c_p is Btu/lbm-°F. Compute the heat required to change the temperature of 1 lbm of this substance from 100 to 200°F.

6-9 If the specific heat of a substance is given by $c = a + bt$, derive an expression for the average specific heat over a range t_1 to t_2. (Be careful!)

6-10(a) Air at 100 psia and 100°F enters a tank through a pipe. How much is the flow work per pound of air?

(b) Water at 100 psia and 100°F enters a tank through a pipe. How much is the flow work per pound of water?

6-11 Assume that a liquid pump is pumping an incompressible liquid with a density $\rho = 60$ lbm/ft³ in an isothermal process from 10 to 40 psia. Relate the flow work into and out of the pump to the work required to do the pumping.

6-12 An ideal gas is compressed from p_1 and T_1 to p_2 isothermally. If the same gas is compressed to p_2 in a process $pv^{1.4} = \text{constant}$, will the final temperature be greater or less? Will the work required be greater or less?

6-13 A steel bar 8 ft long and $\frac{1}{2}$in.² in cross-sectional area is stretched isothermally from a load of $\sigma_1 = 10{,}000$ lbf/in² to $\sigma_2 = 30{,}000$ lbf/in². $E = 30 \times 10^6$ lbf/in.². Show that the work required is given by the expression

$$W = \frac{V}{2E}(\sigma_2^2 - \sigma_1^2)$$

and find the value of W.

6-14 See Example 6.1. Derive an expression for the work required to stretch the band isothermally from L_0 to a new length L. Then using the values given below, compute the work: $L_0 = 6$ in., $A_0 = 0.05$ in.², $K = 0.15$ lbf/in.²-°R, $T = 80$°F, $L_{final} = 12$ in.

6-15 Show that the work done per unit volume in a change of state from 1 to 2 of a paramagnetic substance obeying Curie's law is

$$W = \frac{T}{2C}(M_2^2 - M_1^2) = -\frac{C}{2T}(H_2^2 - H_1^2)$$

6-16 If the equation of state of a dielectric substance is given by $PT = CE$, derive an expression in terms of E_1 and E_2 for the work done in an isothermal process increasing the electric field.

6-17 Steam enters the cylinder of a steam engine at a constant pressure of 100 psia. The bore and stroke are 6 and 8 in., respectively. How much work is done? How much of this is useful to man if atmospheric pressure is 15 psia?

6-18 Heat is added to an ideal gas confined in a cylinder by a piston. Heat is slowly added and the piston is moved so that the gas remains isothermal.

(a) Show the process described on a p-v plot.

(b) Derive an expression for the work done by the gas as it expands from v_1 to v_2.

(c) If $Q = c_v \, \Delta T$ or $Q = c_p \, \Delta T$, then it would appear that no heat is transferred since $\Delta T = 0$. Explain.

6-19 Assume air has a specific heat at constant volume of 0.17 Btu/lbm-°F. A pound of air in a rigid container has heat added to raise its temperature from 70 to 75°F. After cooling back to 70°F, the same air is mixed by a high-speed paddle wheel until the temperature is again 75°F. How does this process illustrate that heat and work are not properties?

6-20 Which of the following systems or processes, if any, would you assume to be adiabatic? Explain briefly.

(a) Water flows through a car radiator.

(b) Water is pumped by a car water pump.

(c) Air passes through a small turbine in an air powered drill.

(d) Steam passes through a large turbine.

(e) A light bulb operates in steady state.

(f) A well-insulated refrigerator operates in steady state.

6-21 For each of the following systems and processes, state whether the heat and work are positive, negative, or zero.

(a) The system is the H_2O in a radiator. The process is that the valve is closed while the H_2O is steam and it all cools to room temperature.

(b) The system is the air in a balloon. The process is that the balloon breaks . . . .

(c) The system is the H_2O and food confined in a pressure cooker. The process is that the stove burner under the cooker is turned on.

(d) The system is the paint in a can. The process is that the paint is stirred with a stick.

6-22(a) A gas system executes a process in which its pressure increases. Show how that process might look on a p-v plot.

(b) What area represents $\int p \, dv$ on the plot?

(c) Under what conditions does $\int p \, dv$ represent work?

(d) What area represents $\int v \, dp$ on the plot?

6-23 Derive an expression for the work done by a van der Waals' gas expanding isothermally in a cylinder from v_1 to v_2.

7

Induction of the Clausius inequality and the first law

Up to this point in our development we have operational definitions for three basic and important terms: thermal equilibrium, work, and heat. We used observation and experiments with thermal equilibrium to lead us to temperature. Now we shall use observations and experiments with work and heat—as well as temperature if appropriate—to induce other important principles. We shall carry out this step here by presenting you with data from which the desired results may be induced. The data will be hypothetical; they are concocted to illustrate the logical process by which the laws are induced.

The names of the relations that we induce are (1) the Clausius inequality and (2) the first law. In Chap. 8, we show how the Clausius inequality leads directly to what is called the second law.

Both of the relations to be discovered here (the Clausius inequality and the first law) are induced from observations of *cyclic* processes. We begin this chapter, therefore, with a discussion of cycles.

7-1 Cycles and properties

A cycle is a process in which the system ends up at the same state at which it started. Schematically a cycle may be represented on a property diagram (e.g., p-v or p-T) as shown in Fig. 7.1.

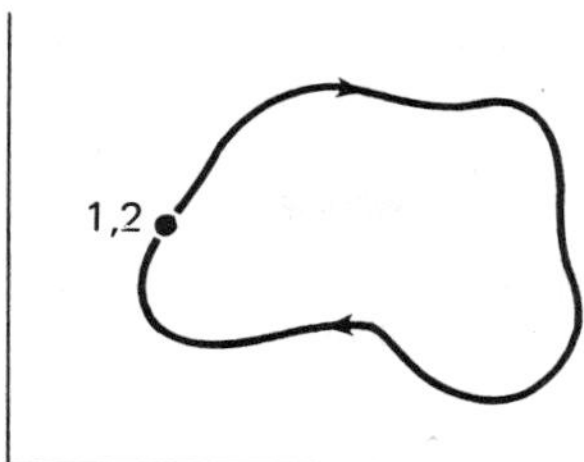

Figure 7.1 Schematic representation of a cyclic process.

Cycle processes are especially important because of their use in energy conversion and because they operate continuously. A cyclic energy converter is defined as a closed system operating in cycles that exchanges heat and work with its environment. A cyclic energy converter that *delivers* net work to its environment is called a *cyclic heat engine*. An example of a cyclic heat engine is the mass of steam that circulates through a steam power plant.

A cyclic energy converter that *receives* net work from its environment is called a *cyclic heat pump* or refrigerator. An example is a mass of Freon circulating in a refrigeration system. In these examples, the equipment in the system (turbines, pumps, etc.) may also be included as part of the cyclic energy converter as long as no mass crosses the boundary of the closed system defined as the energy converter.

It is common practice to represent heat engines and heat pumps schematically as shown in Fig. 7.2. In real energy converters (engines and pumps), heat is usually *rejected* to the *atmosphere* (air, ocean, earth, etc.). The heat *input* to engines is usually from the high temperature products of combustion of a fuel and the heat pumped by refrigeration is from cold spaces such as for food storage boxes and air-conditioned offices.

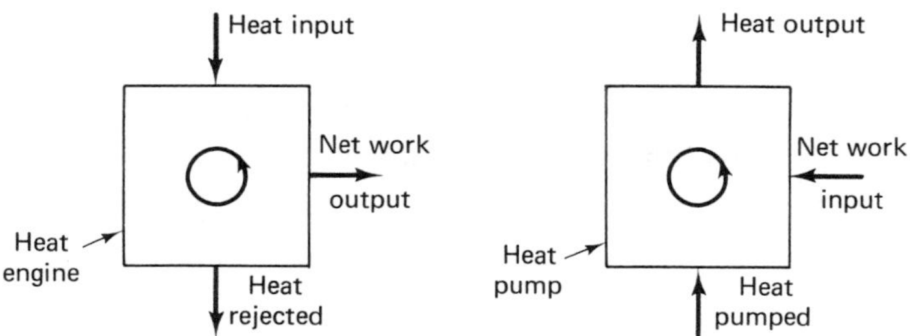

Figure 7.2 Schematic representations of heat engine and heat pump.

There are at least two good reasons for looking at cycles in the search for significant relationships among temperature, work, and heat. One is that cycles are so very common and important in energy conversion. Another is that the cycle, because the system returns to its initial state, enables us to experiment without having to worry about the equations of state of the various substances used. Related to this point is the very important fact that if we can find some parameter (for now, say, X) for which the sum total of all changes in a cycle add to zero, then that parameter is a property. (We shall prove this assertion in the next paragraph.) Thus, if there are new properties to be discovered, they may be discovered in experiments with cyclic processes.

We shall now prove that if the net change in a parameter (X) in a cycle is always zero, then the parameter (X) is a property. Recall that the distinguishing characteristic of a property is that its change of value in a process is dependent only on the end states and not on the process itself. Thus, such quantities as temperature, mass, and volume are properties but heat and work are not. Schematically this identifying or defining characteristic of properties may be shown by the following sketch and notation:

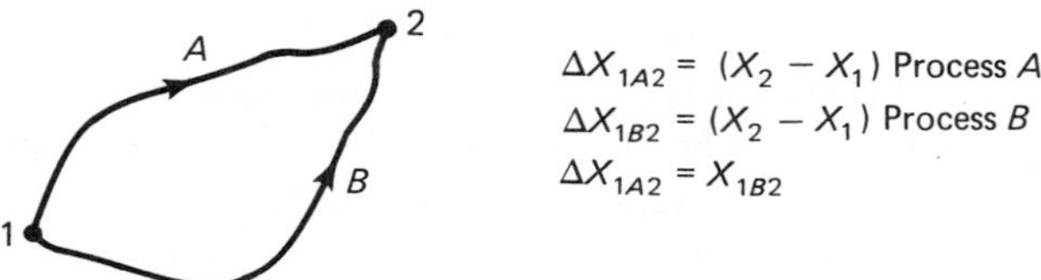

where X stands for the value of a known property.

Now let us consider some quantity Y that is measured as a system undergoes a cycle from 1 to 2 and back to 1.

The nature of Y is to be such that its net change or value in a cycle is always zero. We take the system through two cycles, 1-A-2-B-1 and 1-A-2-C-1, as shown in the sketch and apply the given fact that the changes in Y be zero for each cycle.

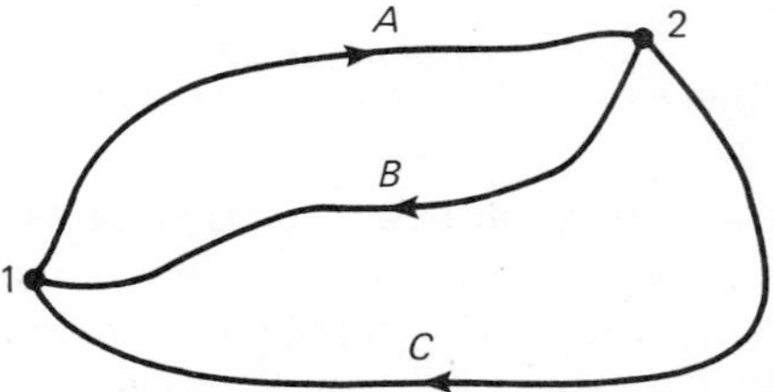

For 1-*A*-2-*B*-1,

$$\Delta Y_{1\text{-}A\text{-}2} + \Delta Y_{2\text{-}B\text{-}1} = 0$$
$$\Delta Y_{1\text{-}A\text{-}2} + \Delta Y_{2\text{-}C\text{-}1} = 0$$

Subtracting these two equations, we find

$$\Delta Y_{2\text{-}B\text{-}1} = \Delta Y_{2\text{-}C\text{-}1}$$

That is, the change in Y is the same along path B as along path C. Since these paths are quite arbitrary, the change in Y in a state change is *independent* of the path or process. Thus, we have shown that if any quantity always sums to zero in a cyclic process, that quantity is a property of the system. This result will be important to us later.

It was noted before but should be recalled again that heat and work are *not* properties of systems. They are interactions between systems and very much dependent on the process.

We can now turn our attention to the major purpose of this chapter—induction of the Clausius inequality and the first law.

7-2 Induction of the Clausius inequality

7-2(a) Why look for a relationship involving heat and temperature?

As noted in Chap. 2, the process of induction or generalization often involves not only careful perception of the data but also hard work, intelligence, and sometimes perhaps genius. Of course, we all make our easy little daily generalizations in order to function reasonably competently in the world. For example, if we know that Mary has been late to work on Monday for each of the last 4 weeks, we simply don't let important plans depend on her being there on time next Monday. If our old car hasn't started on the last 80 rainy or damp mornings, then we don't count on it starting on the next wet day. And if we contracted a bad case of poison ivy from tramping through a particular woods, either we don't tramp through that place again or else we wear high boots. These are all examples of decisions based on rather trivial generalizations of the kind made regularly. Note that none are absolutely certain conclusions. Mary might be there on time next Monday, once in a while the old car does start in the rain, and we might well miss the ivy patch on another walk.

In science, the generalization process is more rigorous and is subjected to careful experiment. We might, for example, decide to be scientific about the behavior of the old car and begin to keep records of when it starts and when it doesn't along with such items as temperature, humidity, the amount of gas in the tank, the day of the week, the hours since the car was driven, etc. But which should we look for? How do we spot the proper generalization?

Our experience helps us to decide what to look for. We note that the car has more trouble on cold wet days than on warm wet days so we would want to watch temperature. We also use other facts that are already known. The level of gas in the tank isn't related to the ignition system so we won't bother with those data. And we may have a theory to help us. We think perhaps that condensation on the ignition wires, plugs, and distributor cap is draining off the current so let's look at the humidity and the temperature of the car itself. Even with all this going for us, a great deal of hard data taking and some genius in seeing the proper relationships among the data would be required.

So to help us find inductive relationships we use our experience, gained in everyday life and in the laboratory, we use what is already known, and we use any theories or hunches that we may have. To induce the Clausius inequality, an important induction of thermodynamics, we are going to have to look at heat and temperature in cycles. More specifically, we are going to have to look at the ratio of heat to temperature in cycles

$$\sum_{\text{cycle}} \frac{Q}{T} \quad \text{or} \quad \oint \frac{d'Q}{T}$$

We have already discussed why cyclic process will be considered but why heat and temperature and, especially, why their ratio?

Heat and temperature are almost always intimately connected. Heat is transferred not only when but also always when there is a temperature difference and the greater the temperature difference, the greater the heat transfer rate. Thus, that some thermodynamically significant relationship between heat and temperature might exist should not seem at all unlikely.

Why their ratio? One rather trivial answer is that their ratio has the same dimensions (heat/temperature) as another important thermodynamic property: specific heat. A more important clue might come from experience with cyclic energy converters—experience that admittedly students do not have.

7-2(b) The Clausius inequality

Let us suppose that for good reasons, it has been decided to study the ratio of heat to temperature in cyclic processes. The results might look like this:

System	*Description of cycle*	$\oint \frac{d'Q}{T}$ (Btu/°R)
Freon-12	Refrigerator	−6.3
Steam	Power plant	−122.8
A ton of bricks	Heat to 500°F in furnace, cool in atmosphere	−5280
Air	Compress in a cylinder, cool to original temperature, expand to original pressure, and heat to original temperature	−1.5
Block and inclined plane	Block slides down, is pushed back up, and then cooled to original temperature	−0.41

What could you induce from such data? Of course, a number of conclusions are possible but the one that experience and more stringent experiments suggests is that the inductive conclusion to be reached is as follows: In a cycle,

$$\oint \frac{d'Q}{T} \quad \text{or} \quad \sum_{\text{cycle}} \frac{Q}{T} \tag{7-1}$$

is always negative. That is,

$$\oint \frac{d'Q}{T} < 0 \tag{7-2}$$

This relationship is known as the *Clausius inequality*.

It is important that students understand the notation here. The equation applies to a closed system in a cycle. $d'Q$, as always, is positive for heat transferred *to* the system and negative for heat transferred from the system. Remember that heat is measured at the system boundaries. The T in the equation is the absolute temperature of the system at the point where the heat is added or removed. (Note that T, being absolute, is therefore always positive.)

EXAMPLE 7.1

1. The working fluid in a cyclic heat engine receives an amount of heat equal to 500 Btu while it is at a temperature of 1000°R. It rejects 400 Btu while it is at a temperature of 500°R. Compute $\oint (d'Q/T)$ for this cycle.
2. An inventor claims to have a refrigerator that operates in cycles taking in 4000 Btu/hr where the engine fluid temperature is 0°F and rejecting 6000 Btu/hr where the engine fluid temperature is 100°F. Does this engine violate the Clausius inequality?

SOLUTION:

1.

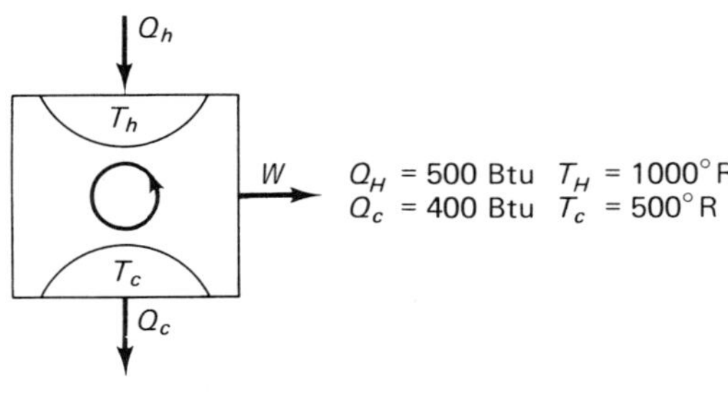

$$\oint \frac{d'Q}{T} = \frac{500}{1000} - \frac{400}{500} = 0.5 - 0.8 = -0.3 \text{ Btu/°R}$$

2.

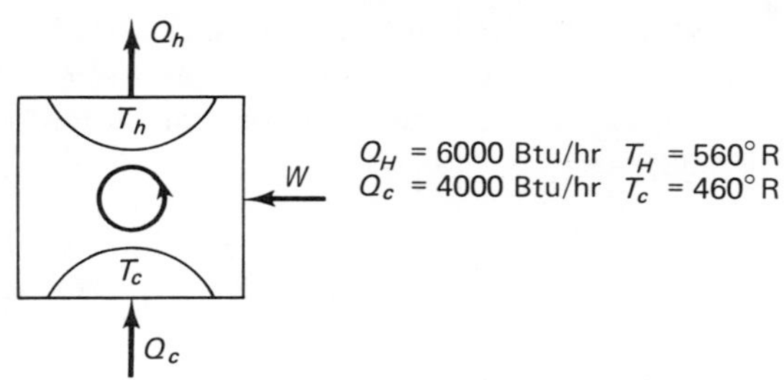

$$\oint \frac{d'Q}{T} = \frac{\dot{Q}_c}{T_c} - \frac{\dot{Q}_H}{T_H} = \frac{4000}{460} - \frac{6000}{560}$$

$$= 8.7 - 10.7 = -2.0 \text{ Btu/hr-°R}$$

The Clausius inequality *is* satisfied because

$$\oint \frac{d'Q}{T} < 0$$

EXAMPLE 7.2. An inventor claims to have a cyclic device that takes heat from the atmosphere into his engine fluid at 5°C, produces work output only, and rejects no heat. Does this violate the Clausius inequality?

SOLUTION:

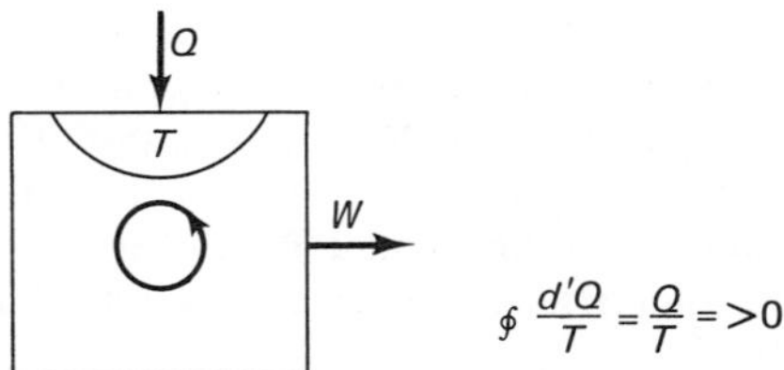

$$\oint \frac{d'Q}{T} = \frac{Q}{T} = >0$$

Yes, the Clausius inequality is violated.

7-3 Induction of the first law

7-3(a)
Why look for a relationship involving heat and work?

Heat is measured by the techniques of calorimetry in units of calories or Btu. Work is measured quite independently and in different units as the product of force and distance. To complicate matters remember that historically there was also a "conservation of caloric" theory. Why, then, bother to look at heat and work in cycles?

One answer lies in experience, again both in life and in the laboratory. Consider the common phenomenon of friction, for example. The work done in sliding a block along is accompanied apparently by a creation of "caloric."

This simple process discredits the caloric theory and gives us an example of work and heat appearing together and apparently related in some way. There is also, again, experience with cyclic energy converters that might be taken into account.

The caloric theory might have been called the principle of conservation of heat. Calorimetry—the measurement of heat—did not really suffer from the death of this theory because the experiments of calorimetry are done with no work and within adiabatic confines, that is, in special processes in which heat *is* conserved. Problems only arose with the theory in such processes as friction where work is also involved and where caloric is apparently created.

Work, too, presented a theoretical problem even though it could be identified and measured. Its measurement required at least the assumption of a conservative (frictionless, idealized) process, that is, a conservation of work process. Such processes must be frictionless or else "work" disappears and "caloric" appears.

The solution to the situation came with the observation that it must be some combination of heat and work that is conserved.

7-3(b) The first law

Consider the following data, all for cyclic processes:

	Cycle Net heat		*Cycle Net work*	
System	Btu	Joules	ft-lbf	Joules
10 lbm air	+1	+1055	+778	+1055
4 lbm H_2O	−2	−2110	−1556	−2110
1 lbm O_2	+10	+10,550	+7780	+10,550
0.3 lbm stone	−5	−5275	−3890	−5275
492 tons brick	+100	+105,500	+77,800	+105,500
6 lbm steel	−10	−1055	−7780	+42,200
0.002 lbm He	+40	+42,200	+31,120	+42,200
9 lbm H_2O	+0.10	+105.5	+77.8	+105.5
6 lbm air	+0.05	+52.7	+38.9	+52.7
1 lbm Al	−0.01	−10.5	−7.8	−10.5

What would *you* conclude from such data?

One could conclude several things, for example, that heat and work always have the same sign in a cycle or that in a cycle heat is never greater than 1000 Btu (105,500 j) and work is never greater than 77,800 ft-lbf (105,500 j). It would be easy with another experiment to show this one is incorrect, however; or you could decide that *in a cycle*

$$Q\,(\text{Btu}) = \frac{W\,(\text{ft-lbf})}{778} \quad \text{or} \quad Q\,(\text{joules}) = W\,(\text{joules}) \tag{7-3}$$

This latter inductive conclusion has been observed to be the case in all such experiments (not including nuclear reactions) and is a statement of the first law. In better notation,

$$\sum_{\text{cycle}} Q = \frac{1}{J} \sum_{\text{cycle}} W \tag{7-4}$$

$$\oint dQ = \frac{1}{J} \oint dW \tag{7-5}$$

$$\oint \left(dQ - \frac{dW}{J} \right) = 0 \tag{7-6}$$

where J is simply a proportionality constant dependent on the units used to measure heat and work. With heat in Btu and work in ft-lbf, $J = 778$ ft-lbf/Btu. Other values for J are given below:

Heat	*Work*	*J*
Btu	ft-lbf	778 ft-lbf/Btu
Btu	kw-hr	$\frac{1}{3412}$ kw-hr/Btu
Btu	hp-hr	$\frac{1}{2544}$ hp-hr/Btu
cal	joules	4.186 j/cal
joules	joules	1

Henceforth in this book we shall omit the J from all equations and assume that Q and W are expressed in the same units or that the appropriate conversion factor will be included by students in their numerical work. In the SI metric units, both heat and work are measured in Newton-meters or joules (1 N-m = 1 j) so that $J = 1$.

Let us return now to the heat engine as an example of the application of the first law. Consider again a simple engine receiving heat Q_h per cycle rejecting Q_c (as demanded by the Clausius inequality per cycle) and doing net work W_n per cycle. With heat and work in the same units, the first law gives

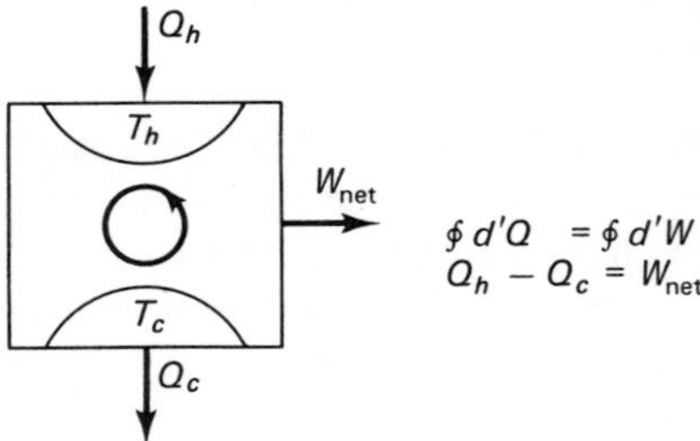

where Q_c and Q_h are both positive as written. The point to notice here is that since Q cannot be zero (or negative) by the Clausius inequality (see Example 7.2 and Prob. 7-1), the net work cannot be as large as the heat input. The first law and the Clausius inequality combine to require that

$$W_{\text{net}} < Q_h \tag{7-7}$$

We can therefore say that it is impossible for a cyclic heat engine to convert heat completely into work. This, of course, begins to sound very much like the concept of energy degradation discussed in Chap. 2. And we shall show in Chap. 9 that the Clausius inequality, the principle of degradation of energy, and the second law all are logically equivalent statements.

EXAMPLE 7.3. You are an adviser to a patent lawyer. A client of the lawyer presents drawings of an engine that he claims produces 10 kw output while taking in 4×10^7 j of heat at 500°C and rejecting 1×10^7 j to the atmosphere at 30°C. Would you advise the lawyer to pursue the patent?

SOLUTION:

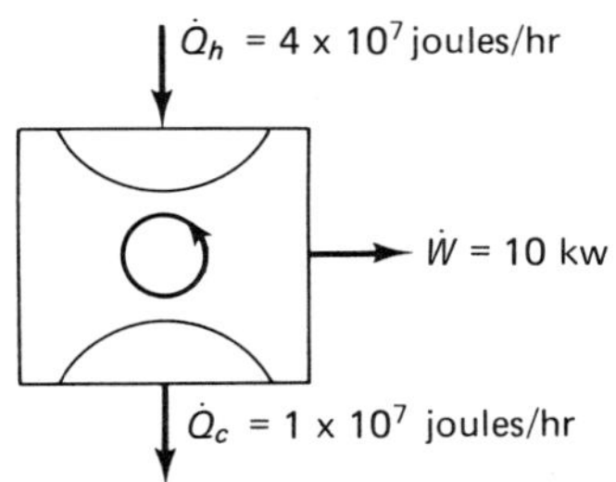

In an hour,

$$\oint \dot{Q} = (4 - 1) \times 10^7 = 3 \times 10^7 \text{ j}$$

$$\oint \dot{W} = 10 \text{ kw-hr} - 10^4 \times 3600 \text{ j} = 3.6 \times 10^7 \text{ j}$$

The engine violates the first law! You had better advise the lawyer to have the inventor check his figures.

EXAMPLE 7.4. On the p-v plot an idealized steam power cycle is shown. The following process heat and work terms are known:

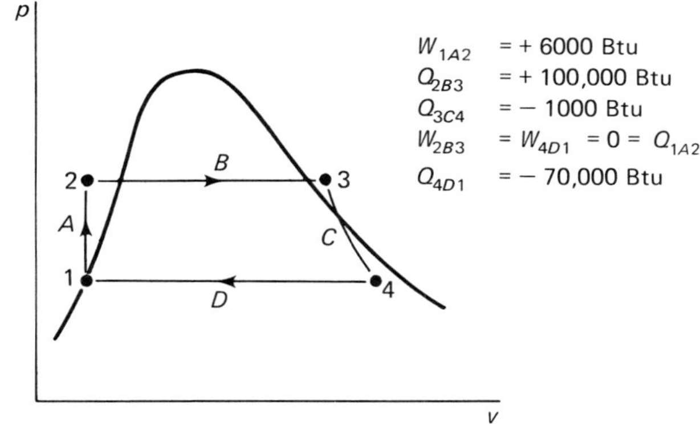

Find W_{3C4} (the work output of the turbine).

SOLUTION:

$$\oint d'Q = \oint d'W$$

$$Q_{1A2} + Q_{2B3} + Q_{3C4} + Q_{4D1} = W_{1A2} + W_{2B3} + W_{3C4} + W_{4D1}$$

$$0 + 100{,}000 - 1000 - 70{,}000 = 6000 + 0 + W_{3C4} + 0$$

$$W_{3C4} = 23{,}000 \text{ Btu}$$

7-4 Summary

Before going further, let us now review the logical development as a whole.

We have taken the operational-inductive approach to development of the laws. This begins with operational definitions and then uses experience and experimental observations to suggest important inductive conclusions. We have considered operational definitions for such basic terms as force, mass, length, and time and for the fundamental thermodynamic terms of thermal equilibrium, heat, and work. From consideration of experiments and observations with mass, we induced the principle of conservation of mass, which we then applied to both open and closed systems. From experiments and observations with thermal equilibrium, we induced the zeroth law and we then went on to develop deductively the "new" property, temperature. In this chapter we have induced both the Clausius inequality and the first law. This was done in a rather hypothetical manner using some made-up data. Students should understand that this was merely to illustrate the concept. Historically, of course, those who first actually induced the laws had much more data but much less organization since they did not know exactly what they were looking for as we did here. One suspects they began to *feel* the laws from experience, both in daily life and in the laboratory, something like your own experience. After that, specific experimental "verification" was not too difficult. Incidentally, the original developments of this type took place in the period from about 1800 to 1850. The big names are Benjamin Thompson (Count Rumford), Joule, Carnot, Kelvin, and Clausius.

In any case, *we* have now imagined ourselves to have induced the "laws" too. Our position is illustrated schematically on the next page.

In this chapter, the Clausius inequality and the first law have been introduced as inductive generalizations from observations and experiments with heat, work, and temperature. The Clausius inequality is

$$\oint \frac{dQ}{T} < 0 \tag{7-8}$$

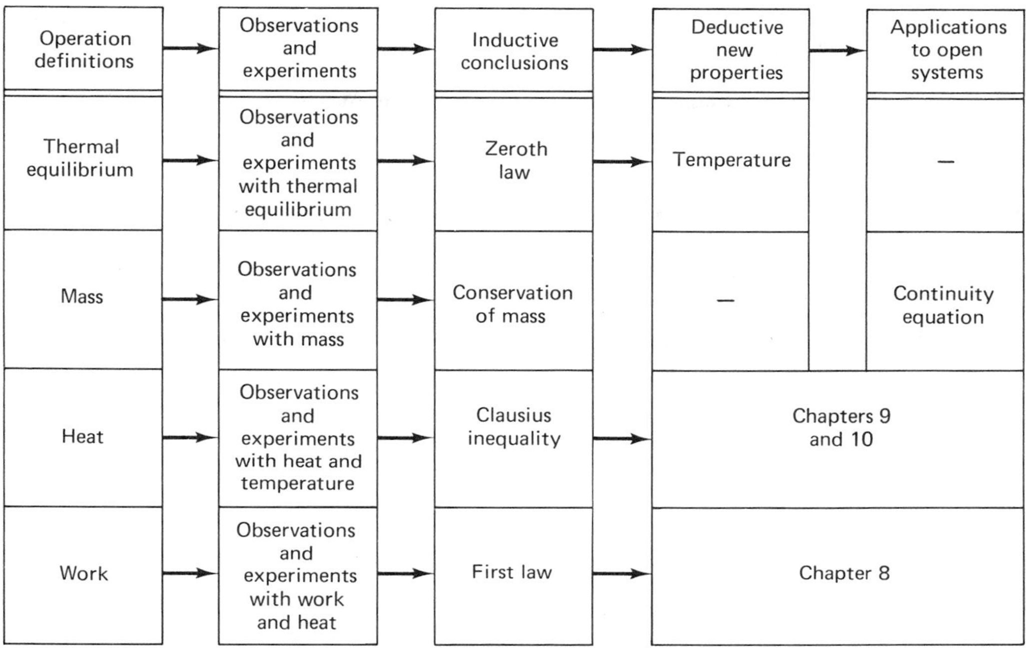

Operation definitions
Observations and experiments
Inductive conclusions
Deductive new properties
Applications to open systems
Thermal equilibrium
Observations and experiments with thermal equilibrium
Zeroth law
Temperature
—
Mass
Observations and experiments with mass
Conservation of mass
—
Continuity equation
Heat
Observations and experiments with heat and temperature
Clausius inequality
Chapters 9 and 10
Work
Observations and experiments with work and heat
First law
Chapter 8

The first law is

$$\oint (d'Q - d'W) = 0 \tag{7-9}$$

In the next chapter, we explore the deductive and practical consequences of the first law. It will lead us to a new property, energy, and to a new equation for open systems called the *energy equation*. In Chap. 9 and 10, we shall explore the consequences of the Clausius inequality and find that it leads us to a new property, entropy, and to a new inequality for open systems called the *entropy inequality*. Then in later chapters we shall combine all that we have learned to restudy relationships among properties of substances and man's efforts at energy conversion.

A very good place to find out more about the way the laws of thermodynamics are developed is to go to your library and read the following article:

"On the History and Exposition of the Laws of Thermodynamics" by Keenan, J. H., and Shapiro, A. S. *Mechanical Engineering*, **69,** 1947, pp. 915–921.

Also, the following book is excellent for students who wish to understand better the way in which all scientific principles are developed:

The Nature of Physical Theory by P. W. Bridgman. Dover Publications, New York, 1936.

Problems

7-1 Show that the net heat rejected (Q_c) from a cyclic heat engine cannot be negative.

7-2 If electricity can be purchased for \$0.02/kw-hr and the interest rate is 10%, how much would you pay for a machine that could take heat from the environment (earth, air, ocean, etc.) and produce 1,000,000 kw-hrs/yr of work?

7-3 A cylinder of air executes the following cycle starting at 1.3 bars absolute and 30°C:

Process A: Isothermal compression to 3.3 bars absolute.
B: Isobaric cooling to 20°C.
C: Constant volume cooling to 1.3 bars absolute.
D: Isobaric heating to 30°C.

(a) Show the cycle on a p-v plot.
(b) Show the area that represents the net work in the cycle. Is the net work positive or negative?
(c) Compute the net work per pound of air.

7-4 The p-v plot shows a concocted steam power cycle. Heat is exchanged in processes B and D only. Compute $d'Q/T$ for the cycle if $p_B = 400$ psia and $p_D = 1$ psia.

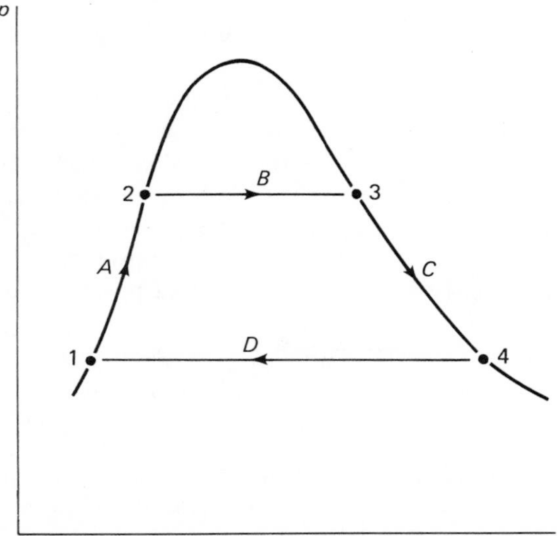

7-5 A student proposes to conceal a small refrigerator in a closet that is kept closed and locked. As a thermodynamicist, comment on this.

7-6 A heat engine is supplied with 100,000 Btu/hr at 800°F and rejects heat at 100°F. What is the maximum horse power this engine can produce?

7-7 A refrigerator receives 1 kw of heat at −40°C and rejects heat at +40°C. What is the least amount of work required to operate the refrigerator?

7-8 A cyclic energy converter operates at the temperatures and with the energy flows as shown below. Is it a refrigerator or an engine?

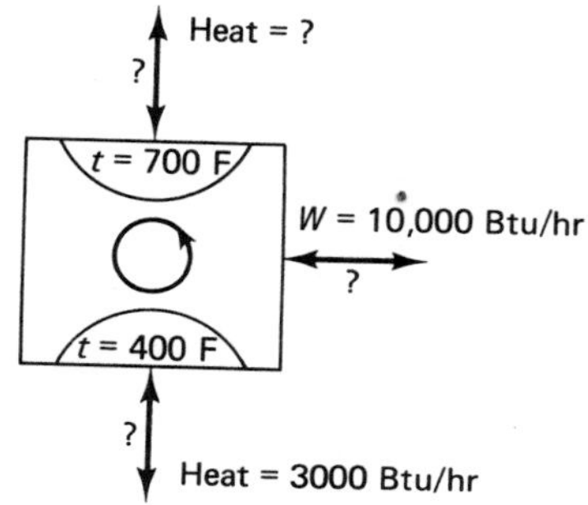

8

Internal energy, the energy equation for open systems and enthalpy

This is a chapter on applying the first law. It answers such questions as What is the property energy? How is it obtained from the first law? How is energy conservation applied in closed and open systems? What is internal energy? Enthalpy? How are internal energy and enthalpy related to specific heat? How are data for internal energy and enthalpy found for real substances? For ideal gases? What kinds of engineering assumptions can be made using the first law?

8-1 Energy and internal energy

8-1(a) Energy

In the last chapter, we showed that if the net change in some quantity is always zero in a cycle, then that quantity is a property. We also showed that $(d'Q - d'W)$ is a property. This new property is called *energy* and is given the symbol E. Symbolically,

$$d'Q - d'W = dE \tag{8-1}$$

Integrating for a process A from state 1 to state 2 gives

$$\sum_{1A2} Q - \sum_{1A2} W = \Delta E_{12} \tag{8-2}$$

or in more simple notation

$$Q_{1A2} - W_{1A2} = \Delta E_{12} \tag{8-3}$$

The notation ΔE_{12} means the change in the property E in the process 1-2. Since E is a property, this change is independent of the process and so the process is not indicated. We often write

$$\Delta E_{12} = E_2 - E_1 \tag{8-4}$$

even though we do not always know, or need to know, the actual values of E_1 and E_2. As an expedient, an arbitrary value of zero is sometimes assigned at some arbitrary state so that other relative values can be tabulated.

Energy, then, is the new property that develops out of the first law.

Incidentally, we do *not* (not *ever*) write ΔQ or ΔW. Such notation would imply that heat and work are properties. Unless you believe in caloric, you can't talk about the change in the amount of heat. We write ΔE because E is a property.

8-1(b) Kinetic, potential, and internal energy

The energy E of a closed mass system is usually divided into three parts because two can be easily identified. Those two are kinetic energy and potential energy. The remainder is called the *internal* energy and is given the symbol U. Thus we write

$$E = U + \text{K.E.} + \text{P.E.} \tag{8-5}$$

To derive equations for the kinetic and potential energies, we apply the first law together with Newton's law in special cases. First, consider an adiabatic system, all of whose properties remain fixed except velocity. Then kinetic energy will be the only property change. Let a horizontal force F be applied to the system to change its horizontal velocity from $\mathcal{V}$ to $\mathcal{V} + d\mathcal{V}$ and its horizontal position from X to $X + dX$.

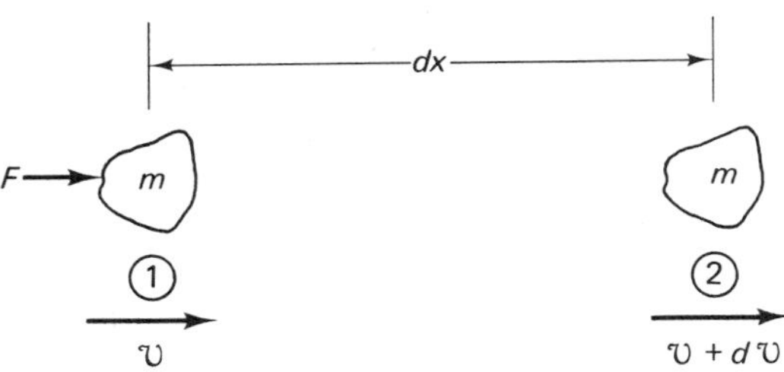

The first law for this process gives

$$Q_{12} - W_{12} = \Delta E_{\text{kinetic}} \tag{8-6}$$

But the process is adiabatic so $Q_{12} = 0$. The work done by the system is

$$W_{12} = -\int_1^2 F\,dX \tag{8-7}$$

Newton's second law gives

$$F = ma = m\frac{d\mathscr{V}}{d\theta} \tag{8-8}$$

Combining,

$$W_{12} = -\int_1^2 m\frac{d\mathscr{V}}{d\theta}\,dX = -\int_1^2 m\frac{dX}{d\theta}\,d\mathscr{V} = -\int_1^2 m\mathscr{V}\,d\mathscr{V} \tag{8-9}$$

$$W_{12} = -\tfrac{1}{2}m(\mathscr{V}_2^2 - \mathscr{V}_1^2) \tag{8-10}$$

And since

$$W_{12} = -\Delta E_{\text{kinetic}}$$

we find

$$\Delta E_{\text{kinetic}} = \tfrac{1}{2}m(\mathscr{V}_2^2 - \mathscr{V}_1^2) \tag{8-11}$$

or in general

$$E_{\text{kinetic}} = \tfrac{1}{2}m\mathscr{V}^2 \tag{8-12}$$

EXAMPLE 8.1. A system executes the process A shown and it is found that $W_{1A2} = +60$ j and that $Q_{1A2} = +80$ j. What is the work in process B if $Q_{1B2} = -10$ j?

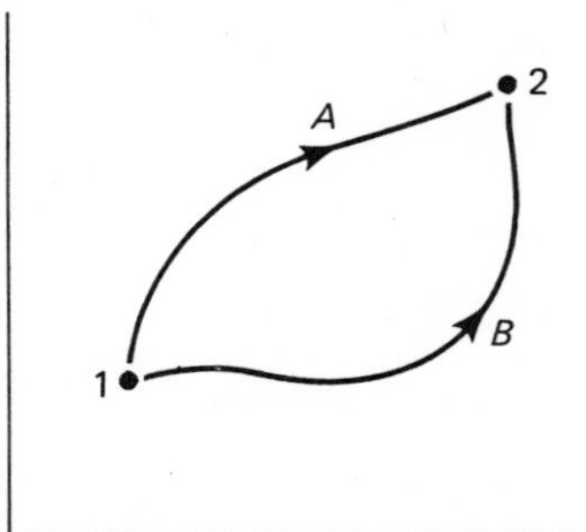

SOLUTION: For process $1A2$,

$$E_{12} = Q_{1A2} - W_{1A2} = 80 - 60 = 20 \text{ j}$$

Since E is a property,

$$(\Delta E_{12})_A = (\Delta E_{12})_B$$

Thus for process $1B2$,

$$\Delta E_{12} = Q_{1B2} - W_{1B2}$$
$$20 = -10 + W_{1B2}$$
$$W_{1B2} = 30\text{ j}$$

For potential energy, consider an adiabatic process in which the only change is the elevation in a gravity field. Then potential energy will be the only property change.

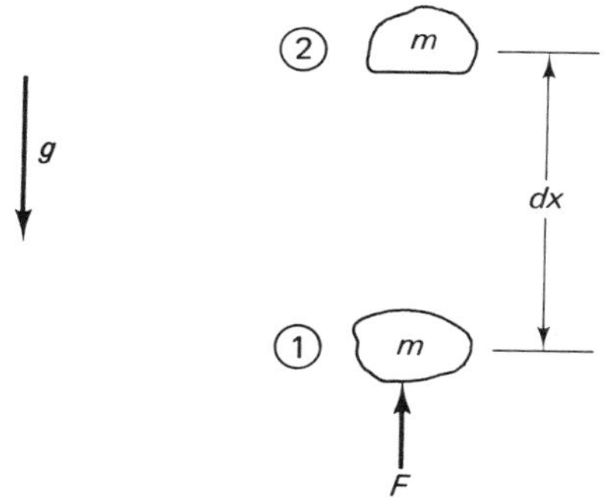

The work done *by* the system is

$$W_{12} = -\int_1^2 F\,dx = -\int_1^2 mg\,dx = -mg(h_2 - h_1) \tag{8-13}$$

$$E_{\text{potential}} = mg(h_2 - h_1) \tag{8-14}$$

or

$$E_{\text{potential}} = mgh \tag{8-15}$$

Thus, if the total energy E is divided as indicated above, we find

$$E = U + E_{\text{kinetic}} + E_{\text{potential}} \tag{8-16}$$

or

$$E = U + E_{\text{kinetic}} + E_{\text{potential}} \tag{8-17}$$

$$E = U + \tfrac{1}{2}m\mathcal{V}^2 + mgh \tag{8-18}$$

If we divide through by the mass of the system and use lowercase letters for the energy per unit mass, the result is

$$e = u + \tfrac{1}{2}\mathcal{V}^2 + gh \tag{8-19}$$

As noted on page 141, e is called the energy (or sometimes *total* energy) and u is called the *internal* energy.

In using the equations above, engineers must be careful to use a consistent set of units. The conversion factors g_0 and J are not shown in the equations but are often needed.

EXAMPLE 8.2. Suppose the internal energy u of a 100-lbm system is known to be 100 Btu/lbm, its velocity is 100 ft/sec, its mass is 10 lbm, its height above a gravity potential reference is 100 ft, and $g = 30$ ft/sec². Compute the total energy E.

SOLUTION:

$$e = u + \text{K.E.} + \text{P.E.}$$

$$\text{K.E.} = \frac{1}{2}\mathcal{V}^2 = \frac{1}{2}(100)^2 \frac{1}{(778)(32.2)} = \frac{(100)^2}{50{,}000} = \frac{10{,}000}{50{,}000} = 0.2 \text{ Btu/lbm}$$

$$(\text{ft}^2/\text{sec}^2)(\text{Btu/ft-lbf})(\text{sec}^2/\text{ft}) \rightarrow \text{Btu/lbm}$$

$$\text{P.E.} = gh = (30)(100)\frac{1}{(32.2)}\frac{1}{(778)} = \frac{3000}{25{,}000} = 0.12 \text{ Btu/lbm}$$

$$(\text{ft/sec}^2)\,(\text{ft})\,(\text{sec}^2/\text{ft})(\text{lbf/lbm})\,(\text{Btu/ft-lbf})$$

$$e = 100 + 0.2 + 0.12 = 100.32 \text{ Btu/lbm}$$

$$E = me = (100)(100.32) = 10{,}032 \text{ Btu}$$

8-2 Generalized energy and open systems

8-2(a) The concept of generalized energy

It is very helpful to think about work, heat, and internal energy together as a kind of *generalized energy* that is conserved. Using the symbol $\mathscr{E}$ for generalized energy, we might use the concept as follows for a closed system process from state 1 to state 2:

$$\mathscr{E}_{\text{initially}} + \mathscr{E}_{\text{added}} = \mathscr{E}_{\text{finally}} + \mathscr{E}_{\text{removed}} \tag{8-20}$$

$$E_1 + Q_{\text{added}} + W_{\text{on System}} = E_2 + Q_{\text{removed}} + W_{\text{by System}} \tag{8-21}$$

Verbally, this says that the initial energy (E_1) plus "energy" added as heat and work equals the final energy (E_2) plus "energy" removed as heat and work. Using the conventional notation where Q stands for net heat *to* the system and W stands for net work *by* the system, Eq. (8-21) easily reduces to the first law equation defining energy:

$$Q_{12} = Q_{\text{added}} - Q_{\text{removed}} \tag{8-22}$$

$$W_{12} = W_{\text{by System}} - W_{\text{on System}} \tag{8-23}$$

$$E_1 + Q_{12} = E_2 - W_{12} \tag{8-24}$$

$$E_2 - E_1 = Q_{12} - W_{12} \tag{8-25}$$

8-2(b) The energy equation for open systems

The development of the first law and its interpretation as conservation of "energy" has been restricted to closed systems so far. Many applications, however, are best handled using an open system (or control volume). Because such applications are so common, a general equation is developed for the first law as it is applied to the control volume or open system. The best way to develop this equation, which for obvious reasons is often called *the energy equation for open systems* or just *the energy equation*, is to use the concept of conservation of generalized energy. This approach is not very rigorous mathematically and students who wish a more precise or theoretical treatment are referred to one of the books mentioned in the Suggested Reading.

Applied to a control volume or open system (which is a region of space defined by a boundary across which material may flow) the principle of conservation of generalized energy is

$$\begin{bmatrix}\text{The rate of}\\ \text{“generalized energy”}\\ \text{flow in}\end{bmatrix} - \begin{bmatrix}\text{the rate of}\\ \text{“generalized energy”}\\ \text{flow out}\end{bmatrix} = \begin{bmatrix}\text{the rate of change of}\\ \text{“generalized energy”}\\ \text{inside}\end{bmatrix} \tag{8-26}$$

Note that this is completely analogous to the development of the continuity equation from conservation of mass where we wrote

$$\begin{bmatrix}\text{Rate of mass}\\ \text{flow in}\end{bmatrix} - \begin{bmatrix}\text{rate of mass}\\ \text{flow out}\end{bmatrix} = \begin{bmatrix}\text{rate of change}\\ \text{of mass inside}\end{bmatrix} \tag{8-27}$$

Generalized energy flow into or out of the open system may consist of three things: work (including flow work), heat, and the energy E of any material flowing across the boundary of the system. The generalized energy inside the system or control volume is only the internal, kinetic, and potential energy of the material inside because heat and work are not stored.

For consistency and simplicity, let us adopt the convention that $\sum_{\text{area}} \dot{Q}$ represents the *net* rate of heat transfer being transferred *into* the region and that $\sum_{\text{area}} \dot{W}$ represents the *net* rate of work being delivered *from* the region. Also, there may be any number of mass flows into or out of the region, each mass flow carrying its own E (or e per unit mass) so that we shall write

$$\sum_{\text{in}} (\dot{m}e)_{\text{in}}$$

for the energy flow into the region associated with the mass flow in and

$$\sum_{\text{out}} (\dot{m}e)_{\text{out}}$$

for the energy flow out associated with mass flow out. Referring to Fig. 8.1

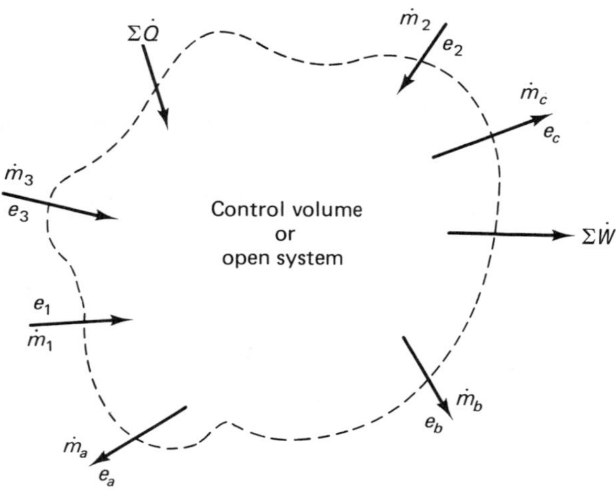

Figure 8.1 Open system for energy equation.

$$\sum_{\text{in}} (\dot{m}e)_{\text{in}} = \dot{m}_1 e_1 + \dot{m}_2 e_2 + \dot{m}_3 e_3 + \cdots \tag{8-28}$$

$$\sum_{\text{out}} (\dot{m}e)_{\text{out}} = \dot{m}_a e_a + \dot{m}_b e_b + \dot{m}_c e_c + \cdots \tag{8-29}$$

Students should note that the energy values (e) are each associated with a particular $\dot{m}$ and are the energies of *those* $\dot{m}$'s as they enter or leave the region. That is, the e's are evaluated *at the boundaries* at the same points where the associated mass flow ($\dot{m}$'s) are determined.

Using this notation, then, the Eq. (8-26) becomes

$$\sum_{\text{area}} Q + \sum_{\text{in}} (\dot{m}e)_{\text{in}} - \sum_{\text{area}} \dot{W} - \sum_{\text{out}} (\dot{m}e)_{\text{out}} = \frac{d}{d\theta}(E_{\text{c.v.}}) \tag{8-30}$$

where $E_{\text{c.v.}}$ is the total energy E of the matter inside the control volume at any time.

Students should now recall or review the concept of flow work presented in Chap. 5. There it was shown that whenever a mass flow crosses the boundary of an open system or control volume, a flow of work also occurs.

Computation showed that the rate of such work is given by

$$\dot{W}_{\text{flow}} = \dot{m}pv \tag{8-31}$$

and that the work flow can be said to be in the direction of mass flow. It is common practice to incorporate this into the Eq. (8-30) by separating the total work $\dot{W}$ into flow work and other work. Symbolically, all nonflow work is represented by $\dot{W}_x$ and since there is flow work associated with *every* $\dot{m}$, we can write

$$\sum_{\text{area}} \dot{W} = \sum_{\text{area}} \dot{W}_x + \sum_{\text{out}} (\dot{m}pv)_{\text{out}} - \sum_{\text{in}} (\dot{m}pv)_{\text{in}} \tag{8-32}$$

(Note that the sign convention has been observed.) With this, conservation of energy for the control volume or open system becomes

$$\sum_{\text{area}} \dot{Q} + \sum_{\text{in}} (\dot{m}e)_{\text{in}} - \sum_{\text{area}} \dot{W}_x - \sum_{\text{out}} (\dot{m}pv)_{\text{out}} + \sum_{\text{in}} (\dot{m}pv)_{\text{in}} - \sum_{\text{out}} (\dot{m}e)_{\text{out}} = \frac{d}{d\theta}(E_{\text{c.v.}}) \tag{8-34}$$

Combining and rearranging gives

$$\sum_{\text{area}} \dot{Q} - \sum_{\text{area}} \dot{W}_x = \frac{d}{d\theta}(E_{\text{c.v.}}) + \sum_{\text{out}} [\dot{m}(e + pv)]_{\text{out}} - \sum_{\text{in}} [\dot{m}(e + pv)]_{\text{in}} \tag{8-35}$$

This equation is called the *energy equation*. It is the first law applied to a control volume or open system. The analogous equation for a control mass or closed system is

$$Q - W = \Delta E \tag{8-36}$$

The term $(e + pv)$ that appears in Eq. (8-35) can be expanded as follows

$$e + pv = u + \tfrac{1}{2}\mathscr{V}^2 + gz + pv \tag{8-37}$$

$$= (u + pv) + \tfrac{1}{2}\mathscr{V}^2 + gz \tag{8-38}$$

8-3 Enthalpy

The expression $(u + pv)$ consists only of properties and hence may itself be considered a property. Because it always appears in the energy equation, it is given a name *enthalpy* and a symbol of its own, h:

$$\text{Enthalpy} = h = u + pv \tag{8-39}$$

or

$$H = U + pV = m(u + pv) \tag{8-40}$$

Since problems involving the flow of fluids are so common, enthalpy is usually tabulated in addition to internal energy—and sometimes even in preference to it.

Students often ask for a physical interpretation or "meaning" for enthalpy. It is not really possible to give this without being a little imprecise but it may be noted that enthalpy is the sum of the internal energy *plus* the flow work. Since the flow work is *always* present when a substance crosses an open system boundary, it represents an energy flow as much associated with the material as is the internal energy u. We might say, therefore, that enthalpy h is to an open system what internal energy u is to a closed system. But to say this, one must note that such a statement is rather nonrigorous. Enthalpy is internal energy plus flow work.

Using enthalpy in the notation, the energy equation becomes

$$\sum_{\text{area}} \dot{Q} - \sum_{\text{area}} \dot{W}_x = \frac{d}{d\theta}(E_{\text{c.v.}}) + \sum_{\text{out}} \left[\dot{m}\left(h + \frac{\mathscr{V}^2}{2} + gz\right)\right]_{\text{out}} - \sum_{\text{in}} \left[\dot{m}\left(h + \frac{\mathscr{V}^2}{2} + gz\right)\right]_{\text{in}} \tag{8-41}$$

8-4 The steady flow energy equation

Many problems involve *steady* flow; that is, conditions at a given point in space are not a function of time. In these cases, the total energy inside the control volume $E_{\text{c.v.}}$ does not change with time, so

$$\frac{d}{d\theta}(E_{\text{c.v.}}) = 0 \tag{8-42}$$

and the steady flow energy equation is

$$\sum_{\text{area}} \dot{Q} - \sum_{\text{area}} \dot{W}_x = \sum_{\text{out}} \left(h + \frac{\mathscr{V}^2}{2} + gz\right)\dot{m}_{\text{out}} - \sum_{\text{in}} \left(h + \frac{\mathscr{V}^2}{2} + gz\right)\dot{m}_{\text{in}} \tag{8-43}$$

Another common special case of the energy equation occurs when, in steady flow, there is only a single inlet and a single outlet. Then the continuity equation gives

$$\dot{m}_{\text{in}} = m_{\text{out}} = \dot{m} \tag{8-44}$$

and the steady flow energy equation reduces to

$$\sum_{\text{area}} \dot{Q} - \sum_{\text{area}} \dot{W}_x = \dot{m}\left[\left(h + \frac{\mathscr{V}^2}{2} + gz\right)_{\text{out}} - \left(h + \frac{\mathscr{V}^2}{2} gz\right)_{\text{in}}\right] \tag{8-45}$$

Dividing through by $\dot{m}$ puts this equation to a basis of a unit mass passing through:

$$\frac{\dot{Q}}{\dot{m}} - \frac{\dot{W}_x}{\dot{m}} = \left(h + \frac{\mathscr{V}^2}{2} + gz\right)_{\text{out}} - \left(h + \frac{\mathscr{V}^2}{2} + gz\right)_{\text{in}} \tag{8-46}$$

For convenience, we usually simply write

$$Q - W_x = \left(h + \frac{\mathscr{V}^2}{2} + gz\right)_{\text{out}} - \left(h + \frac{\mathscr{V}^2}{2} + gz\right)_{\text{in}} \tag{8-47}$$

EXAMPLE 8.3. A kilogram of steam enters a turbine at 50 bars absolute and 600°C and leaves as saturated vapor at 1.5 bars absolute. During its pass through the turbine its elevation falls 3 m and its velocity is reduced from 3 to 0.3 m/sec. Compute its change in energy, Δe.

SOLUTION:

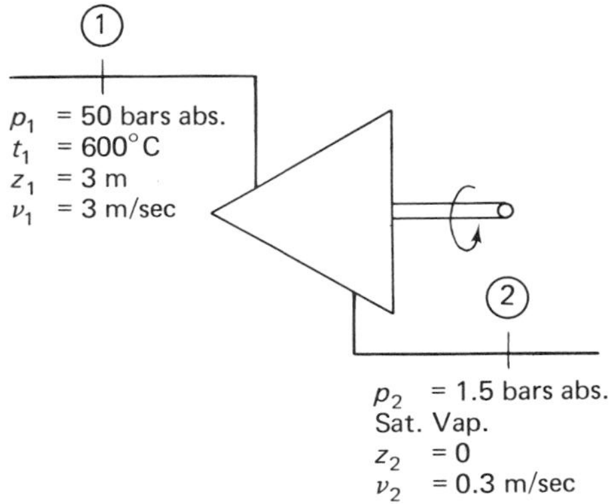

From the steam tables (Appendix C),

$$u_1 = 3273.0 \text{ j/gm}$$
$$v_1 = 78.69 \text{ cm}^3/\text{gm}$$
$$u_2 = 2519.7 \text{ j/gm}$$
$$v_2 = 1159.3 \text{ j/gm}$$
$$\Delta e = \Delta(u + \tfrac{1}{2}\mathscr{V}^2 + gz) = \Delta u + \Delta\tfrac{1}{2}\mathscr{V}^2 + \Delta gz$$
$$\Delta u = u_2 - u_1 = 2519.7 - 3273.0 = -753.3 \text{ j/gm}$$
$$= -753{,}300 \text{ j/kg}$$
$$\Delta\tfrac{1}{2}\mathscr{V}^2 = \tfrac{1}{2}(\mathscr{V}_2^2 - \mathscr{V}_1^2) = -\tfrac{1}{2}(9 - 0.09)1 = -4.455 \text{ j/kg}$$

Units:

$$m^2/\text{sec}^2 \times 1(\text{N-sec}^2/\text{kg-m}) \rightarrow \text{N-m/kg} \rightarrow \text{j/kg}$$
$$\Delta gz = g(z_2 - z_1) = -9.8(3 - 0)1 = -29.4 \text{ j/kg}$$

Units:

$$\text{m/sec}^2 \times \text{m} \times 1(\text{N-sec}^2/\text{kg-m}) \rightarrow (\text{N-m/kg}) \rightarrow \text{j/kg}$$
$$\Delta e = -753{,}300 - 4.5 - 29.4 = -753{,}334 \text{ j/kg}$$

Note: Students should note the unimportance of the kinetic and potential energy changes relative to Δu.

EXAMPLE 8.4. We shall see later in this chapter that the internal energy and enthalpy of an ideal gas are given by

$$du = c_v \, dT$$
$$dh = c_p \, dT$$

Suppose that a closed rigid container of 5 lb of this gas has heat added at a rate of 5 Btu/min. If the initial pressure and temperature of the gas were 15 psia and 80°F, how long will it be until the pressure is 150 psia? Assume c_p and c_v are constants at 0.30 and 0.23 (Btu/lbm-°R), respectively.

SOLUTION: Translation into diagrams and symbols:

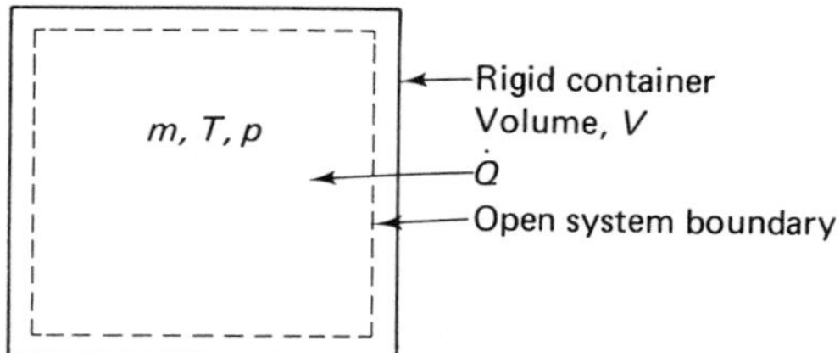

System: Open as shown.
Equation of state: Ideal gas: $pV = MRT$.
Process: Constant volume heat addition.
Fundamentals: Conservation of mass: $m =$ constant.
Energy equation:

$$\sum_{\text{area}} \dot{Q} - \sum_{\text{area}} \dot{W}_x = \frac{d}{d\theta}(E_{\text{c.v.}}) + \sum_{\text{out}} \left(h + \frac{\mathscr{V}^2}{2} + gz\right)\dot{m}_{\text{out}} - \sum_{\text{in}} \left(h + \frac{\mathscr{V}^2}{2} + gz\right)\dot{m}_{\text{in}}$$

Note that although there is no mass flow crossing the boundary, we must find a *rate* of change of the properties and hence need a *rate* equation. There is no work so $\dot{W}_x = 0$. Neglect P.E. and K.E. so $E_{\text{c.v.}} = (mu)_{\text{c.v.}}$. There is no flow across the boundary of the open system so

$$\dot{m}_{\text{in}} = \dot{m}_{\text{out}} = 0$$

Thus,

$$\sum \dot{Q} = m \frac{du_{\text{c.v.}}}{d\theta}$$

Now since

$$du = c_v\, dT \qquad \dot{Q} = mc_v \frac{dT}{d\theta}$$

Applying the equation of state $pv = RT$ and noting that the volume is constant, we find

$$\frac{P_1}{T_1} = \frac{P_2}{T_2} \quad \text{or} \quad T_2 = T_1 \frac{P_2}{P_1} = 540 \frac{150}{15} = 5400°\text{R}$$

Integrating the reduced energy equation gives

$$\int_0^\theta d\theta = \frac{mc_v}{\dot{Q}} \int_{T1}^{T2} dT$$

$$\theta = \frac{(5)(0.23)}{(5)} (5400 - 540) = (0.23)(4860)$$

$$\theta = 1118 \text{ min} = 18.6 \text{ hr}$$

EXAMPLE 8.5. A thin copper sheet that has been heated to $t_1 = 500°\text{F}$ is to be cooled in air and we wish to know the rate at which its temperature will fall. As a first approximation, assume that the heat transfer coefficient from the sheet to the air is constant at $h = 2$ Btu/hr-ft²-°F difference between the sheet and air. Also assume c_p for copper is 0.1 Btu/lbm-°F and that the air temperature is 70°F.

SOLUTION:

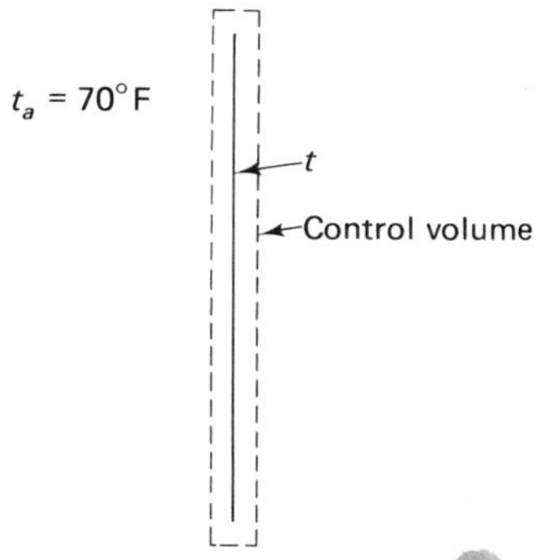

t_a = 70° F

t

Control volume

Energy equation:

$$\sum_{\text{area}} \dot{Q} - \sum_{\text{area}} \dot{W}_x = \frac{d}{d\theta}(E_{\text{c.v.}}) + \sum \left(h + \frac{\mathscr{V}^2}{2} + gz\right)\dot{m}_{\text{out}} - \sum{}' \left(h + \frac{\mathscr{V}^2}{2} + gz\right)\dot{m}_{\text{in}}$$

There is no mass flow across the boundary so $\dot{m}_{\text{in}} = \dot{m}_{\text{out}} = 0$. Also, no work crosses the boundary so $\dot{W}_x = 0$. The energy equation thus reduces to

$$\sum_{\text{area}} \dot{Q} = \frac{d}{d\theta}(E_{\text{c.v.}})$$

The *rate* of heat transfer is given by

$$\sum_{\text{area}} \dot{Q} = -hA(t - t_a)$$

where A is the total surface area of the sheet and the minus sign is needed to account for the fact that the heat is transferred out of the control volume ($t > t_a$). The internal energy may be taken as the energy of the sheet and expressed as

$$\frac{d}{d\theta}(E_{\text{c.v.}}) = \frac{d}{d\theta}[mc_p(t - t_0)] = mc_p\frac{dt}{d\theta}$$

where m is the mass of copper and t_0 is a constant arbitrary reference temperature. Combining these results into the energy equation gives

$$-hA(t - t_a) = mc_p\frac{dt}{d\theta}$$

$$-\frac{hA}{mc_p}\int_0^\theta d\theta = \int_{t_1}^t \frac{dt}{t - t_a}$$

$$-\frac{hA}{mc_p}\theta = \ln\frac{t - t_a}{t_1 - t_a}$$

$$\frac{t - t_a}{t_1 - t_a} = e^{-(hA/mc_p)\theta}$$

$$t = t_a + (t_1 - t_a)e^{-(hA/mc_p)\theta}$$

Checking this result shows that, at $\theta = 0$, $t = t_1$ and that, at $\theta = \infty$, $t = t_a$. In between, t falls exponentially.

8-5 Internal energy, enthalpy, and specific heats

In Chap. 5, we introduced the concept of specific heat (c) and gave it an operational definition. Briefly, it is the quantity of heat required to raise the temperature of a unit mass by 1° in a specified process. We can now show the relationship of specific heat to internal energy (u) and enthalpy (h).

Consider adding heat to one unit mass ($m = 1$) of a substance in a *no-work constant volume* process. The first law gives

$$d'Q = dU = m\,du = du \tag{8-48}$$

$$d'Q = du$$

But, from the definition of c_v, $d'Q$ also equals

$$d'Q = mc_v\,dT = C_v\,dT \tag{8-49}$$

Therefore

$$du = c_v\, dT \tag{8-50}$$

Now if we take T and v to be the independent properties, we may write for a pure substance [see Sec. 4-1(c)]

$$u = u(T, v) \tag{8-51}$$

$$du = \left(\frac{\partial u}{\partial T}\right)_v dT + \left(\frac{\partial u}{\partial v}\right)_T dv \tag{8-52}$$

But for the process in question, $dv = 0$, we obtain by comparison with Eq. (8-50)

$$c_v = \left(\frac{\partial u}{\partial t}\right)_v \tag{8-53}$$

This is the definition of specific heat at constant volume.

For a constant *pressure* heat addition, we obtain

$$d'Q - d'W = du \tag{8-54}$$

In this case, however, $d'W$ is not zero because the system will expand and hence do ($p\, dv$)-type work on its environment. Thus.

$$d'Q - p\, dv = du \tag{8-55}$$

$$d'Q = du + p\, dv \tag{8-56}$$

And also

$$h = u + pv \tag{8-57}$$

$$dh = du + p\, dv + v\, dp \tag{8-58}$$

For a constant pressure process, $dp = 0$, so

$$dh = du + p\, dv \tag{8-59}$$

Hence

$$d'Q = dh = c_p\, dT \tag{8-60}$$

Taking T and p as independent, we write for a pure substance

$$h = h(T, p) \tag{8-61}$$

$$dh = \left(\frac{\partial h}{\partial T}\right)_p dT + \left(\frac{\partial h}{\partial p}\right)_T dp \tag{8-62}$$

Remembering $dp = 0$ in the case in question and comparing with Eq. (8-60),

$$c_p = \left(\frac{\partial h}{\partial T}\right)_p \tag{8-63}$$

This is the definition of specific heat at constant pressure.

EXAMPLE 8.6. A steam turbine is supplied with a steady flow of 10 lbm/min of steam at 800 psia and 800°F. Exhaust steam is saturated vapor at 20 psia. Heat transfer from the turbine to the environment is 12,000 Btu/hr. Compute the horsepower output of the turbine.

SOLUTION:

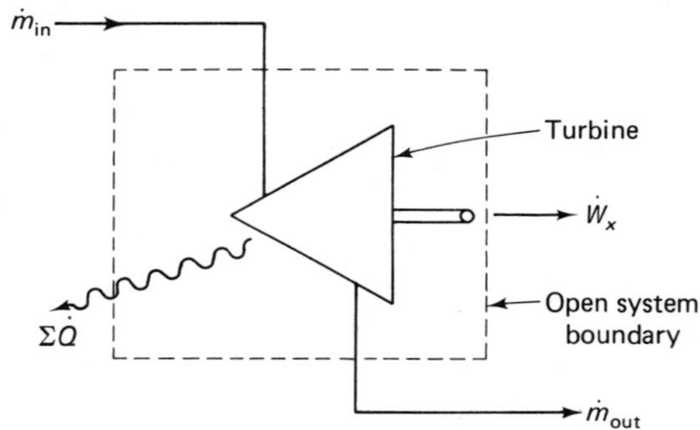

Given:

$$\dot{m}_{in} = 10 \text{ lbm/min} \qquad \text{State 2: Saturated vapor}$$
$$p_1 = 800 \text{ psia} \qquad p_2 = 20 \text{ psia}$$
$$t_1 = 800°\text{F}$$
$$\sum \dot{Q} = 12{,}000 \text{ Btu/hr (direction as shown)}$$

System: Open as shown.
Equation of state: Steam tables.
Process: Steady flow.
Fundamentals:
Continuity:

$$\dot{m}_{in} = \dot{m}_{out} = \dot{m}$$

Energy equation:

$$\sum \dot{Q} - \sum \dot{W}_x = \sum_{out}\left(h + \frac{\mathscr{V}^2}{2} + gz\right)\dot{m}_{out} - \sum_{in}\left(h + \frac{\mathscr{V}^2}{2} + gz\right)\dot{m}_{in}$$

Neglect K.E. and P.E. terms.

$$\dot{Q} - \dot{W}_x = \dot{m}(h_2 - h_1)$$

From steam tables,

$$h_1 = 1398.6$$
$$h_2 = 1156.3$$

$$-\frac{12{,}000}{6} - \dot{W}_x = 10(1156.3 - 1398.6) = -2432 \text{ Btu/min}$$

$$-200 - \dot{W}_x = -2423$$

$$\dot{W}_x = 2223 \text{ Btu/min}$$

$$\dot{W}_x = 2223 \times \frac{60}{2544} = 52.4 \text{ hp}$$

EXAMPLE 8.7. The data shown are from the steam tables. Use them to estimate c_p for steam at 100 psia and 420°F.

p psia	T °F	h Btu/lbm
100	400	1227.5
100	420	1238.1
100	440	1248.5

SOLUTION: By definition,

$$c_p = \left(\frac{\partial h}{\partial T}\right)_p$$

Estimating this as

$$c_p \simeq \left(\frac{\Delta h}{\Delta T}\right)_p \simeq \left(\frac{h_{440} - h_{400}}{40}\right)_{p = 100}$$

$$c_p \simeq \frac{1248.5 - 1227.5}{40} = \frac{21.0}{40.0}$$

$$c_p \simeq 0.525 \text{ Btu/lbm-°F}$$

8-6 Internal energy and enthalpy as properties

Both internal energy and enthalpy are properties. Thus, for a pure substance, their values are fixed and determined whenever any two independent properties are fixed. That is, if p and T are the independent properties, we may write

$$u = u(p, T) \qquad h = h(p, T) \tag{8-64}$$

$$du = \left(\frac{\partial u}{\partial p}\right)_T dp + \left(\frac{\partial u}{\partial T}\right)_p dT \qquad dh = \left(\frac{\partial h}{\partial p}\right)_T dp + \left(\frac{\partial h}{\partial T}\right)_p dT \tag{8-65}$$

Analogous equations can be written in case other properties (e.g., p and v) are the independent ones.

In a number of tables and graphs of property data, values of both u and h are not given because the computation of one from the other is so simple using the definition $h = u + pv$. One must take care to use consistent units, however. (See the computation in Example 8.2.)

8-6(a) Tabulated values

Internal energy and/or enthalpy are commonly given in tables of properties. Most property tables are designed with p and T as the most convenient input variables. Any table can, or course, be worked "backward" as a following example will show. Given the values of two independent properties, tables can be used to find the values of all the other properties listed.

It should be noted that the subscripts f, g, i, fg, etc., applied to u and h have the same meaning as when applied to specific volume. That is, h_f means the enthalpy of saturated liquid at the given conditions.

Also, the equations involving quality x that were derived for specific volume also apply to specific enthalpy and specific internal energy. That is, for a two-phase mixture of liquid and vapor,

$$h = h_f + xh_{fg} = (1 - x)h_f + xh_g \tag{8-66}$$

$$u = u_f + xu_{fg} = (1 - x)u_f + xu_g \tag{8-67}$$

8-6(b) Ideal gases

It can be shown—and we shall show it in a later chapter—that if $pv = RT$, then internal energy (u) and enthalpy (h) are functions *only* of temperature. That is, symbolically,

$$u = u(T) \quad \text{if} \quad pv = RT \tag{8-68}$$

$$h = h(T) \quad \text{if} \quad pv = RT \tag{8-69}$$

This does not in any way change the fact that two independent properties are required to determine the state—it merely states that u, h, and T are *not* independent for ideal gases. This should make it possible to develop tables of properties for ideal gases with only one independent input variable (usually temperature) and this in fact is what has been done. The gas tables by Keenan and Kaye enable one to look up the internal energy and/or enthalpy *of substances for which* $pv = RT$ if only the temperature is known.

From the definition of specific heat at constant volume

$$c_v = \left(\frac{\partial u}{\partial T}\right)_v \tag{8-70}$$

and from the fact that for an ideal gas ($pv = RT$), $u = u(T)$ only, we obtain

$$c_v = \frac{du}{dT} \quad \text{or} \quad du = c_v\, dT \tag{8-71}$$

Analogously we find

$$c_p = \frac{dh}{dT} \quad \text{or} \quad dh = c_p\, dT \tag{8-72}$$

Equations (8-71) and (8-72) hold for *all* processes *of an ideal gas*. There is some tendency for new thermodynamics students to think that $du = c_v\, dT$ can only be used in constant volume ideal gas problems and that $h = c_p\, dT$ applies only in constant pressure ideal gas problems. Not so. Both equations apply in both processes—and any other processes—*provided* the system is an ideal gas obeying $pv = RT$. The equations follow simply from the definitions $c_v = (\partial u/\partial T)_v$ and $c_p = (\partial h/\partial T)_p$ and from the fact that u and h are not functions of p and/or v for ideal gases.

It should be noted that c_p and c_v are not necessarily constant but may vary with temperature, and usually do. In fact, it is the variation of the specific heats with temperature that is accounted for in the gas tables. Otherwise u and h would just be linear functions of T.

Portions of the gas tables are found in Appendix E.

An important result for ideal gases can be obtained from going back to the definition of h and combining it with some of the results above as follows:

$$h = u + pv \tag{8-73}$$

Differentials

$$dh = du + d(pv) \tag{8-74}$$

If $pv = RT$,

$$dh = c_p\, dT \qquad du = c_v\, dT \qquad d(pv) = R\, dT \tag{8-75}$$

Thus

$$c_p\, dT = c_v\, dT + R\, dT \tag{8-76}$$

Finally

$$c_p = c_v + R \tag{8-77}$$

or

$$R = c_p - c_v \tag{8-78}$$

Sometimes over small temperature ranges or where the specific heats are "nearly" constant, c_p and c_v are assumed constant for computation purposes. In these cases, integration gives

$$u_2 - u_1 = c_v(T_2 - T_1) \tag{8-79}$$

$$h_2 - h_1 = c_p(T_2 - T_1) \tag{8-80}$$

It should be noted that the gas constant R has dimensions similar to those of specific heat: energy/mass-degree. For specific heat, we usually use units of Btu/1bm-°F. But, in $pv = RT$, R is usually ft-lbf/lbm-°R. The Fahrenheit/Rankine difference has no effect here because they appear in the denominator (*per* degree) and the degree *size* is the same on both scales. But the ft-lbf and Btu differ by a factor of 778, so when using Eq. (8-79) and (8-80) in numerical work, care must be taken with units.

Another concern with dimensions and units regarding the gas constant and specific heats has to do with the mass basis. Though engineers generally use the pound-mass (lbm), sometimes the pound-*mole* (lbmole) is more convenient and is used. On a pound-mole basis, the gas constant is a universal constant symbolized by $\bar{R}$ and equal to 1545 ft-lbf/lbmole-°R, or $\frac{1545}{778} \simeq$ 2 Btu/lbmole-°R, or 8312 j/kgmole-°K. Students will remember that we obtain the gas constant R for a particular gas by dividing $\bar{R}$ by the molecular weight of the gas:

$$R = \frac{\bar{R}}{MW} \tag{8-81}$$

Thus for air

$$R = \tfrac{1545}{29} = 53.3 \text{ ft-lbf/lbm-°R} \cong 0.07 \text{ Btu/lbm-°R} \tag{8-82a}$$

$$R = \tfrac{8312}{29} \cong 287 \text{ j/kg-°K} \tag{8-82b}$$

EXAMPLE 8.8. An adiabatic steady flow turbine produces 200 Btu/lbm of work output and the enthalpy of the steam leaving is 1150.0 Btu/lbm. What is the temperature of the steam entering if its pressure is 600 psia?

SOLUTION:

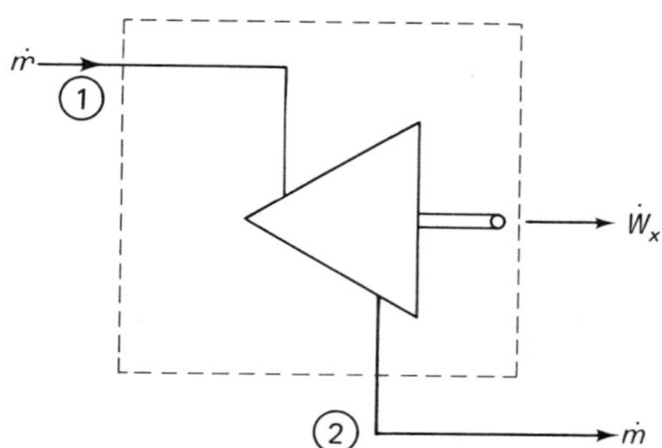

Applying the steady flow energy equation to the open system shown gives, with K.E. = P.E. = 0,

$$-\dot{W}_x = \dot{m}(h_2 - h_1)$$

$$h_1 = h_2 + \frac{\dot{W}_x}{\dot{m}}$$

$$h_1 = 1150.0 + 100 = 1250.0 \text{ Btu/lbm}$$

In the tables, at 600 psia, we find

$$t = 500°\text{F} \qquad h = 1216.2 \text{ Btu/lbm}$$
$$t = 600°\text{F} \qquad h = 1289.5 \text{ Btu/lbm}$$

Thus, we know the temperature is between 500 and 600°F. We can find more accurate results, if we wish, by interpolation as follows:

$$T = 500 \qquad h = 1216.2$$
$$T = T_x \qquad h = 1250$$
$$T = 600 \qquad h = 1289.5$$

$$\frac{T_x - 500}{600 - 500} = \frac{1250 - 1216.2}{1289.5 - 1216.2} = 0.461$$

$$T_x = 500 + 46.1 = 546.1°\text{F}$$

Note that this interpolation assumes a *linear* relationship between h and T, which is not very accurate over such a large temperature range. Using a more detailed table, the result is 542.7°F.

EXAMPLE 8.9. Steam leaves a steady flow turbine at 1.5 bars absolute and a quality of 0.90. How much heat per pound must be removed from the steam in the condenser if the liquid leaving the condenser is saturated liquid?

SOLUTION: From the steady flow energy equation applied to the condenser, we obtain

$$\frac{\dot{Q}}{\dot{m}} = h_2 - h_1$$

1

2

where h_2 and h_1 refer to the state of the H_2O leaving and entering the condenser. From the tables, we find at 1.5 bars

$$h_f = 467.11 \text{ j/gm}$$
$$h_g = 2693.6 \text{ j/gm}$$

Now $h_2 = h_f$. To find h_1,

$$h_1 = h_f + xh_{fg} = h_f + x(h_g - h_f)$$
$$= 467.1 + 0.90(2693.6 - 467.1)$$
$$h_1 = 2471 \text{ j/gm}$$

Then

$$\frac{\dot{Q}}{\dot{m}} = 467 - 2471 = 2004 \text{ j/gm}$$

EXAMPLE 8.10. The following data are taken from the gas tables for air.

T °R	h Btu/lbm	u Btu/lbm
200	47.67	33.96
210	50.07	35.67
220	52.46	37.38
520	129.06	92.04
550	131.46	93.76
560	133.86	95.47

Estimate c_p and c_v for air at 210°R and at 550°R.

SOLUTION: At 210°R,

$$c_p \simeq \frac{\Delta h}{\Delta t} = \frac{4.79}{20} = 0.24 \text{ Btu/lbm}$$

$$c_v \simeq \frac{\Delta u}{\Delta t} = \frac{3.42}{20} = 0.171 \text{ Btu/lbm}$$

At 550°R,

$$c_p \simeq \frac{4.80}{20} = 0.24 \text{ Btu/lbm}$$

$$c_v \simeq \frac{3.43}{20} = 0.1715 \text{ Btu/lbm}$$

Note: Similar calculations at other temperatures give these results:

T(°R)	c_p	c_v
1000	0.250	0.181
2000	0.278	0.210

EXAMPLE 8.11. Air at 40 psia and 80°F in a closed insulated rigid container is subjected to violent stirring until its temperature is 100°F.

1. How much heat is transferred to the air?
2. Compute Δh and Δu for the air if c_p and c_v are constant at 0.24 Btu/lb-°F and 0.17 Btu/lb-°F.
3. Air at 40 psia and 80°F is heated in a constant pressure cylinder to 100°F. Compute Δh and Δu if c_p and c_v are the same as in part 2.

SOLUTION:

1. Zero
2. $\Delta h = c_p\,\Delta T = 0.24(20) = 4.8$ Btu/lbm
   $\Delta u = c_v\,\Delta T = 0.17(20) = 3.4$ Btu/lbm
3. Same as part 2.

8-7 Processes at constant internal energy or enthalpy

Processes occur at constant internal energy when $Q = W$ for the process. Thus, if heat is added to an expanding gas at just the right rate, the heat will just equal the work done, and the process will take place at constant internal energy. This occurs occasionally but not often in engineering work.

One special constant u process occurs when $Q = W = 0$. This is the expansion of a system into a vacuum as shown:

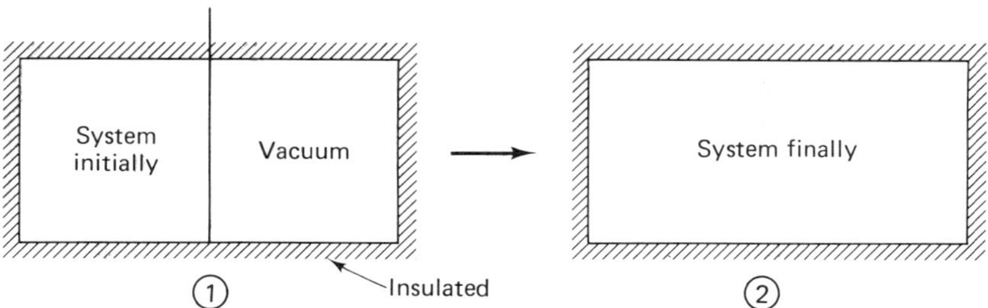

In this process, heat is zero either because of the insulation or because of the rapid speed of the process. Work is zero because expansion into a vacuum involves no force. Therefore, from the first law, internal energy must be constant. This process is of more theoretical than practical value to most engineers.

Unlike u = constant processes, constant enthalpy processes are quite common in engineering applications. They usually occur in steady flow situations in which work is zero and the remaining terms other than enthalpy are either negligible or cancel themselves out.

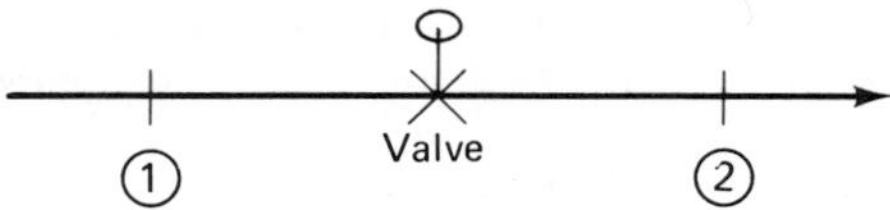

The steady flow energy equation on a per unit mass basis is

$$Q - W_x = h_2 - h_1 + \frac{\mathscr{V}_2^2 - \mathscr{V}_1^2}{2} + g(Z_2 - Z_1) \tag{8-83}$$

Heat transfer is usually negligible in a valve per unit mass flowing unless heating or cooling is overt, height changes are zero or minute, and changes in kinetic energy are often quite small (in Btu/lb) or actually zero. The equation thus becomes

$$h_1 = h_2 \tag{8-84}$$

The same assumptions often apply to flow through capillary tubes and some other devices with no moving parts. Care must be exercised, however, to ensure that the assumptions are valid in every application. Cases in which there is a compressible substance and a significant pressure drop may cause important velocity changes.

EXAMPLE 8.12. The ratio of specific heats, $\gamma = c_p/c_v$, for all diatomic gases under conditions where they closely obey the ideal gas equation of state is about 1.40. Compute $\bar{c}_p$ and $\bar{c}_v$ (on a per mole basis) for diatomic gases.

SOLUTION: Given

$$\gamma = \frac{c_p}{c_v} = 1.4$$

Remembering

$$\bar{c}_p - \bar{c}_v = \bar{R} = \frac{1545}{778} \simeq 2.0 \text{ Btu/lbmole-°R}$$

Combining the two equations above to solve for $\bar{c}_p$ and $\bar{c}_v$ gives

$$\bar{c}_p = 7.0 \quad \text{and} \quad \bar{c}_v = 5.0 \text{ Btu/lbmole-°R}$$

In metric units the results are

$$\gamma = \frac{c_p}{c_v} = 1.4$$

$$\bar{c}_p - \bar{c}_v = \bar{R} = 8312 \text{ j/kgmole-°K}$$

Solving simultaneously gives

$$\bar{c}_p = 29{,}090 \text{ j/kgmole-°K}$$

$$\bar{c}_v = 20{,}780 \text{ j/kgmole-°K}$$

8-8 Energy and kinetic theory

Students who wish to continue their study of the kinetic theory of ideal gases may now finish Chap. 17 starting with Sec. 17-3.

8-9 Making assumptions in applying the first law

In Chap. 1, a methodology for problem solving in macroscopic thermodynamics was outlined and this outline has been followed in most of the problem examples in the book. Problem solving in any engineering field almost always involves the creation of an idealized model of the real situation. Usually the real situation is far too complex for the tractable application of principles—so we simplify. The success of engineering has been that our models are simple enough to solve but complex enough to give useful results. To date the biologists and social scientists (the economists have tried hardest) have had no such luck.

In formulating models, it is essential that appropriate assumptions be made. We have already discussed some assumptions associated with the equations of state in Chap. 4 and some associated with heat and work in Chap. 5. Now that we have the first law and some new properties we can work some complex problems that often involve additional assumptions. Some of those frequently made are listed below for your consideration. Some are repeated from earlier discussions. Please, don't take this as a list of rules because there are always exceptions, special cases, and new developments that void or change our usual way of doing things.

1. Liquids, expecially H_2O, are highly incompressible and except for "very large" Δp's their density can be considered constant.
2. Gases obey $pv = RT$ quite well at temperatures about twice (or higher) their critical temperatures and/or at pressures about one-tenth (or less) their critical pressures.
3. Specific heats of gases are only moderate functions of temperatures and thus for "ordinary" changes in temperature (e.g., up to several hundreds of degrees for air) the use of constant specific heats is valid and the gas tables are not needed. Note, for example, two calculations are required to heat 1 lb of air at constant volume from 200 to 500°F.

$$\text{Constant } c_v = 0.171 \text{ Btu/lbm-°F}$$

$$Q = (1)(0.171)(300) = 51.3 \text{ Btu}$$

From the gas tables,

$$Q = 1(u_f - u_i) = 85.20 - 33.96 = 51.24 \text{ Btu}$$

4. The enthalpy of liquids not compressed to extremely high temperatures can be approximated by the enthalpy of saturated liquid at the same temperature.
5. The work done by a closed system in expanding into a vacuum is zero. Though the process of filling a vacuum is seldom of practical importance, it is nevertheless a common occurrence in classical thermodynamic textbooks. Students should note that work is force times distance and since the vacuum presents no resistance to the motion of the invading substance, there is no force present and no work is done. *Caution:* This is not to say that work cannot be done by or on the system in other places but only that the work associated with the part moving into the vacuum is zero.
6. Heat transfer is a "slow" rate process and hence usually negligible in processes that take place very "rapidly." There is no way to be absolute in this statement because *slow* and *rapid* are relative terms. Nevertheless, processes that occur in a matter of seconds can *usually* be assumed to be adiabatic.
7. Processes that occur slowly allow time for heat transfer and are usually assumed to be isothermal. This is simply a corollary of item 6. The adiabatic and isothermal processes are often opposite extremes—in reality processes fall in between.
8. The work done by a constant pressure environment on a control mass that it expels is equal to the pressure of the environment times the volume expelled.
9. Changes in kinetic and potential energy are usually small and negligible compared with internal energy changes in processes involving large exchanges of heat and/or work. To illustrate this, note the change in velocity required to be equivalent to 1 Btu/lbm:

$$1 = \frac{\Delta(\mathscr{V}^2)}{2g_0 J} - \frac{\Delta(\mathscr{V}^2)}{50{,}000}$$

$$\Delta(\mathscr{V}^2) \simeq 50{,}000 \text{ ft}^2/\text{sec}^2$$

$$\Delta\mathscr{V}^2 \simeq 224 \text{ ft/sec}$$

And also note the height change required to be equivalent to 1 Btu/lbm at $g = 32.2 \text{ ft/sec}^2$:

$$1 = \frac{g\,\Delta(z)}{g_0\,J} = \frac{\Delta(z)}{778} \qquad \Delta z = 778 \text{ ft}$$

In metric units, the change in velocity required to equal 2300 j/kg (about 1 Btu/lbm) is

$$2300 = \frac{\Delta(\mathscr{V}^2)}{2}$$

$$\Delta(\mathscr{V}^2) = 4600 \text{ m}^2/\text{sec}^2$$

$$\Delta\mathscr{V} = 68 \text{ m/sec}$$

The height change required to equal the same 2300 j/kg at $g = 9.8$ m/sec^2 is

$$2300 = g\,\Delta z$$

$$\Delta z = \frac{2300}{9.8} = 235 \text{ m}$$

Of course, kinetic energy changes cannot be neglected in nozzles or diffusers, which are devices intended to change kinetic energy, but in most ordinary piping situations to turbines, pumps, heat exchangers, etc., kinetic energy changes are of secondary importance.

10. Except when valves or other devices that are intended to reduce pressure are present, neglecting pressure drops in piping is a good first approximation in thermodynamic studies. In a heat exchanger, for example, pressure drops in the piping will be most important to the final design but one can do a reasonably good energy analysis by first neglecting them. (Note that this gives a $p =$ constant process equation.)

11. The gas that remains inside a confined space in a "restrained" blowdown process executes a "slow," adiabatic expansion in which $W = p\,dv$ is a good assumption. [The "blowdown process" occurs when a tank of compressed gas is vented through a valve or nozzle to the atmosphere. The assumption stated here is that *the gas that remains inside the tank* expands slowly enough to allow $W = p\,dv$ but quickly enough to be adiabatic. Of course, if the blowdown is *very* slow, the gas inside will remain isothermal (see item 6) but most blowdown experiments or processes take place rather quickly.]

The list of assumptions above is certainly not all-inclusive but it covers a good many common assumptions made to solve real as well as textbook problems. *Students must recognize and be responsible for making assumptions appropriately*. You can't assume the problem away but neither can you solve infinitely complex problems, and you should know better than to spend days and dollars computing an answer to $\pm 0.1\%$ when all that is needed is $\pm 1\%$, or $\pm 10\%$.

8-10 Summary

In this chapter, we have seen how experiments with closed systems in cycles involving heat and work lead inductively to the first law,

$$\oint (d'Q - d'W) = 0 \tag{8-85}$$

and how the fact that

$$\oint (dQ - dW) = 0$$

leads to a new property, energy (E):

$$dQ - dW = dE \qquad Q_{1A2} - W_{1A2} = \Delta E = E_2 - E_1 \tag{8-86}$$

where

$$E = U + \tfrac{1}{2}m\mathscr{V}^2 + mgh \tag{8-87}$$

Heat, work, and the energy E can now be generalized into a concept of "energy" that is conserved for a closed system. When this principle of conservation of "generalized energy" is applied to an open system or control volume, the result is given by the energy equation

$$\sum_{\text{area}} \dot{Q} - \sum_{\text{area}} \dot{W}_x = \frac{d}{d\theta}(E_{\text{c.v.}}) + \sum_{\text{out}} \left(h + \frac{\mathscr{V}^2}{2} + gz\right)\dot{m}_{\text{out}} - \sum_{\text{in}} \left(h + \frac{\mathscr{V}^2}{2} + gz\right)\dot{m}_{\text{in}} \tag{8-88}$$

where $h = u + pv$ and W_x *ex*cludes flow work. Equation (8-88) is applicable by reduction to a variety of special cases (e.g., steady flow) and students *must* be competent in its use.

In this chapter, we also developed more formal definitions for c_v and c_p:

$$c_v = \left(\frac{\partial u}{\partial T}\right)_v \qquad c_p = \left(\frac{\partial h}{\partial T}\right)_p \tag{8-89}$$

Finally, we discussed the use of internal energy and enthalpy as properties—the new properties that come out of our study of the first law.

Now we shall turn our attention to the second law to see how it is developed from the Clausius inequality and to how the new property, entropy (s), is deduced from the law.

Students interested in additional reading on the contents of this chapter should try the following Suggested Reading:

The Dynamics and Thermodynamics of Compressible Fluid Flow, Vol. I, by A. S. Shapiro. Ronald Press, New York, 1954. (See especially Chapters 1 and 2.)

Problems

8-1 One kilogram of saturated H_2O liquid at 6 bars absolute is enclosed in a cylinder that is arranged to maintain a constant pressure. Heat is added to the H_2O in the cylinder until the contents are half liquid and half vapor *by mass*. Atmospheric pressure is 1 bar. Compute the heat added.

8-2 Where steam exists in the two-phase region in a pipeline, it is impossible to determine the state of the steam by the usual method of measuring the pressure and temperature. The most common way of overcoming this problem is to use a device called a *calorimeter*. (See Prob. 8-5.) A bright (?) young engineer has proposed a different scheme, however, that makes use of an evacuated bottle. Steam from the main is allowed to fill the evacuated bottle and the temperature in the bottle is recorded as soon as the pressures are equalized.

Suppose the pressure and temperature in a steam main are found to be 100 psia and 327.8°F, respectively. An evacuated bottle is used as described above and when the pressure in the bottle is 100 psia, the temperature is found to be 350°F. What is the enthalpy of the steam in the main?

8-3 The heating radiator in an office is supplied with steam at 1.5 bars absolute and 150°C. A trap ensures that only liquid leaves the radiator and the temperature of the liquid leaving is found to be 90°C. Compute the heat transfer per pound of steam.

8-4 Steam at 80 psia and 400°F is allowed to expand in steady flow through a nozzle to a pressure of 40 psia. The outlet temperature of the steam is found to be 300°F. What throat area is required if the flow rate of steam is to be 1 lbm/sec?

8-5 Steam in a main is measured to be at a pressure of 140 psia and a temperature of 353°F. A calorimeter as shown is connected to the main and a small steady flow of steam is bled through the calorimeter to the atmosphere. The pressure and temperature in the calorimeter are measured to be 20 psia and 300°F, respectively. What is the quality (x) of the steam in the main?

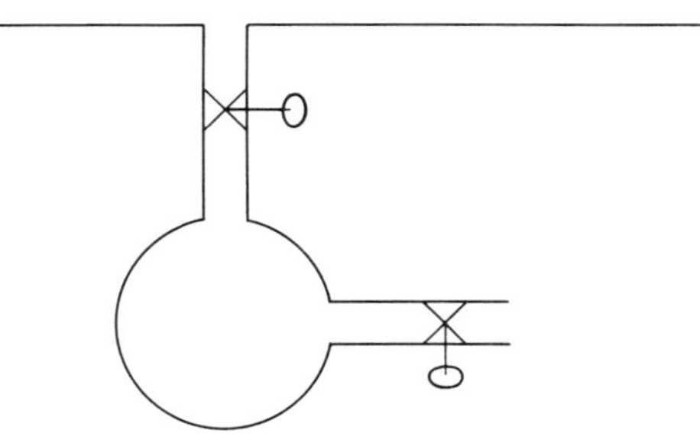

8-6 A flash tank is a device with no moving parts that is sometimes used in a refrigeration system. The tank is kept at a lower pressure than the refrigerant supplied to it and also serves as a separator for the liquid and vapor phases.

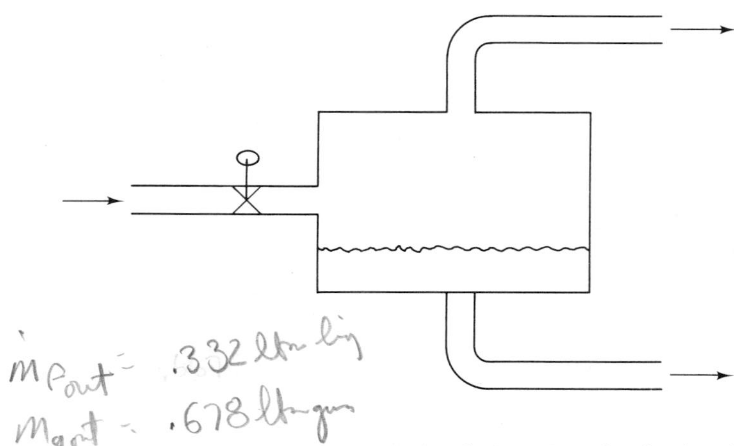

A system using Freon-12 and a flash tank maintained at 12 psia is supplied with refrigerant at 50 psia and an enthalpy of 50.97 Btu/lbm. Compute the flow rates of liquid and vapor leaving the tank per pound of Freon entering.

8-7 A process known as the Joule–Thomspon process can be used to liquefy gases. (See sketch below.) Air is compressed to 150 atm and then cooled to atmospheric temperature (290°K) and supplied to the device at (state 1). Next it passes through a counterflow heat exchanger and then through an expansion or throttling valve to atmosphere (state 3). A two-phase mixture (hopefully) leaves the valve and is separated, the liquid product being drained off and the gaseous portion passed back through the heat exchanger.

Find the liquid yield of this system per kilogram supplied at state 1. Also show state 1 through 6 on the pressure–enthalpy chart on p. 169 that may be used for obtaining data on air as needed.

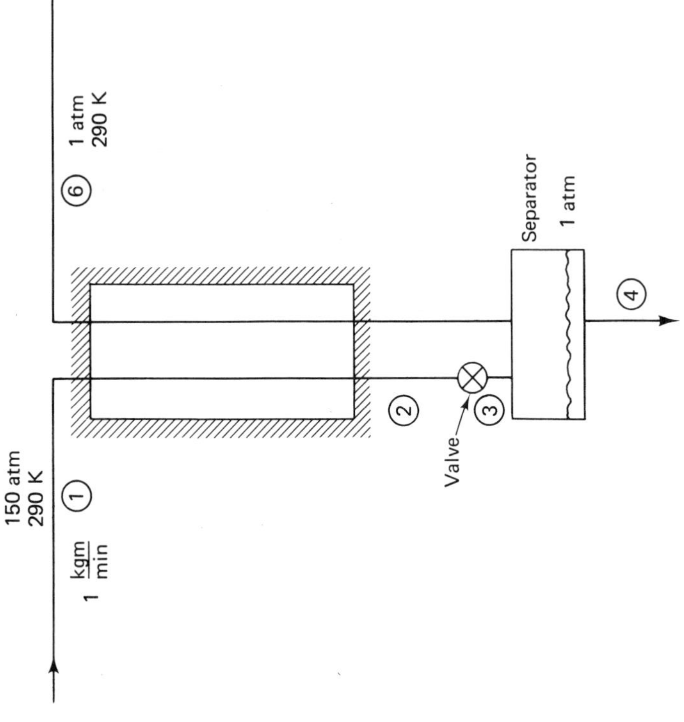

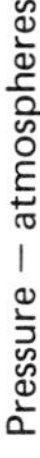

Approximate *p-h* chart for air

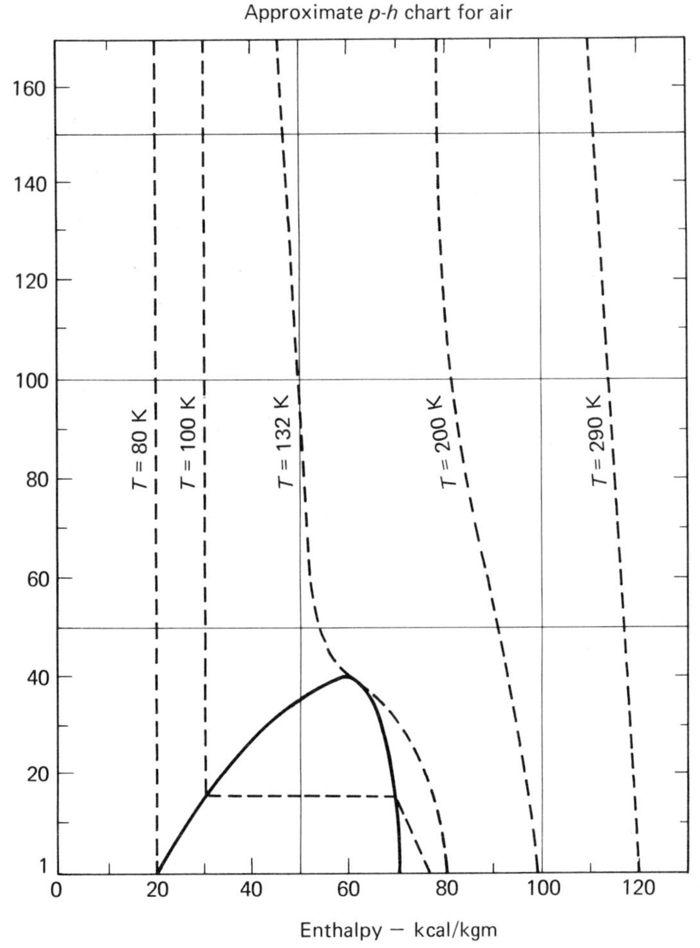

8-8 One-tenth cubic meter of nitrogen gas at 2 bars absolute is allowed to expand isothermally to 1 bar absolute. Find the heat transferred to the nitrogen during this process.

8-9 The system shown below is an automatic reclining backrest devised by engineers at Siesta Industries, Inc. When the company president sits down for his afternoon nap, the backrest slowly moves to the left. After a time, the bracket has moved 6 in. to the left and the whole system is in equilibrium. As he rests, the company president wonders how much work has been done on the air in the cylinder. He knows that the initial pressure of the air was atmospheric (15 psia), that the spring constant is 50 lb/in., that the spring was initially relaxed, that the piston area is 10 in.^{2}, and that the initial volume is 60 in.^{3}.

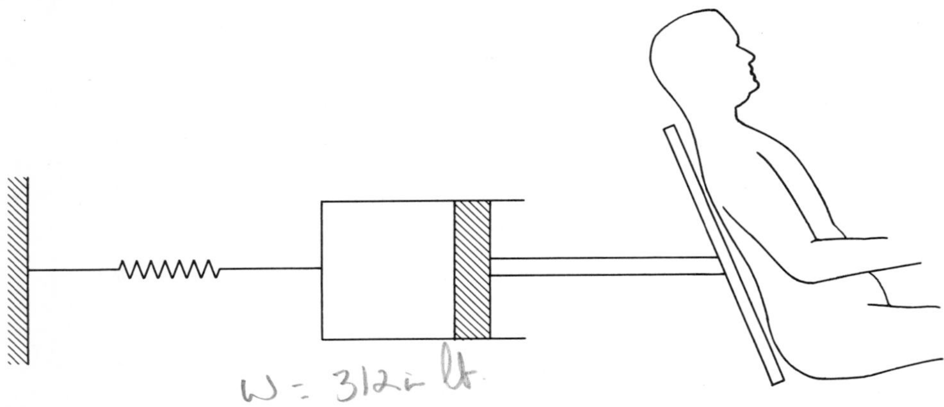

8-10 A tank initially at 70°F and atmospheric pressure (15 psia) is to be filled *today* from a constant pressure line at 200 psia so that the pressure in the tank *tomorrow* will be 150 psia. To what pressure should the tank be filled today? Room temperature remains constant at 70°F.

8-11 Air flows steadily into an insulated nozzle as shown. The air enters at 1.4 bars absolute and 25°C and leaves at 1 bar absolute and 15°C. The outlet area of the nozzle is 1 cm^2. Find the mass rate of flow of the air.

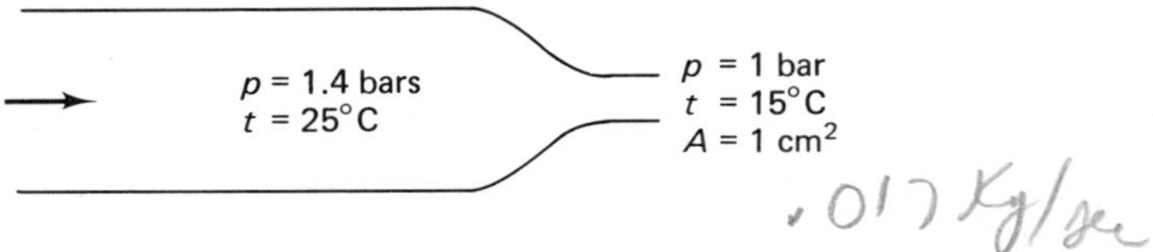

8-12 The following scheme has been proposed for maintaining a constant temperature in a space capsule that contains electronic equipment. Initially pressurized, the capsule is to be provided with a small but adjustable opening for controlled air leakage. It is contended that the leakage of air will remove the energy added by the dissipation of heat from the electrical equipment, thereby maintaining a constant temperature inside the capsule.

Suppose a capsule with a volume of 0.1 m³ has an internal heat source of 10.0 watts. What should be the rate of flow of air from the capsule to maintain the temperature constant at 27°C?

8-13 A 20-ft³ tank is supplied with a constant flow of 7 lb/hr of air at 30 psia and 100°F. It is also equipped with an electric heater that adds heat at a steady rate of 500 Btu/hr. In order to hold the tank pressure constant at 25 psia, air is allowed to escape through a variable and controlled opening to the atmosphere at 15 psia. If the tank is initially at 80°F, what must be the initial rate of escape of air?

8-14 A 0.5-m³ tank is evacuated but contains a small leak. Air leaks in so slowly that the temperature in the tank remains essentially constant at the surrounding room temperature, 20°C. Atmospheric pressure is 1 bar. When the pressure in the tank has reached 0.5 bar, how much heat will have been transferred to the air that has leaked into the tank?

8-15 For monatomic gases in their ideal range, the ratio of specific heat is found to be 1.67. Compute $\bar{c}_p$ and $\bar{c}_v$ for an ideal monatomic gas. Also compute the specific heat of argon and helium on a per pound basis.

8-16 Steam is flowing steadily through an insulated pipeline 3.0 in.² in area in which there is a pressure drop due to friction. At a certain section in the pipe the steam pressure is 100 psia and the temperature is 500°F. At another section, downstream from the first, the pressure is 60 psia and the temperature is 400°F. Find the rate of flow.

8-17 In a refrigeration system, two steady flow streams of Freon-12 are to be simultaneously mixed and heated in an insulated tank. The heating is accomplished by passing air through pipes that pass through the tank. One Freon stream has a flow rate of 100 lbm/min and enters as saturated liquid at 12 psia. The other Freon flow rate is 50 lbm/min and it enters at 12 psia as saturated vapor. The Freon mixture leaves the tank at 5 psia and 20°F. The air enters at 80°F and leaves at 50°F. Compute the required air flow rate. For air, $c_p = 0.24$, $c_v = 0.17$ Btu/lbm-°R.

8-18 Air flows steadily and adiabatically in a constant area duct with a standing discontinuity (or shock) as shown. The problem is to determine the pressure and temperature of the air after the shock.

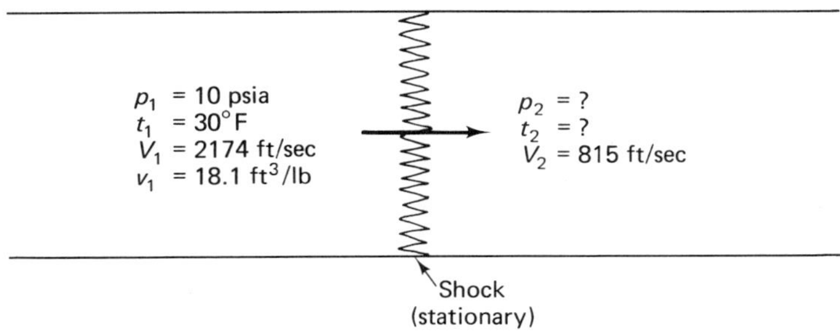

8-19 One-tenth cubic foot of air at 30 psia and 80°F is allowed to expand isothermally to 15 psia. Find the heat transferred to the air during this process.

8-20 Saturated steam vapor enters a compressor at 1 bar absolute and leaves at 1.5 bars absolute and 150°C in steady flow. The heat transferred from the compressor is 10^6 j/hr. Steam flow rate is 40 kg/hr. Compute the power required to drive the compressor.

8-21 Steam enters the nozzle of a turbine with a low velocity at a pressure of 400 psia and $t = 600$°F. It leaves at 260 psia and a velocity or 1540 ft/sec. The rate of steam flow is 3000 lbm/hr. Compute the quality or temperature of the steam leaving the nozzle and the exit area of the nozzle. State assumptions.

8-22 An ideal gas is confined at 80°F (27°C) and 2 atm pressure in the left side of an insulated container. The right side, of equal volume, is evacuated. The partition is broken and the gas fills the container. What will be the new temperature and pressure of the gas?

8-23 Put numbers into the results obtained in Example 8.5 and compute the time required for the copper sheet to reach a temperature where you could handle it without gloves.

8-24 A 5-in. diameter aluminum billet with a mass of 50 lbm is to be preheated prior to extrusion to a temperature of 1000°F. The desired temperature of the billet for extrusion is 700°F. How long will it take the billet to cool? For aluminum, $c_p = 0.1$ Btu/lbm-°F. You may assume the heat transfer coefficient is approximately constant at $h = 1.5$ Btu/hr-ft^2-°F.

9

The concept of reversibility and the second law

In Chap. 7, we induced two general principles. The first law

$$\oint d'Q - d'W = 0 \tag{9-1}$$

and the Clausius inequality

$$\oint \left(\frac{d'Q}{T}\right) < 0 \tag{9-2}$$

In the last chapter, we dealt with the consequences of the first law and deduced a new property, energy. Now we embark on a process starting with the Clausius inequality that will lead to the second law and to another new property, entropy. We can then show how this new law and property relate to the concept of energy degradation that we described earlier.

As a part of the development, we shall describe a limiting, ideal process called *reversible* for which we shall find

$$\oint \left(\frac{d'Q}{T}\right)_{\text{reversible}} = 0 \tag{9-3}$$

You may recall how we showed earlier that if the cyclic integral of "something" is zero, then the "something" is a property. Therefore $(d'Q/T)_{\text{reversible}}$

is a property. We call it *entropy* and give it the symbol s. That is,

$$ds = \left(\frac{d'Q}{T}\right)_{\text{reversible}} \tag{9-4}$$

We shall show that this new property is *not* conserved but always has a net increase. We should not be surprised to find a property that is not conserved from the second law because we already noted from our experience that the second law must have something to do with the observation that processes only go in one direction. Hence we might expect a property that also only goes in one direction. From our concept of energy degradation we might expect the property to always have a net decrease instead of a net increase as entropy does. We shall see that entropy appears with a negative sign in the expression for work potential, however, a fact that corrects for the direction and confirms our expectation.

Though the Clausius inequality and the second law appear at first to be concerned with heat, actually the major application has to do with work or, more specifically, the cyclic (continuous) conversion of heat to work. As we have noted previously, this is extremely important to mankind because man (especially technological man) needs work, whereas he primarily finds heat as a source of energy. Let us go on now to the development so we can see just how this law limits us.

9-1 The concept of reversibility

9-1 (a) Reversible processes

A process executed by a system is called *reversible* if the system *and its environment* can be restored to their initial states and leave no other effects anywhere. Another term for reversible might be *completely restorable*. The definition requires that work and heat exchanged between a system and its environment in a reversible process can be restored to each in exactly the same form so that both are returned completely to their initial states.

To illustrate this concept, consider a closed system that changes by process R from state 1 to state 2 while receiving heat Q_{1R2} and doing work W_{1R2}. If the system can be returned to state 1 along path R while restoring Q_{1R2} to

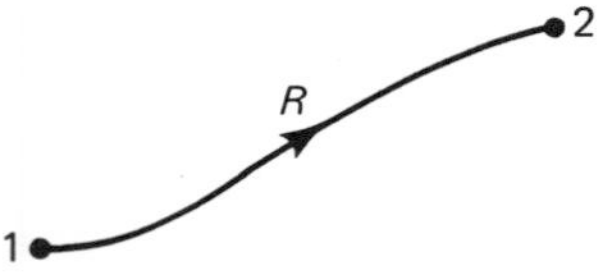

the environment and receiving W_{1R2} from the environment and while restoring the environment to its initial state (with no other effects anywhere), then process R may be called *reversible*.

One way of thinking about the reversible process is to note that it can be *completely* erased leaving no net effects at all on the physical world.

We have already discussed, in Chap. 2, the observation that since real processes occur in one direction only, no real processes are reversible. But with care we can make some processes *almost* reversible. As an example, consider a system that is to be raised in temperature from T_1 to T_2. If we had available a collection of constant temperature reservoirs at T_1, $T_1 + \Delta T$, $T_1 + 2\,\Delta T, \ldots, T_2$, where ΔT is quite small, we could accomplish the temperature rise of the system from T_1 to T_2 in a series of very short temperature steps. (Because we shall refer to it later, please notice that only very small temperature gradients will ever exist within the system because only a very small ΔT is ever imposed on it.) Now we can restore the system to its original temperature using the same reservoirs. When this is done, we shall have all the reservoirs also restored to their initial conditions, except for the ones at T_1 and T_2. The T_1 reservoir will have gained a small amount of heat [enough to lower the system from $(T_1 + \Delta T)$ to T_1] and the T_2 reservoir will have lost a small amount [to raise the system from $(T_2 - \Delta T)$ to T_2]. We can't restore these two reservoirs but by making ΔT extremely small we can almost obtain a reversible situation because only these two reservoirs are not restored, and even they are *nearly* back to their initial point. The transfer of heat, therefore, is always irreversible but by reducing the ΔT involved it can be made nearly reversible.

Another almost reversible case can be illustrated with the case of a closed system expanding against a cylinder. Suppose we put a series of very small weights, each Δm, on the piston one at a time, thereby compressing and doing work on the gas inside. Let us say we compress the gas from p_1 to p_2 in this fashion. Now if we remove the weights one by one, we receive back almost

all the work done as the remaining weights are lifted again. In fact, each weight except the first and last ones on are restored to their initial height. The environment is thus restored (as well as the system) except for a small ΔZ for these two weights. Again, by making the Δm smaller and smaller, we can reduce the unrestored event in the environment to a very small amount but never quite to zero. The compression, or expansion, of a gas therefore is never a reversible process but by reducing the speed of the process we can make it nearly reversible. Referring to our discussion in Chap. 6, students should note that the smaller is Δm, the more nearly will $p' = p$ at all times during the almost reversible process. In other words, there are never more than very small pressure gradients within the system.

From the preceding discussion, it should be apparent that only by transferring heat infinitely slowly using an infinitesimal dT can we obtain a completely reversible heat transfer process. Similarly, only by compressing or expanding a system infinitely slowly using an infinitesimal dm (or dp) can we obtain a completely reversible compression or expansion process. Said in another way, if heat is transferred at a *finite* rate or with a finite ΔT, the process is *not* reversible. And a compression or expansion that takes place at *finite* speed with a finite Δm or Δp is also *not* reversible. We therefore can identify real processes that involve heat transfers or ΔT's, or finite speed motions (Δm's or Δp's), as *not* reversible. Of course, we know that *no real* process is reversible but it is often useful to identify the physical processes themselves, for as we shall see, some processes are more nonreversible than others.

Other sources of nonreversibility can also be identified. In the interest of brevity, these will only be listed here. We already have the first two.

1. Finite rate heat transfer.
2. Finite rate expansion or compression of a fluid.
3. Friction, whether between solids or within a fluid or solid.
4. Electrical resistance.
5. Magnetic hysteresis.
6. Expansion of a fluid to a region of lower pressure (finite Δp).
7. Chemical reactions.
8. Mixing of different substances.
9. Osmosis.
10. Diffusion.

9-1 (b) Internal reversibility

In the two examples of almost reversible processes just discussed, the *system* is completely restored to its initial state but the *environment* is not quite restored, depending on the size of ΔT or Δm. We noted that there were never more than very small temperature or pressure gradients within the systems in these almost reversible processes. When this is the case, we say that the *system* executes an *internally reversible* process. There are many situations when systems or subsystems execute internally reversible processes even though the total situation (system plus environment) is not reversible or not even almost reversible. The compression or expansion of a gas in a cylinder is an example. We have noted that unless the motion of the piston is nearly the speed of sound in the gas, then $p \simeq p'$ and there are only negligible pressure gradients in the gas. Then regardless of external happenings, the gas itself undergoes an almost reversible process. We say that it is *internally reversible*.

As another example, many systems can be raised in temperature without appreciable internal temperature gradients, even though gradients outside the system are quite large. This occurs when the system has a very high thermal conductivity or when heat is added relatively slowly or when the system is a fluid or gas that is kept well mixed. It occurs also when heat is added to or removed from an equilibrium two-phase mixture that, as we know, stays at constant temperature until one phase is gone. Note too that when we speak of a reservoir, we assume that it is internally reversible to heat transfer as a part of its definition because it is so large that finite heat exchanges don't affect its temperature.

The key idea in identifying internally reversible systems is the *absence of gradients* of potential *within* the system. We have used temperature and pressure (i.e., force) potentials as examples. The principle applies to chemical, electrical, and magnetic potentials as well.

Another way to think about the internally reversible process is that it is one in which equilibrium very nearly exists in all parts of the system at all times. Since there are "no" gradients of potential, "no" forces for disequilibrium exist. An internally reversible system thus is always in a state of equilibrium, meaning that if we suddenly leave it alone, it will stay exactly as it was at the moment of our departure.

9-1(c) Reversible cycles

A cycle made up of reversible processes is called a *reversible cycle*. Such a cycle can be reversed with no changes except in the direction of the heat and work quantities exchanged with the environment. As a schematic illustration of this, consider a system of engine working fluid operating as a heat engine receiving Q_h, rejecting Q_c, and producing work W as shown in Fig. 9.1.

In the diagram, the components labeled *reservoir* are assumed to be constant temperature sources or sinks. The natural environment (atmosphere, ocean, earth) serves as such a reservoir for many of our processes. But there are other source and sink reservoirs. The hot gases in a combustion chamber are a source for the steam in the boiler tubes; the air inside a refrigerator is a

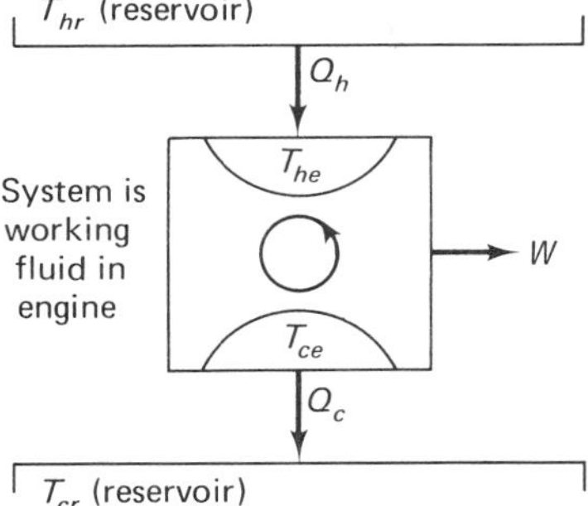

Figure 9.1 Schematic diagram of cyclic heat engine.

source for the refrigerant in the tubes of the refrigeration system; and so on.

In the diagram, at the time and place where Q_h is transferred to the system (the engine working fluid), the system temperature is T_{he}. Also, at the time and place where Q_c is transferred from the system, the system temperature is T_{ce}. If this system is executing a reversible cycle, then it is possible to reverse the direction of process resulting in a heat pump with Q_h, Q_c, and W changed only in direction (not magnitude) as shown in the sketch.

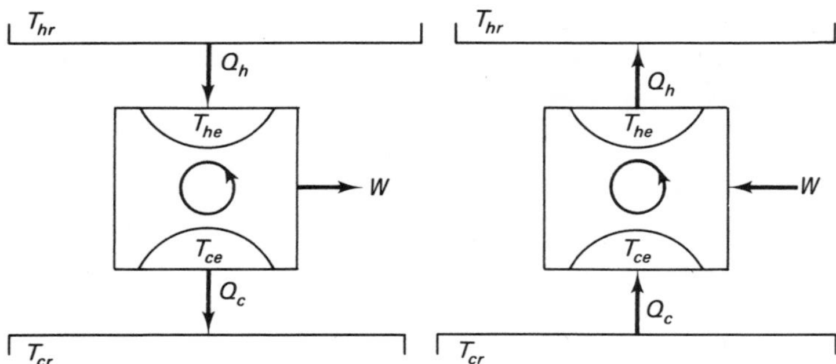

The working fluid system described is said to be *internally reversible*. If, in addition, the heat exchanges between the working fluid and reservoirs are also reversible (that is, if $T_{hr} = T_{he}$ and $T_{cr} = T_{ce}$), then the whole process is reversible, not just the engine itself. We would then call this an *externally reversible engine*, meaning it is not only reversible inside itself but also reversible in the way it interacts with its environment. An externally reversible system is completely reversible.

Students should bear in mind that in real situations the temperatures of the reservoirs and working fluid are not always constant throughout the various heat exchangers. When phase changes occur in the working fluid, of course, T_{he} and/or T_{ce} may be essentially constant temperature but often there is also some superheating or subcooling after the phase change is complete so even in these cases the assumption of a constant temperature for T_{he} and T_{ce} is something of an idealization. The constant temperature model greatly helps our theoretical discussion here, however, and illustrates the principles that apply to the more complicated, realistic situations.

EXAMPLE 9.1. Suppose a very large body at 1000°R transfers heat Q to another at 900°R. Both are large enough to remain practically isothermal. Since each body is internally reversible by the definitions of *reservior* and *internally reversible*, is this a reversible process?

SOLUTION: No. A significant temperature gradient exists in the space between the reservoirs.

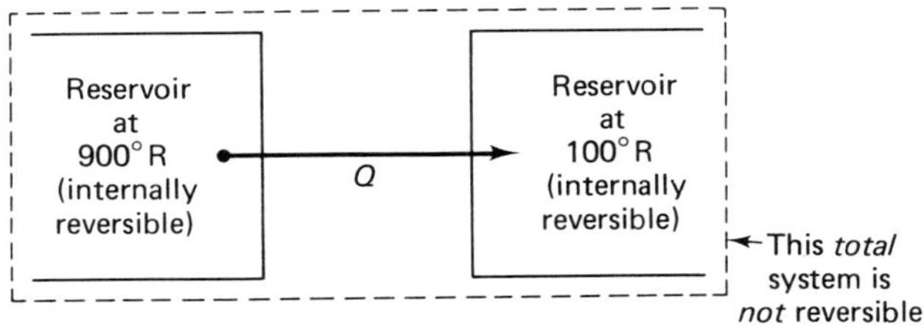

EXAMPLE 9.2. Which, if any, of the following systems would you say are internally reversible in the processes described?

1. The air compressed initially in a bottle that escapes when the cork is pulled.
2. The fly spray *that remains in the can* during the spraying of a fly.
3. The fly spray that leaves the can during the spraying of a fly.
4. A 1-in. diameter copper bar that is cooled from 200°F to room temperature by natural convection.
5. A 25-cm diameter copper billet that is quenched in water from 500 to 30°C.

SOLUTIONS: Systems 1, 3, and 5 are not internally reversible. Systems 2 and 4 are internally reversible, at least *almost*.

EXAMPLE 9.3. Consider a simple steam energy conversion system consisting of three elements:

1. A heat exchange (usually a boiler) in which hot gases at 1000°F transfer 10,000 Btu/min to the working fluid at 800°F.
2. The cyclic energy converter.
3. A heat exchanger (usually a condenser) in which the working fluid at 100°F transfers 8000 Btu/min to cooling water from a stream at 60°F.

Sketch this situation schematically and label the various temperatures, heat flows, and work (T_{hr}, $\dot{Q}_h$, W, etc.) numerically.

SOLUTION:

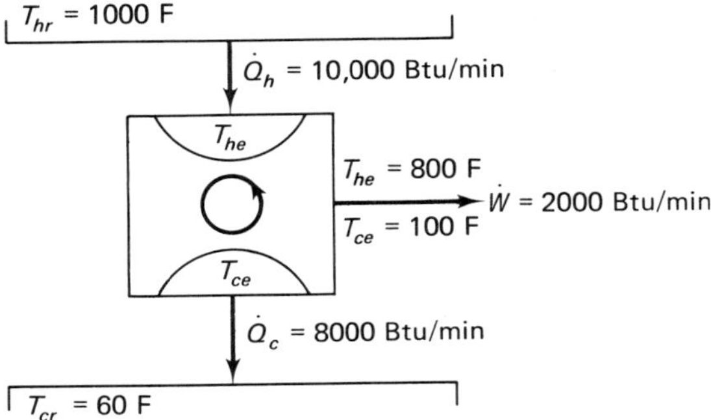

EXAMPLE 9.4. Consider a simple refrigeration system consisting of three elements:

1. A heat exchanger (usually an evaporator) in which 10^6 j/min from the cooled space at $-10°C$ is transferred to the working fluid at $-20°C$.
2. The heat pump itself.
3. A heat exchanger (usually a condenser) in which 3×10^6 j/min of heat is transferred from the working fluid at 40°C to cooling water at 30°C.

Sketch this situation schematically and label the various temperatures and heat flows.

SOLUTION:

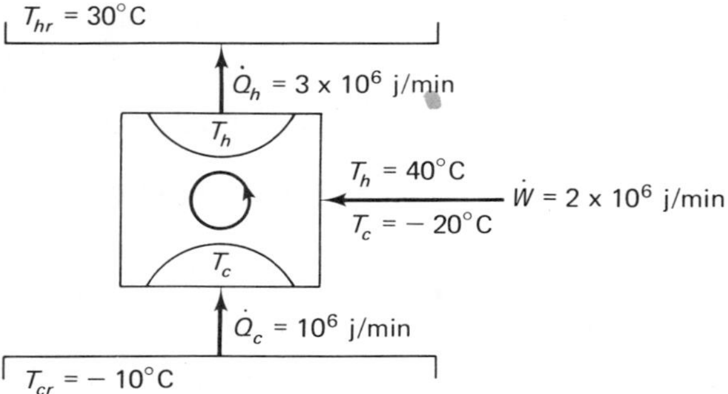

9-2 The second law

Consider again the totally reversible cyclic heat engine and cyclic heat pump shown in the preceding section. If we compute $\oint d'Q/T$ for these cycles, we obtain

Heat Engine:

$$\oint_E \frac{d'Q}{T} = \left(\frac{Q_h}{T_h}\right)_E - \left(\frac{Q_c}{T_c}\right)_E \tag{9-5}$$

Heat Pump:

$$\oint_P \frac{d'Q}{T} = \left(\frac{Q_c}{T_c}\right)_P - \left(\frac{Q_h}{T_h}\right)_P \tag{9-6}$$

Since in the totally reversible case the heat flows and temperatures are the same,

$$\left(\frac{Q_h}{T_h}\right)_E = \left(\frac{Q_h}{T_h}\right)_P \quad \text{and} \quad \left(\frac{Q_c}{T_c}\right)_E = \left(\frac{Q_c}{T_c}\right)_P \tag{9-7}$$

Therefore, adding Eq. (9-5) and (9-6) we find

$$\oint_E \frac{d'Q}{T} + \oint_P \frac{d'Q}{T} = 0 \tag{9-8}$$

But by the Clausius inequality, *both* terms of Eq. (9-8) must be negative if the cycles are to conform to nature as we have observed it. Therefore, we must conclude that reversible cycles and hence reversible processes are not possible.

We note, however, that if *both* terms *equal* zero,

$$\oint_E \frac{d'Q}{T} = \oint_P \frac{d'Q}{T} = 0 \tag{9-9}$$

then Eq. (9-8) is satisfied. We conclude therefore that the limiting case of $\oint d'Q/T = 0$ applies to reversible cases. That is,

$$\oint \left(\frac{d'Q}{T}\right)_{\text{reversible}} = 0 \tag{9-10}$$

Equation (9-10) may be taken as a statement of the second law.

We are in a slightly embarrassing situation here for having just *de*duced a law when we have said earlier that laws are always *in*duced. The source of this apparent difficulty is that the names of these equations and concepts have been established historically and are in such common use that we must conform. It will help somewhat if we henceforth combine Eq. (9-10) with the Clausius inequality into one relationship and call it all the second law:

$$\oint \frac{d'Q}{T} \le 0 \tag{9-11}$$

where it is understood that the inequality applies to real cycles and the equality sign to reversible processes only.

In recapitulation, we first induced the Clausius inequality

$$\oint \frac{d'Q}{T} < 0$$

from observation, experience, and experiment. Then we defined the concept of reversibility as a completely restorable process—slow, frictionless, no potential gradients, etc. We extended the concept of reversibility to cycles. Combining the concept of reversibility with the Clausius inequality then led to the result that the limiting case—when $\oint d'Q/T$ approaches zero— is the reversible situation. And we agreed to combine this result with the Clausius inequality into the single relation, Eq. (9-11), and call it the *second law.*

This law might also be stated verbally by noting that *reversible cycles are not possible.* This statement ties in closely with the somewhat loose statement of the second law that we developed in Chap. 2 based on everyday experience. There we said that "without outside influence, processes occur in only one direction." In our new terms, that just means that there are no reversible processes in real life.

The development has been presented here using illustrations with only a single high temperature source for Q_h and a single low temperature sink for Q_c. For the sake of simplicity, we shall only state but not show that all the conclusions hold for multiple sources and sinks as well.

Now we can go on to learn how all this relates to work and the degradation of energy, for the second law places strict limitations on the performance of cyclic heat engines.

9-3 The second law and cycles

9-3(a) Limitations on heat engines

Consider a simple engine as shown below receiving heat at the rate of $\dot{Q}_h$ at a constant temperature T_{he} and rejecting heat at a rate of $\dot{Q}_c$ at a constant temperature T_{ce}. The reservoir temperatures are T_{hr} and T_{cr}.

(It should be noted that there are no real engines or refrigerators that actually receive and/or reject heat at a single temperature of the working fluid (T_{he}, T_{ce}). In this discussion, then, we are talking about hypothetical systems (except Sec. 10-6 on the Carnot cycle). The general qualitative conclusions are not affected by this assumption, however, which greatly simplifies both the logic and the arithmetic. It may be convenient here for students to think of T_{he} and T_{ce} to be appropriate *average* temperatures of the working fluid during heat addition and heat rejection.)

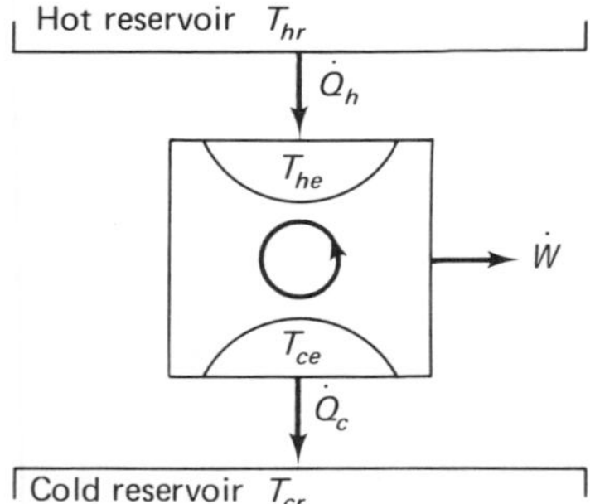

The efficiency, as always, is a measure of the useful output divided by the costly input. In the case of a heat engine, the useful output is the net work $\dot{W}$ and the costly input is the heat $\dot{Q}_h$. Thus,

$$\eta_{\substack{\text{heat}\\\text{engine}}} = \frac{\text{net work output}}{\text{costly heat input}} = \frac{\dot{W}}{\dot{Q}_h} \tag{9-12}$$

From the first law, we have

$$\dot{W} = \dot{Q}_h - \dot{Q}_c \tag{9-13}$$

So

$$\eta_{\substack{\text{heat}\\ \text{engine}}} = \frac{\dot{W}}{\dot{Q}_h} = \frac{\dot{Q}_h - \dot{Q}_c}{\dot{Q}_h} = 1 - \frac{\dot{Q}_c}{\dot{Q}_h} \tag{9-14}$$

For a given time period covering an integral number of engine cycles, this becomes

$$\eta_{\substack{\text{heat}\\ \text{engine}}} = 1 - \frac{Q_c}{Q_h} \tag{9-15}$$

The first law prohibits an efficiency greater than unity but allows an efficiency equal to unity. The second law, however, will *not* allow such a high efficiency. To prove this, consider the imaginary engine shown below that has $\eta = W/Q = 1$.

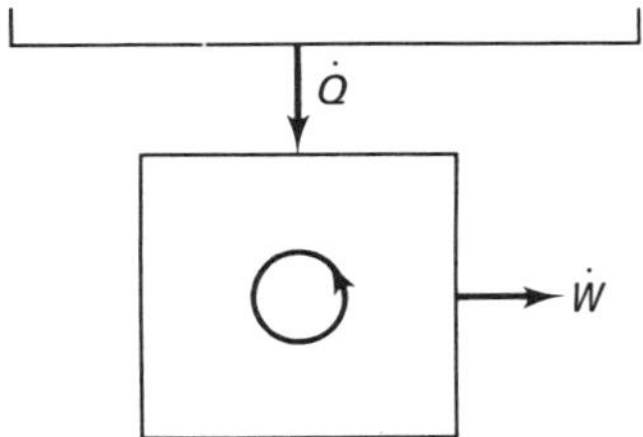

Applying the second law to the working fluid in the engine, we find

$$\oint \frac{d'Q}{T} \leq 0 \quad \text{or} \quad \frac{Q}{T} \leq 0 \tag{9-16}$$

T is absolute temperature and is therefore positive. Q is heat supplied and is also positive. Thus, the left side cannot be either negative or zero and the assumed engine, whether real or reversible, is not possible by the second law.

The fact that it is not possible to construct a cyclic heat engine with an efficiency of unity is often called the *Kelvin–Planck statement of the second law.* We view it here as simply a deductive consequence of our previous work.

Students should note that the impossibility of a cyclic heat engine with an efficiency of unity can be stated as a requirement that all real cyclic heat engines must exchange heat with at least two reservoirs—one a source and one a sink. Some typical sources and sinks that students will recognize are the combustion products and river water for a steam power generating system, the sun and the earth's atmosphere for a solar engine, and the combustion products and the earth's atmosphere (via the radiator) in an automobile. It is possible to use one part of the natural environment as a source and another part at a lower temperature as a sink, as with the solar engine, but you can't just use some one part of nature as a source and let it go at that. You also have to find another place to reject the heat (Q_c) that cannot be converted to work, hence our problems of environmental thermal pollution.

We have shown that a cyclic heat engine cannot have an efficiency of unity. But can we approach unity as a limit? If not, how large an efficiency can we obtain?

Refer again to the engine in Fig. 9.1. For the engine working fluid, we have

$$\oint \frac{d'Q}{T} \leq 0 \tag{9-17}$$

or

$$\frac{Q_h}{T_{he}} - \frac{Q_c}{T_{ce}} \leq 0 \tag{9-18}$$

Rearrangement gives

$$\frac{Q_h}{Q_c} \geq \frac{T_{ce}}{T_{he}} \tag{9-19}$$

where the inequality sign applies to real cases and the equality sign to reversible cases. Combining this with the expressions for efficiency, we find

$$\eta_{\substack{\text{heat}\\ \text{engine}}} = 1 - \frac{Q_c}{Q_h} \tag{9-20}$$

or

$$\leq 1 - \frac{T_{ce}}{T_{he}} \tag{9-21}$$

Now since the efficiency is net work output divided by heat input, we see that

$$\frac{W_{\text{net output}}}{Q_{\text{in}}} \leq 1 - \frac{T_{ce}}{T_{he}} \tag{9-22}$$

What then is the work potential (that is, the maximum possible useful work that can be obtained from Q_{in} in an engine working between T_h and T_c)? Obviously, the equals sign gives the maximum so

$$\text{Work potential of } Q = Q\left(1 - \frac{T_c}{T_h}\right) \tag{9-23}$$

The second law has other important theoretical and practical implications. Let us rewrite

$$\eta_{\substack{\text{heat}\\\text{engine}}} \leq 1 - \frac{T_{ce}}{T_{he}} \tag{9-24}$$

First, they show that a reversible engine has the best possible efficiency when operating between the same temperatures, T_{he} and T_{ce}, and they show what that efficiency is:

$$\eta_{\substack{\text{reversible}\\\text{heat}\\\text{engine}}} = 1 - \frac{T_{ce}}{T_{he}} = \frac{T_{he} - T_{ce}}{T_{he}} \tag{9-25}$$

Also since the development said nothing at all about what kind of reversible engine was being considered, the equality sign results apply to *all* reversible engines; hence all reversible engines have the same efficiency as given by Eq. (9-25).

Furthermore, by inspecting Eq. (9-25), we can learn that to improve the efficiency of a reversible engine we should either raise T_{he} or lower T_{ce} or both.

Now let us see what all this means in terms of energy degradation.

We have said that a loss of work potential—the ability to do useful work—is energy degradation. We can now calculate the work potential of a quantity of heat. So let's first of all see what happens when heat is simply transferred from one system to another, which must obviously be at a lower temperature. Suppose 1000 Btu is transferred from a system at 2500°R to one at 2000°R. What is the loss of work potential; that is, what is the degradation?

Assuming an available environment at 500°R, the work potential of 1000 Btu at 2500°R is

$$A_{2500} = 1000(1 - \tfrac{500}{2500}) = 800 \text{ Btu}$$

But the work potential of 1000 Btu at 2000°R, after it has transferred, is

$$A_{2000} = 1000(1 - \tfrac{500}{2000}) = 750 \text{ Btu}$$

Thus, the loss of work potential, or the degradation, is 50 Btu.

Now let's suppose we have a heat engine working at $T_{he} = 2000$°R and $T_{ce} = 600$°R and taking in 1000 Btu per cycle. If that engine has an efficiency of 40%, how much degradation per cycle is being done?

$$\eta_{\substack{\text{heat}\\\text{engine}}} = \frac{W_{\text{net out}}}{Q_{\text{in}}} = 0.40 = \frac{W_{\text{net out}}}{1000}$$

$$W_{\text{net}} = 400 \text{ Btu}$$

From conservation of energy, if the heat input is 1000 Btu per cycle, the net work output is 400 Btu per cycle, then the heat rejected at T_{ce} must be 600 Btu. The work potential of this heat is

$$A_{600} = 600(1 - \tfrac{500}{600}) = 100 \text{ Btu}$$

Thus, the work potential input to the engine is 750 Btu; whereas the output is 400 + 100 = 500. (The work potential of 400 Btu of work is 400 Btu.) So the degradation in the engine, per cycle, is 250 Btu.

Now if the rejected heat (600 Btu at 600°R) is transferred to the atmosphere at 500°R, there will be more degradation equal to 100 Btu since the work potential of the energy at 500°R is zero. Thus, the total degradation in this whole process is 50 + 250 + 100 = 400 Btu, as you might expect, because the work potential of 1000 Btu at 2500°R is 800 Btu, and 400 Btu of net work is delivered.

To illustrate a point, let's do the example above again assuming the engine is internally reversible. Temperatures are the same. *Now* what is the degradation in the engine itself?

If the engine is internally reversible, its efficiency will be

$$\eta_{\substack{\text{reversible}\\\text{heat}\\\text{engine}}} = 1 - \frac{T_{ce}}{T_{he}} = 1 - \frac{600}{2000} = 0.70$$

Therefore its work output will be

$$W_{\text{net out}} = 0.70(1000) = 700 \text{ Btu}$$

The work potential of the heat input is still 800 Btu and the work potential of the heat output is still 100 Btu so there is *no* loss of work potential. This, of course, is true of all reversible engines: They do not degrade energy. But, alas, they also do not exist! And since the second law requires all real engines to have efficiences less than reversible engines, all real engines do degrade energy. Hence we also refer to the second law as the *law of degradation of energy*.

Sometimes the second law is stated in other terms. One common version is called the *Kelvin–Planck statement* and it states that it is impossible to construct a heat engine that is more efficient than a reversible engine working between the same reservoirs. Another version is called the *Clausius statement* and it states that it is impossible for net heat to be transferred continuously or by a cyclic process from a lower to a higher temperature leaving no other effects. Students of this book should recognize that both of these statements are covered by the Clausius inequality. The engine more efficient than reversible will have a $\oint d'Q/T$ *greater* than zero and so will any device able to take a quantity of heat from a lower to higher temperature. As a proof of the latter, consider the situation shown in the sketch:

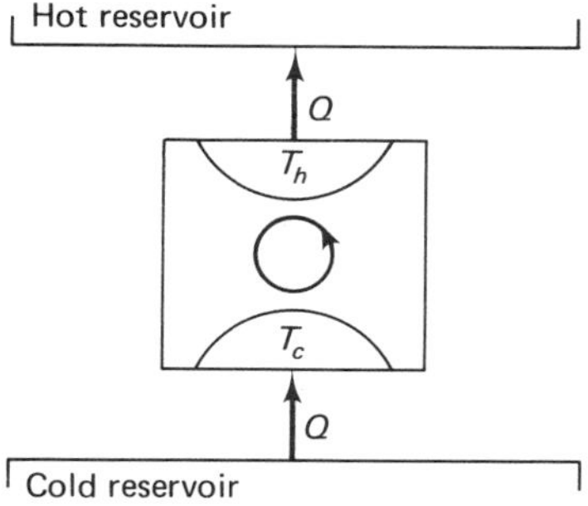

For the cyclic device,

$$\oint \frac{d'Q}{T} = \frac{Q}{T_c} - \frac{Q}{T_h} = +$$

since $T_h > T_c$. Also see Prob. 9-8.

9-3(b) The thermodynamic temperature scale

It is possible to define a temperature scale in terms of the heat transfers with reversible engines. We note that the work obtainable from a reversible cyclic heat engine is given by

$$W_{\substack{\text{reversible}\\ \text{engine}}} = Q_h\left(1 - \frac{T_{ce}}{T_{he}}\right) \qquad \text{Reversible only} \tag{9-26}$$

Using the first law to eliminate W in favor of the Q's and rearranging gives

$$W_{\substack{\text{net}\\ \text{out}}} = Q_h - Q_c = Q_h\left(1 - \frac{T_{ce}}{T_{he}}\right) \qquad \text{Reversible only} \tag{9-27}$$

$$1 - \frac{Q_c}{Q_h} = 1 - \frac{T_{ce}}{T_{he}} \qquad \text{Reversible only} \tag{9-28}$$

$$\frac{Q_c}{Q_h} = \frac{T_{ce}}{T_{he}} \qquad \text{Reversible only} \tag{9-29}$$

EXAMPLE 9.5. A system executes an internally reversible cycle while exchanging heat with three reservoirs as shown. Compute the direction and magnitude of heat exchange Q_3.

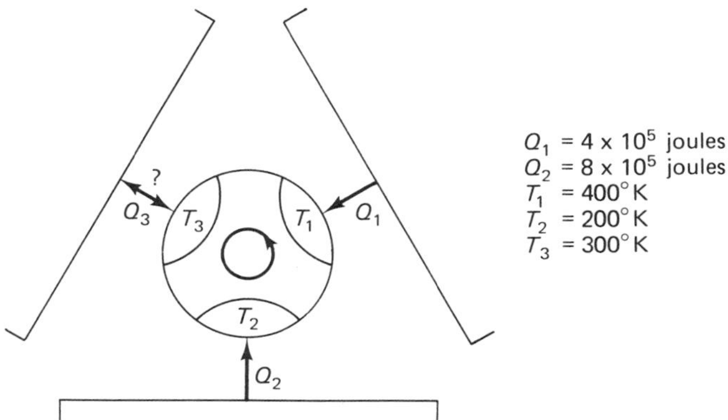

SOLUTION: Since the cycle is internally reversible,

$$\frac{Q_1}{T_1} + \frac{Q_2}{T_2} + \frac{Q_3}{T_3} = 0$$

$$\frac{4 \times 10^5}{400} + \frac{8 \times 10^5}{200} + \frac{Q_3}{300} = 0$$

$$Q_3 = -300(5 \times 10^3) = -15 \times 10^5 \text{ j}$$

EXAMPLE 9.6. A heat engine (see sketch on p. 182) is supplied with 10,000 Btu/hr from a reservoir at 3000°R while the working fluid is at 1000°F. The sink is at 500°R and the working fluid is at 140°F where heat is rejected at a rate of 8000 Btu/hr.

1. What is the actual efficiency of the engine?
2. What fraction is this of the internally reversible efficiency?
3. What fraction is this of the completely reversible efficiency?

SOLUTION:

1. $\eta_{\text{actual}} = \frac{W_{\text{out}}}{Q_{\text{in}}} = \frac{Q_{\text{in}} - Q_{\text{out}}}{Q_{\text{in}}} = \frac{10{,}000 - 8000}{10{,}000} = 0.20$

2. $\eta_{\substack{\text{internal}\\\text{reversible}}} = 1 - \frac{T_{ce}}{T_{he}} = 1 - \frac{600}{1460} = 0.59$

$\frac{0.20}{0.59} = 0.34$ (fraction of internally reversible efficiency)

3. $\eta_{\substack{\text{external}\\\text{reversible}}} = 1 - \frac{T_{cr}}{T_{hr}} = 1 - \frac{500}{3000} = 0.83$

$\frac{0.20}{0.83} = 0.24$ (fraction of externally reversible efficiency)

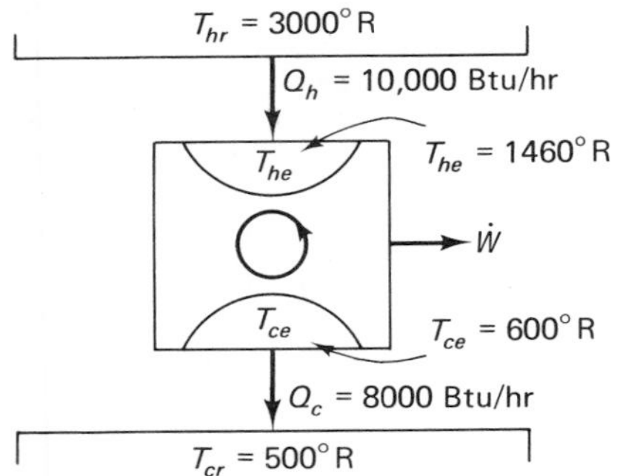

Now imagine a reversible engine operating between the conventional steam and ice points. We know how to measure heat flows so that if we add to the equations above an equation that defines the *size* of the degree (T_{steam}

$- T_{\text{ice}} = 100$ for Celsius/Kelvin scales and 180 for Fahrenheit/Rankine scales), we have two equations in two unknowns, T_{steam} and T_{ice}. The resulting temperature scale is called the *thermodynamic temperature scale*. It is of little practical importance because there are no reversible engines and even if there were, this would be a very difficult thermometer to use. This result is of great importance theoretically, however, because the thermodynamic scale *can* be shown to correspond exactly to the absolute scales determined with the constant volume—zero pressure gas thermometer. Thus, this puts our use of temperature on a firm theoretical, as well as practical, basis and should relieve the uneasy feeling we have had about temperature since Chap. 3.

9-3(c) Limitations on heat pumps

The preceding discussion on heat engines can be extended to heat pumps (refrigerators). The usual function of a heat pump is to remove heat from something called the *source*. Typical sources are the inside of the refrigerator box, the water in an ice maker, or the room air circulating through an air conditioner. Heat is rejected to a sink, usually the atmosphere.

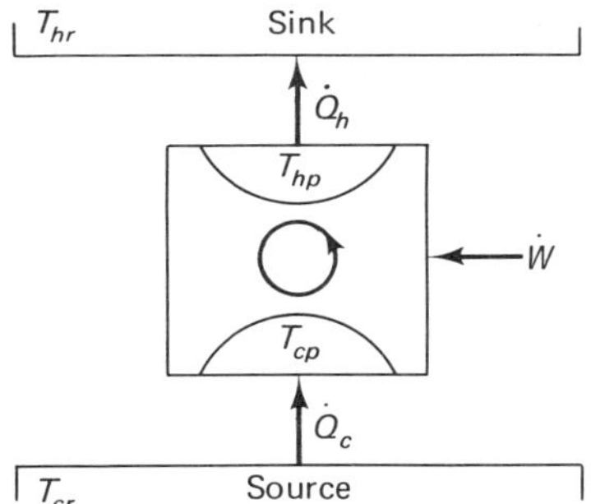

The performance of such a device is defined as the useful output divided by the costly input. In this case the useful output is $\dot{Q}_c$ because removing heat from the source is the purpose of the device. The costly input, or course, is the work we must do, $\dot{W}$. The performance measure is not called "efficiency" because it can sometimes be greater than unity for a heat pump. So it is called the *coefficient of performance* (COP).

$$\text{COP} = \frac{\text{useful output}}{\text{costly input}} = \frac{\dot{Q}_c}{\dot{W}} \tag{9-30}$$

Applying the first law,

$$\dot{Q}_c + \dot{W} = \dot{Q}_h \tag{9-31}$$

Thus,

$$\text{COP} = \frac{\dot{Q}_c}{\dot{Q}_h - \dot{Q}_c} \tag{9-32}$$

Or, for a given integral number of engine cycles,

$$\mathrm{COP} = \frac{Q_c}{Q_h - Q_c} \tag{9-33}$$

By methods analogous to those used for heat engines it can be shown that reversible heat pumps have the best possible COP for given source and sink temperatures, that all reversible heat pumps have the same efficiency when operating between the same temperatures, and that COP of a reversible heat pump is

$$\mathrm{COP}_{\substack{\text{reversible}\\ \text{heat}\\ \text{pump}}} = \frac{T_c}{T_h - T_c} \tag{9-34}$$

Equation (9-36) can be used to argue (a little circuitously) that absolute zero is unattainable in the absence of absolutely perfect heat insulation. If any substance could reach absolute zero, it would be cooler than its surroundings and so unless impossibly perfect insulation were used, it would have *some* heat transferred to it. But at $T_c = 0$, the COP of the best possible refrigerator is zero. It would thus take infinite work to remove such heat. Hence absolute zero could not be maintained. Actually it cannot be attained either because the last bit of heat to be removed in getting there would also require infinite work. The conclusion is that absolute zero is unattainable. This is what is meant by the statement that "you can't get out of the game." If we could obtain absolute zero, then heat engines could be run at efficiencies of unity so we could then at least break even. But it would take negative temperatures for us to be, even theoretically, able to win. But since we cannot get to negative temperatures, or even to zero, we have to stay in the game where the second law says we must always lose not energy but work potential. Wait until Charlie Brown hears about this! Sigh.

9-4 Limitations on processes

In the last chapter, we showed that when operating between the same temperatures, reversible engines have the best possible efficiency. We also showed that all such reversible engines have the same efficiency. Now we want to extend the discussion to *processes* and show that, for given end states and input, a reversible process produces the maximum work and that all reversible processes produce the same work.

The proof can be easily stated. Suppose there existed some irreversible process that produced more work from the same input than a reversible process between the same end states. We could construct (or imagine constructing) a reversible cycle using the reversible process in question. Then if we replace the reversible process with the irreversible process just described, leaving the rest of the cycle the same, we shall have an irreversible cycle that produces more work than a reversible one with the same inputs. But we have just shown that this is not possible because we have proved that reversible cycles have the best possible efficiency. We must conclude, therefore, that the irreversible process described as better (i.e., produces more work) than the reversible one is not possible.

This argument can be made more concise using the following sketch. A

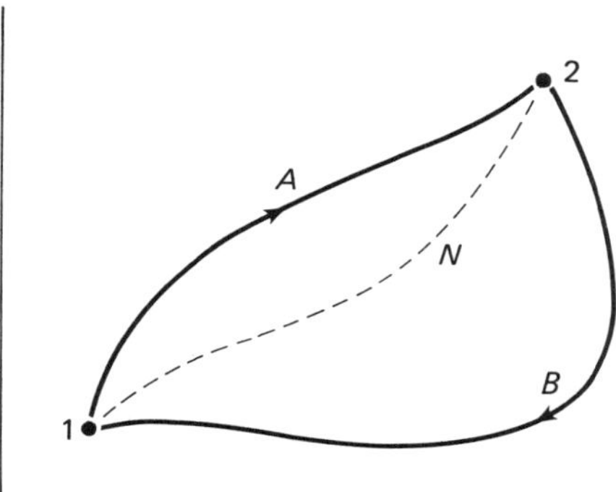

reversible cycle $1A2B1$ is shown. Assume the work per cycle is given by W, where $W = W_{1A2} + W_{2B1}$. Replace process $1A2$ with a nonreversible process $1N2$ such that $W_{1N2} > W_{1A2}$. Now the nonreversible cycle $1N2B1$ will do more work $(W_{1N2} + W_{2B1})$ than the reversible cycle $(W_{1A2} + W_{2B1})$. But this we have already shown is impossible when the two are operating between the same temperatures. We can only conclude therefore that our assumption of a nonreversible process N producing more work than a reversible one operating between the same temperatures is not allowed.

When work utilizing processes are considered, it is similarly easy to show that the reversible process requires the least or minimum possible input of work.

As always, the moral: Reversible processes and cycles are best. In fact, it's imperative: Always strive for reversibility as closely as possible.

9-5 Summary

A process executed by a system is called *reversible* if the system and its environment can be restored to their initial states and leave no other effects. No real processes are completely reversible, though some processes (e.g., compression, expansion, heat transfer) can approach reversibility if they are executed very slowly. The absence of gradients within a system identifies a process as *internally reversible* for that system only.

A cycle made up of reversible (or internally reversible) processes is called a *reversible* (or internally reversible) cycle. An *externally reversible* cycle is one that is internally reversible *and* all heat and work exchanges with the environment are also reversible.

The Clausius inequality, induced in Chap. 7, is that for a system executing a real cycle:

$$\oint \frac{d'Q}{T} < 0 \tag{9-35}$$

But, for an internally reversible cycle,

$$\oint \left(\frac{d'Q}{T}\right)_{\text{reversible}} = 0 \tag{9-36}$$

The second law is defined by the Eq. (9-36) and (9-37).

The efficiency of a cyclic heat engine is defined as

$$\eta_{\substack{\text{heat}\\\text{engine}}} = \frac{\text{net work output}}{\text{costly heat input}} = \frac{\dot{W}}{\dot{Q}_h} \tag{9-37}$$

$$= 1 - \frac{\dot{Q}_c}{\dot{Q}_h} = 1 - \frac{Q_c}{Q_h} \tag{9-38}$$

From the second law, it can be shown that

$$\eta_{\substack{\text{heat}\\\text{engine}}} \leq 1 - \frac{T_{ce}}{T_{he}} \tag{9-39}$$

and that

$$\frac{Q_c}{Q_h} \geq \frac{T_{ce}}{T_{he}} \tag{9-40}$$

where the subscript *e* refers to the temperature of the engine working fluid at the time and place the respective heat is received or rejected. The equality sign refers to reversible processes only. The inequality sign applies to all real processes. Thus the second law limits the maximum efficiency of cyclic heat engines.

The second law also limits the performance of cyclic refrigerators (heat pumps). The coefficient of performance is defined as

$$\text{COP} = \frac{\text{useful output}}{\text{costly input}} = \frac{\dot{Q}_c}{\dot{W}} = \frac{Q_c}{W} \tag{9-41}$$

$$= \frac{Q_c}{Q_h - Q_c} \tag{9-42}$$

The second law gives

$$\text{COP} \leq \frac{T_c}{T_h - T_c} \tag{9-43}$$

The second law makes it possible to define a theoretical temperature scale. For reversible cycles,

$$\frac{Q_c}{Q_h} = \frac{T_c}{T_h} \qquad \text{Reversible only} \tag{9-44}$$

Equation (9-44), together with a definition of degree size, establishes the thermodynamic temperature scale in terms of heat flows to and from a reversible engine or heat pump.

Though apparently dealing exclusively with heat, the second law's most important application has to do with work. The work potential (maximum useful work obtainable by man) of a quantity of heat available at T_h in an environment to serve as a sink at T_c is

$$\text{Work potential of } Q = Q\left(1 - \frac{T_c}{T_h}\right) \tag{9-45}$$

Interested students will find an interesting discussion of thermodynamics in the following suggested reading:

The Nature of Thermodynamics by P. W. Bridgman, Harvard University Press.

Problems

9-1 Estimate the best possible efficiency that could be obtained for engines operating with the reservoirs stated.
(a) The sun and the ocean.
(b) A combustion chamber burning oil or gas and a river.
(c) The top of the ocean and the bottom of the ocean.

9-2 Estimate the best possible coefficient of performance of refrigerators operating between the reservoirs stated.
(a) Inside a household refrigerator and the kitchen.
(b) Forty feet down in the earth and the atmosphere near the earth.
(c) Inside an air-conditioned office and the atmosphere outside.

9-3 A rule of thumb on the performance of actual air-conditioning units is 1 hp/ton where a ton is 12,000 Btu/hr. What coefficient of performance does this represent?

9-4(a) If the H_2O in a power plant cycle receives heat at $T_{he} = 650°C$ and rejects heat at $T_{ce} = 40°C$ and its actual operating efficiency is 30%, what is its internal performance compared with the maximum possible?
(b) If the combustion chamber is at 1100°C and the river cooling water at 27°C what is the plant's overall performance compared with the maximum possible?

9-5 How could you make the process of a block sliding down a plane almost reversible?

9-6 Which, if any, of the following systems (italicized) may be said to be internally reversible in the processes described?

(a) A *copper block* is raised in temperature by sliding down a plane.
(b) The *cooling water* flowing through the condenser of a power plant is raised in temperature from 27 to 35°C.
(c) The *earth* is heated by the sun.
(d) *Water* at 32°F is frozen in a freezer at 0°F.
(e) The *air* that enters an evacuated bottle when the cork is removed.

9-7 Why is there thermal pollution?

9-8 Prove that the Kelvin–Planck statement of the second law is logically equivalent to the Clausius inequality.

9-9(a) Derive an expression for the minimum work required to pump a quantity of heat Q from T_c to the environmental temperature T_0 ($T_0 > T_c$).

(b) Derive an expression for the minimum work required to pump a quantity of heat Q to T_h from T_c in an environment at T_0 ($T_h > T_0 > T_c$).

9-10 In a book on automobile engines, the following is written about steam engines: "Another new system has been rumored to be in development that does not require a condenser such as we have previously mentioned. Instead, the steam exhausting from the engine is sent through a reheating device that, in effect, restores to the steam the energy it lost in the engine. Thus, steam continues to circulate between the engine and the reheater and, according to reports, little heat is lost so this should make the rumored engine quite efficient. It will take time to build and test prototype engines to determine whether or not the arrangement is feasible."

Is it feasible? Explain.

9-11 List the sources of energy degradation in the following systems. Indicate for each system which you feel are the major ones.

(a) An automobile engine.
(b) A gas fired home heating system.
(c) A Coke bottle filling machine.
(d) A power lawn mower in operation.
(e) A radio.
(f) A blast furnace.

10

Entropy, the entropy inequality, and the $Td\,S$ equations

The zeroth law led us to a new property: temperature. The first law led us to a new property: energy. The second law also leads us to a new property: entropy. Temperature and energy are concepts most of us become aware of at a very early age. They are common words in everyday language. They are discussed in high school physics and chemistry. There is some confusion between temperature and heat but nevertheless neither temperature nor energy are entirely new to you.

The same cannot be said about entropy. If you have heard of it, it was likely shrouded in mystery and cloaked by some discussion of order and disorder. Entropy is an abstract concept. Beginning with our experience and some elemental words, we have had to build up to this point through temperature, heat, work, energy, energy conversion, etc. Now we are ready to introduce entropy but students should not expect to obtain an instant physical feeling for this new property. Working with the new property will help. And we shall *try* to explain in the book not only how it is used but also what it means.

We have pointed out that the second law was expected to give us a second property to go with the first law and energy. And we have noted that we should expect the new property to be a directional or one-way kind of property because the second law is a directional or one-way law. We should also expect the new property to have something to do with work potential and degradation. Let us then see how our expectations work out.

10-1 Entropy

10-1 (a) Definition of entropy

In Chap. 4 we showed that if a physical parameter is always zero in a cycle, then that parameter is a property. In Chap. 9, we also showed that *in a **reversible** cycle*,

$$\oint \left(\frac{d'Q}{T}\right)_{\text{reversible}} = 0 \tag{10-1}$$

Thus, it follows that $(d'Q/T)_{\text{reversible}}$ is a property. We call it *entropy* and give it the symbol s.

$$ds = \left(\frac{d'Q}{T}\right)_{\text{reversible}} \tag{10-2}$$

$$\Delta s_{12} = s_2 - s_1 = \int_1^2 \left(\frac{d'Q}{T}\right)_{\text{reversible}} \tag{10-3}$$

Students must note carefully and explicitly that the subscript reversible in writing $(d'Q/T)_{\text{reversible}} = ds$ is *essential*. In general, $(d'Q/T)$ is *not* a property because $\oint d'Q/T$ is *not* zero. Only in those special ideal limiting cases called *reversible* is $\int d'Q/T$ equal to $\int ds$ and a property. We shall see, however, that though defined in imaginary reversible processes the new property is very real, very important, and very useful.

Students must be quite conscious that entropy s *is* always a property but that $d'Q/T$ is *not a property except in reversible processes.* To further illustrate this, consider a system changing from 1 to 2 as shown along possible paths A, B, and R.

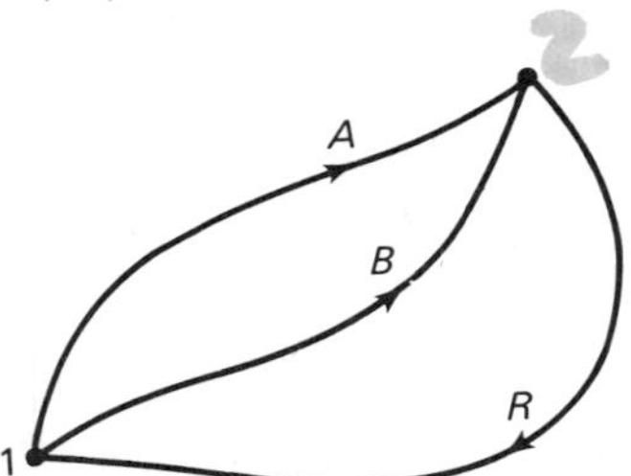

A and B are real or irreversible processes. R is a reversible (imaginary) process. *Regardless of which process is used,* Δs_{12} *is the same* because s is a property and hence its change is not a function of the process but only the end states.

We can use $\int_1^2 d'Q/T$ to compute Δs_{12} only for process R.

$$\int_{1R}^{2} \left(\frac{d'Q}{T}\right)_{\text{reversible}} = \Delta s_{12} \tag{10-4}$$

For the other two processes, computing $\int_1^2 d'Q/T$ is a meaningless exercise. But note that by imagining a reversible process from 1 to 2, we can compute Δs_{12} from $\int_1^2 (d'Q/T)_{\text{reversible}}$ even though the actual process may be A or B or something else. This is one manner in which entropy changes are computed for real processes, that is, by imagining a reversible process between the end states and then computing $\int_1^2 d'Q/T$ for that reversible process.

It should also be noticed at this point that if a system executes an *internally* reversible process, then we may use $\int d'Q/T$ to compute the entropy change for that system in that process even though other external systems involved may not execute reversible processes.

10-1(b) The principle of increasing entropy

Earlier we noted that one reason a second law is necessary is because the first law does not say anything about process directions. We have also said that the property that we expected to result from the second law would reflect this directional character of nature. We shall now show that this indeed is the case with entropy.

To derive this principle, consider a closed system executing two cycles, 1-A-2-B-1 and 1-A-2-C-1, as shown.

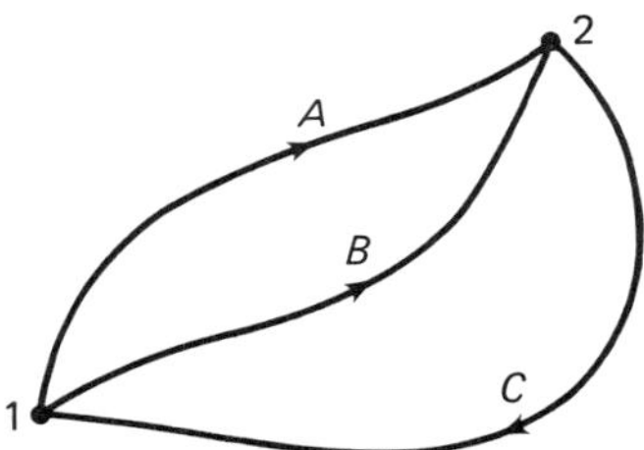

Consider that 1-A-2-B-1 is a reversible cycle and 1-A-2-C-1 is not reversible. Then, of course, processes A and B are reversible and process C is not reversible, though in other respects the processes are arbitrary. Now for 1-A-2-B-1 we have

$$\oint \frac{d'Q}{T} = 0 \quad \text{or} \quad \int_1^2 \left(\frac{d'Q}{T}\right)_A + \int_2^1 \left(\frac{d'Q}{T}\right)_B = 0 \tag{10-5}$$

and for 1-A-2-C-1 we have

$$\oint \frac{d'Q}{T} < 0 \quad \text{or} \quad \int_1^2 \left(\frac{d'Q}{T}\right)_A + \int_2^1 \left(\frac{d'Q}{T}\right)_C < 0 \tag{10-6}$$

Subtracting Eq. (10-5) from Eq. (10-6), we find

$$\int_1^2 \left(\frac{d'Q}{T}\right)_B > \int_1^2 \left(\frac{d'Q}{T}\right)_C \tag{10-7}$$

And since process B is reversible,

$$\int_1^2 \left(\frac{d'Q}{T}\right)_B = \Delta s_{12} \tag{10-8}$$

Hence,

$$\Delta s_{12} > \int_1^2 \left(\frac{d'Q}{T}\right)_C \tag{10-9}$$

Or, in general, $ds > d'Q/T$. The result above is extremely important and students of thermodynamics should remember it as they remember $d'Q - d'W = dE$ or $F = ma$. It applies to *all* real processes.

Now if a system is *adiabatic*, then $dQ = 0$, and $ds_{\text{adiabatic}} > 0$. This shows that, *in an adiabatic system*, the entropy increases in any real process or change of state that occurs. This is called the *principle of increase of entropy*.

Students must take careful note of the fact that the principle of increase of entropy applies *only* to adiabatic systems.

A common student error is to assume that the entropy of ***everything*** *always* goes up. This is not true. In $ds > d'Q/T$, if $d'Q$ is negative, then ds *may* be negative and the entropy decreases. It happens all the time. What the principle of increasing entropy says is that in such cases the entropy of something else, included within some larger boundaries such that the whole affair is adiabatic, will go up *more*. It is *only* the entropy of an *adiabatic system* that *always* goes up.

10-1(c) Constant entropy processes

We have shown that in a reversible process, $ds = (dQ/T)_{\text{reversible}}$. Now if the process is also adiabatic ($d'Q = 0$), we find $ds = 0$ or $s =$ constant. Reversible adiabatic processes are constant entropy. Though seemingly rather specialized, such processes are approached closely enough in practice to be very important. We shall use this fact often after we have studied entropy as a property in Sec. 10-4.

The word *isentropic* is used to denote a constant entropy process.

EXAMPLE 10.1. When 1000 Btu of heat are transferred from a system at 2500°R to a system at 2000°R, what is the change in entropy of each system? Of both systems taken together?

SOLUTION: For the 2500°R reservoir, we assume it executes an internally reversible isothermal process transferring out 1000 Btu. Therefore $ds = d'Q/T$.

$$\Delta s_{12} = \int_1^2 \frac{d'Q}{T} = \frac{Q}{T} = \frac{(-1000)}{2500} = -0.40 \text{ Btu/°R}$$

For the 2000°R reservoir, we assume it executes an internally reversible isothermal process receiving 1000 Btu.

$$\Delta s_{12} = \int_1^2 \frac{d'Q}{T} = \frac{Q}{T} = \frac{(+1000)}{2000} = +0.50 \text{ Btu/°R}$$

Together

$$\Delta s = -0.40 + 0.50 = +0.10 \text{ Btu/°R}$$

Note: Students are referred to Sec. 9.3. In the example there, the degradation in the process above was found to be 50 Btu or $T_0 \, \Delta s$ ($=500 \times 0.10$). That should be a clue to the significance of entropy and its relationship to work potential and degradation.

EXAMPLE 10.2. An aluminum block ($c_p = 400$ j/kg-°K) with a mass of 5 kg is initially at 40°C in room air at 20°C. It is cooled reversibly by transferring heat to a completely reversible cyclic heat engine until the block reaches 20°C. The 20°C room air serves as a constant temperature sink for the engine. Compute

1. The change in entropy for the block.
2. The change in entropy for the room air.
3. The work done by the engine.

SOLUTION:

1. For the block in the reversible process, $ds = (d'Q/T)_{\text{reversible}}$ where $d'Q$ is the heat transferred *to* the block. Also, $d'Q = mc_p \, dT$. Thus,

$$(s_2 - s_1)_{\text{block}} = \int_1^2 \frac{d'Q}{T} = mc_p \int_1^2 \frac{dT}{T} = mc_p \ln \frac{T_2}{T_1}$$

$$= (5)(400) \ln \frac{293}{313} = -(2000)(0.067) = -134 \text{ j/°K}.$$

2. Since the process is completely reversible, if we take the block, the engine, and the room air as a system, then that system is adiabatic. For an adiabatic system in a reversible process, $ds = 0$. Therefore, for the room air,

$$(s_2 - s_1)_{\text{air}} = +134 \text{ j/°K}$$

3. To get the work done, we set up the transient situation as shown:

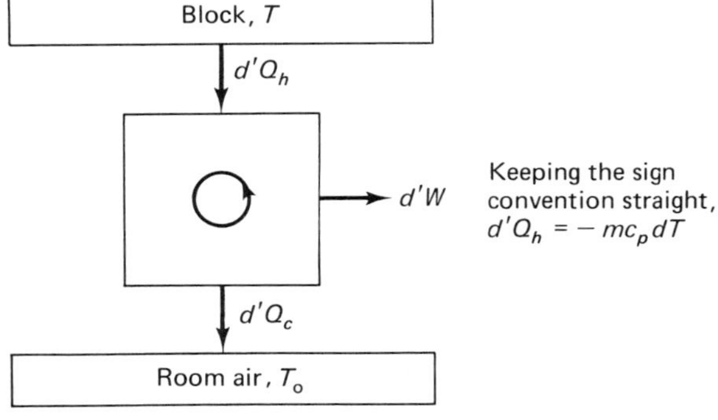

Keeping the sign convention straight, $d'Q_h = -mc_p dT$

The efficiency of the completely reversible engine can be written as

$$\eta_{\text{reversible}} = 1 - \frac{T_0}{T} \quad \text{and} \quad \eta = \frac{d'W}{d'Q_h} = -\frac{d'W}{mc_p\, dT}$$

Equating these gives

$$1 - \frac{T_0}{T} = -\frac{d'W}{mc_p\, dT} \quad \text{or} \quad d'W = mc_p T_0 \frac{dT}{T} - mc_p\, dT$$

Integrating for the total work as T changes from $T_1 = 313°\text{K}$ to $T_2 = 293°\text{K}$ gives

$$W = mc_p T_0 \ln \frac{T_2}{T_1} - mc_p(T_2 - T_1)$$

$$W = (5)(400)(293)(-0.067) - (5)(400)(-20)$$

$$= -39{,}260 + 40{,}000 = 740 \text{ j}$$

EXAMPLE 10.3. Suppose the aluminum block of Example 10.2 is allowed to cool by natural convection to room temperature. Compute

1. The change in entropy for the block.
2. The change in entropy for the room air.
3. The net change in entropy for the universe.

SOLUTION:

1. The block executes an internally reversible process and so we can use $ds = dQ/T$ for it. The result is the same as in Example 10.2. That is,

$$(s_2 - s_1)_{\text{block}} = -134 \text{ j/°K}$$

2. The room air executes an internal reversible process. The heat transferred to it is the heat removed from the block.

$$Q_{\text{to room}} = Q_{\text{from block}} = mc_p(T_1 - T_2) = (5)(400)(20) = 40{,}000 \text{ j}$$

For Δs,

$$(s_2 - s_1)_{\substack{\text{room}\\ \text{air}}} = \int_1^2 \frac{d'Q}{T_0} = \frac{Q}{T_0} = \frac{40{,}000}{293} = 136.5 \text{ j/°K}$$

3. The net change in entropy for the universe is thus

$$\Delta s_{\text{universe}} = -134 + 136.5 = 2.5 \text{ j/°K}$$

Students may note that the work *not* done when the block cools by itself is 740 j and that

$$T_0 \, \Delta s_{\text{universe}} = (293)(2.5) \cong 740 \text{ j}$$

This should give another clue to the meaning of entropy and degradation.

10-2 Generalized entropy and open systems

10-2(a) The concept of generalized entropy

After the first law and the property energy ($\mathscr{E}$) were developed, we showed how the concept of generalized energy made easy the development of the energy equation for open systems. We can generalize the concept of entropy in a similar way in order to apply the second law to open systems. The result, of course, will not be an equation but an inequality. We call it the *entropy inequality for open systems*.

In the generalized concept of energy, the energy $\mathscr{E}$ is a form of energy associated with material, whereas heat Q and work W are forms of energy in transition or energy flux. The generalized energy $\mathscr{E}$ is a generic term embodying both forms.

By direct analogy we may generalize the concept of entropy. The entropy s is associated with material and dQ/T is a kind of entropy flux.

Generalized energy is conserved. Thus, we could write for a closed system in a process.

$$\mathscr{E}_{\text{initial}} + \mathscr{E}_{\text{in}} = \mathscr{E}_{\text{final}} + \mathscr{E}_{\text{out}} \tag{10-10}$$

where $\mathscr{E}_{\text{initial}}$ and $\mathscr{E}_{\text{final}}$ are stored terms and $\mathscr{E}_{\text{in}}$ and $\mathscr{E}_{\text{out}}$ are the flux terms, heat and work. We showed how this leads to the standard first law equation,

$$Q_{12} - W_{12} = \Delta\mathscr{E}_{12} \tag{10-11}$$

Generalized entropy is *not* conserved. From the principle of increasing entropy, we see that it is created. Using $\mathscr{S}$ for generalized entropy, we may therefore write for a closed system process

$$\mathscr{S}_{\text{initial}} + \mathscr{S}_{\text{in}} < \mathscr{S}_{\text{final}} + \mathscr{S}_{\text{out}} \tag{10-12}$$

In Eq. (10-12), $\mathscr{S}_{\text{initial}}$ and $\mathscr{S}_{\text{final}}$ are stored terms and $\mathscr{S}_{\text{in}}$ and $\mathscr{S}_{\text{out}}$ are flux terms, dQ/T. The inequality sign is directed to express the fact that entropy is generated or created in the process. To show that this expression is equivalent to our previous ones, we substitute s and dQ/T for $\mathscr{S}$:

$$s_1 + \int_{1_{\text{in}}}^{2} \left(\frac{d'Q}{T}\right)_{\text{in}} < s_2 + \int_{1_{\text{out}}}^{2} \left(\frac{d'Q}{T}\right)_{\text{out}} \tag{10-13}$$

Using the regular sign convention to handle Q_{in} and Q_{out} gives

$$\int_1^2 \frac{d'Q}{T} = \int_{1_{\text{in}}}^{2} \left(\frac{d'Q}{T}\right)_{\text{in}} - \int_{1_{\text{out}}}^{2} \left(\frac{d'Q}{T}\right)_{\text{out}} \tag{10-14}$$

and thus we obtain

$$s_2 - s_1 > \int_1^2 \frac{d'Q}{T} \tag{10-15}$$

which is the same result obtained in Sec. 10-1(c).

10-2(b) The entropy inequality for open systems

Just as generalized energy made it easy to apply the first law to open systems, so too generalized entropy makes it easy to apply the second law to open systems. For the first law, we wrote

$$\begin{bmatrix}\text{Rate of } \mathscr{E} \\ \text{flow in}\end{bmatrix} - \begin{bmatrix}\text{rate of } \mathscr{E} \\ \text{flow out}\end{bmatrix} = \begin{bmatrix}\text{rate of change} \\ \text{of } \mathscr{E} \text{ inside}\end{bmatrix} \tag{10-16}$$

Taking account of $\mathscr{E}$ coming in, leaving, or stored inside with material; work (including flow work); and heat leads us directly to the energy equation

$$\sum \dot{Q} - \sum \dot{W}_x = \frac{d\mathscr{E}_{\text{c.v.}}}{d\theta} + \sum \left(h + \frac{\mathscr{V}^2}{2} + gz\right)\dot{m}_{\text{out}} - \sum \left(h + \frac{\mathscr{V}^2}{2} + gz\right)\dot{m}_{\text{in}} \tag{10-17}$$

We then wrote this more elegantly as

$$\oint_A d'Q - \oint_A d'W_x = \frac{\partial}{\partial\theta}\int_V \rho e \, dV + \oint \left(h + \frac{\mathscr{V}^2}{2} + gz\right)\rho\vec{\mathscr{V}} \cdot \vec{n}\, dA \tag{10-18}$$

To obtain the corresponding second law expressions, we say

$$\begin{bmatrix}\text{Rate of } \mathscr{S} \\ \text{flow in}\end{bmatrix} - \begin{bmatrix}\text{rate of } \mathscr{S} \\ \text{flow out}\end{bmatrix} < \begin{bmatrix}\text{rate of change} \\ \text{of } \mathscr{S} \text{ inside}\end{bmatrix} \tag{10-19}$$

Note that the principle of increasing $\mathscr{S}$ is adhered to. Substituting s and dQ/T appropriately for $\mathscr{S}$ and referring to Fig. 10.1, we find

$$\begin{bmatrix}\text{Rate of } \mathscr{S} \\ \text{flow in}\end{bmatrix} = s_1\dot{m}_1 + s_2\dot{m}_2 + s_3\dot{m}_3 + \frac{\dot{Q}_a}{T_a} + \frac{\dot{Q}_b}{T_b} \tag{10-20}$$

$$\begin{bmatrix}\text{Rate of } \mathscr{S} \\ \text{flow out}\end{bmatrix} = s_a\dot{m}_a + s_b\dot{m}_b + \frac{\dot{Q}_1}{T_1} + \frac{\dot{Q}_2}{T_2} \tag{10-21}$$

$$\begin{bmatrix}\text{Rate of change} \\ \text{of } \mathscr{S} \text{ inside}\end{bmatrix} = \frac{d}{d\theta}(S_{\text{c.v.}}) \tag{10-22}$$

where $S_{\text{c.v.}}$ is the total entropy associated with the material inside the control volume. Generalizing Eq. (10-20) to (10-22) in the following fashion,

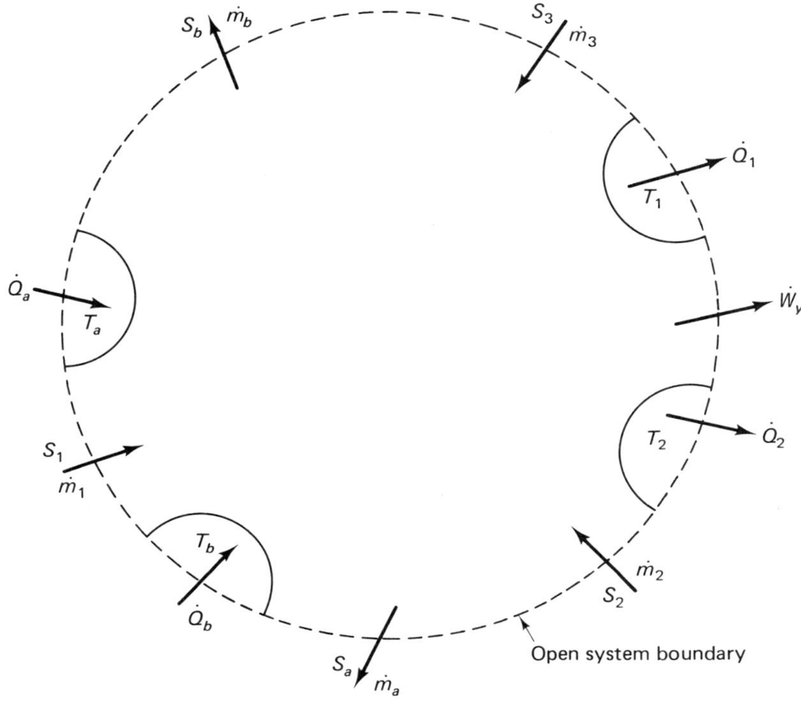

Figure 10.1 Open system for entropy inequality.

$$\begin{bmatrix} \text{Rate of } \mathscr{S} \\ \text{flow in} \end{bmatrix} > \sum_{\text{in}} (\dot{m}s)_{\text{in}} + \sum_{\text{in}} \left(\frac{\dot{Q}}{T}\right)_{\text{in}} \tag{10-23}$$

leads to the complete inequality as

$$\frac{\partial}{\partial \theta}(S_{\text{c.v.}}) > \sum_{\text{in}} (s\dot{m})_{\text{in}} + \sum_{\text{in}} \left(\frac{\dot{Q}}{T}\right)_{\text{in}} - \sum_{\text{out}} (s\dot{m})_{\text{out}} - \sum \left(\frac{\dot{Q}}{T}\right)_{\text{out}} \tag{10-24}$$

Rearranging and letting

$$\sum_{\text{area}} \left(\frac{\dot{Q}}{T}\right) = \sum_{\text{in}} \left(\frac{\dot{Q}}{T}\right)_{\text{in}} - \sum_{\text{out}} \left(\frac{\dot{Q}}{T}\right)_{\text{out}} \tag{10-25}$$

results in

$$\sum_{\text{area}} \frac{\dot{Q}}{T} < \frac{\partial}{\partial \theta} (S_{\text{c.v.}}) + \sum_{\text{out}} (\dot{m}s)_{\text{out}} - \sum_{\text{in}} (\dot{m}s)_{\text{in}} \tag{10-26}$$

Or, in more elegant notation,

$$\oint_A \frac{d'Q}{T} < \frac{\partial}{\partial\theta}\int_V \rho s\, dV + \oint \rho s \vec{\mathscr{V}} \cdot \vec{n}\, dA \tag{10-27}$$

10-3 The $T\,ds$ equations

With the equation $ds = (d'Q/T)_{\text{reversible}}$ we are now in a position to derive two equations that are among the most important of all thermodynamics equations, the so-called $T\,ds$ equations. To make the derivation, we consider a closed system executing a reversible process R such that the only work done is $(p\,dv)$-type work (and $p = p'$), $ds = d'Q/T$, and $E = u$. Writing the first law, we find

$$d'Q - d'W = dE \quad \text{or} \quad T\,ds - p\,dv = du \tag{10-28}$$

Rearranging gives

$$T\,ds = du + p\,dv \tag{10-29}$$

Now, though Eq. (10-29) has been derived in a very restricted set of circumstances, we note that it involves only properties. Remembering that properties are functions only of the end states, not upon the path of process, we conclude that the equation must apply to *any* process between given end states. That is, ($T\,ds = du + p\,dv$) is a completely general equation. In terms of the sketch, we are saying that we used process R in the derivation but because the resulting equation involves only properties, it must apply to any process N as well.

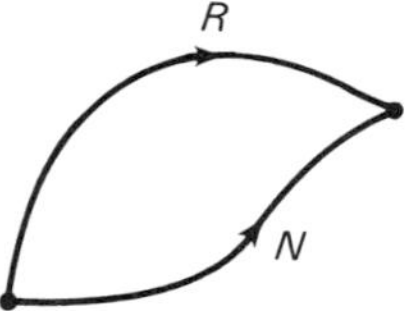

We obtain a second $T\,ds$ equation from the definition of enthalpy, $h = u + pv$. Differentiating,

$$dh = du + p\,dv + v\,dp \tag{10-30}$$

Thus,

$$du + p\,dv = dh - v\,dp \tag{10-31}$$

And so

$$T\,ds = dh - v\,dp \tag{10-32}$$

More will be said about these equations later; they will be used extensively in problems, to find other important relations among properties, and in other derivations.

EXAMPLE 10.4. Compute the work per pound required to pump liquid water from 15 psia and 80°F to 90 psia in a reversible adiabatic steady flow process.

SOLUTION: Translation:

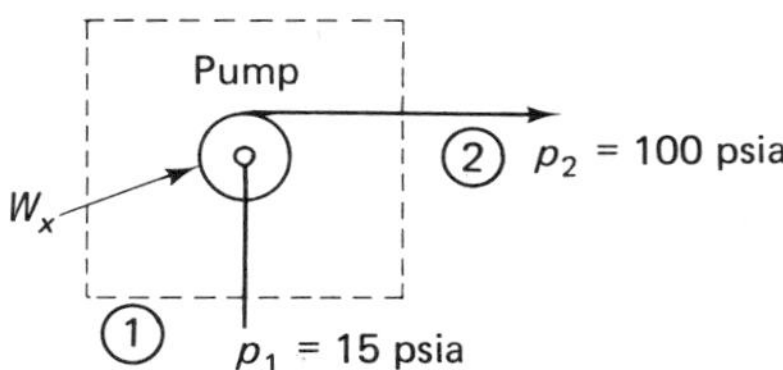

System: Open as shown.
Equation of state: Steam tables (note that liquid H_2O is nearly incompressible.)
Process equation: Reversible adiabatic means s = constant.
Principles:
Continuity:

$$\dot{m}_1 = \dot{m}_2 = \dot{m}$$

Energy:

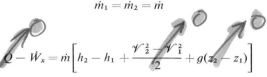

$$\dot{Q} - \dot{W}_x = \dot{m}\left[h_2 - h_1 + \frac{\mathscr{V}_2^2 - \mathscr{V}_1^2}{2} + g(z_2 - z_1)\right]$$

Neglect K.E. and P.E., and $Q = 0$ (given).
Therefore,

$$-\dot{W}_x = \dot{m}(h_2 - h_1) \quad \text{or} \quad W_x = h_2 - h_1$$

Relation among properties:
T ds equations:

$$T\,ds = dh - v\,dp$$

Since $ds = 0$ in this case,

$$dh = v\,dp \quad \text{or} \quad h_2 - h_1 = \int_1^2 v\,dp$$

Combining with energy equation:

$$-W_x = h_2 - h_1 = \int_1^2 v\,dp$$

As liquid H_2O is nearly incompressible, v is constant.

$$-W_x = v(p_2 - p_1)$$

From the tables, for saturated liquid at 80°F, $v_f = 0.016$ ft³/lbm.

$$-W_x = (0.016)(75)(144) = 172 \text{ ft-lbf/lbm} = 0.22 \text{ Btu/lbm}$$

10-4 Work in reversible processes

10-4(a) Work in a reversible closed system

The amount of work produced in a reversible process for a closed system can be found by writing the first law for this process:

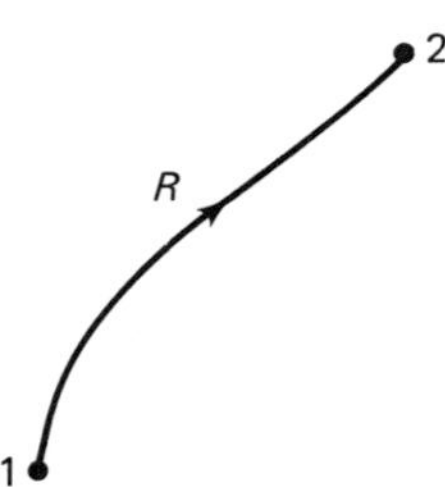

$$Q_{1R2} - W_{1R2} = E_2 - E_1$$

$$= (U_2 + \tfrac{1}{2}m\mathscr{V}_2^2 + mgZ_2) - (U_1 + \tfrac{1}{2}m\mathscr{V}_1^2 + mgZ_1) \qquad (10\text{-}33)$$

and the second law:

$$Q_{1R2} = \int_1^2 T\,ds \qquad (10\text{-}34)$$

and the $T\,ds$ equation:

$$\int_1^2 T\,ds = U_2 - U_1 + \int_1^2 p\,dV \qquad (10\text{-}35)$$

Combining gives

$$W_{1R2} = E_1 - E_2 + Q_{1R2} \qquad (10\text{-}36)$$

$$= E_1 - E_2 + \int_1^2 T\,ds \qquad (10\text{-}37)$$

$$= E_1 - E_2 + (U_2 - U_1) + \int_1^2 p\,dV \tag{10-38}$$

$$= \int_1^2 p\,dV - \tfrac{1}{2}m(\mathscr{V}_2^2 - \mathscr{V}_1^2) - mg(Z_2 - Z_1) \tag{10-39}$$

In the absence of significant kinetic and potential energy changes, this reduces to

$$W_{1R2} = \int_1^2 p\,dV \quad \text{or} \quad dW_{\text{reversible}} = p\,dV \tag{10-40}$$

Bear in mind that the development above applies to *closed system* reversible processes. Since we know that $W_{\substack{\max \\ \text{poss}}}$ is produced in a reversible process, this result is also

$$W_{\substack{\max \\ \text{poss} \\ 1\text{-}2}} = \int_1^2 p\,dV - \tfrac{1}{2}m(\mathscr{V}_2^2 - \mathscr{V}_1^2) - mg(Z_2 - Z_1) \tag{10-41}$$

or

$$dW_{\substack{\max \\ \text{poss}}} = p\,dV - m\mathscr{V}\,d\mathscr{V} - mg\,dZ \tag{10-42}$$

Students should note that this corresponds with the results of our discussion earlier on the work in a very slow compression or expansion process.

10-4(b) Work in reversible steady slow open system processes with single inlet and outlet

Though the case considered in Sec. 10-4(a) is highly specialized, it is useful in many real situations. Most conventional energy conversion systems operate in steady state and contain many devices that have but a single inlet and outlet (e.g., turbines, pumps, etc.). To obtain an expression for the reversible work in these situations, we begin with the steady flow energy equation for single inlet–outlet cases:

$$\sum_{\text{area}} \dot{Q} - \sum_{\text{area}} \dot{W}_x = \dot{m}\left(h_2 + \frac{\mathscr{V}_2^2}{2} + gZ_2\right) - \dot{m}\left(h_1 + \frac{\mathscr{V}_1^2}{2} + gZ_1\right) \tag{10-43}$$

where the subscripts 1 and 2 refer to the inlet and outlet, respectively. By dividing through by $\dot{m}$, we put this on a per pound basis:

$$Q - W_x = \left(h_2 + \frac{\mathscr{V}_2^2}{2} + gZ_2\right) - \left(h_1 + \frac{\mathscr{V}_1^2}{2} + gZ_1\right) \tag{10-44}$$

From the second law for the reversible case we have

$$d'Q = T\,ds \quad \text{or} \quad Q = \int_1^2 T\,ds \tag{10-45}$$

and from the *T ds* equations, we obtain

$$T\,ds = dh - v\,dp \quad \text{or} \quad \int_1^2 T\,ds = h_2 - h_1 - \int_1^2 v\,dp \tag{10-46}$$

Combining these gives

$$\int_1^2 T\,ds - W_{x_{\text{reversible}}} = \left(h_2 + \frac{\mathscr{V}_2^2}{2} + gZ_2\right) - \left(h_1 + \frac{\mathscr{V}_1^2}{2} + gZ_1\right) \tag{10-47}$$

$$h_2 - h_1 - \int_1^2 v\,dp - W_{x_{\text{reversible}}} = \left(h_2 + \frac{\mathscr{V}_2^2}{2} + gZ_2\right) - \left(h_1 + \frac{\mathscr{V}_1^2}{2} + gZ_1\right) \tag{10-48}$$

$$-W_{x_{\text{reversible}}} = \int_1^2 v\,dp + \frac{\mathscr{V}_2^2 - \mathscr{V}_1^2}{2} + g(Z_2 - Z_1) \tag{10-49}$$

In the absence of significant kinetic and potential energy changes, this reduces to

$$W_{x_{\text{reversible}}} = \int_1^2 v\,dp \tag{10-50}$$

Bear in mind that Eq. (10-50) applies to given system steady flow single inlet–single outlet reversible processes. As expressed above, the work is on a per pound basis. If the flow rate $\dot{m}$ is returned, we find

$$\dot{W}_{x_{\text{reversible}}} = \dot{m}\int_1^2 v\,dp \tag{10-51}$$

The work $\dot{W}_x$ is, by the sign convention, the work done *by* the system.

10-5 Entropy as a property

10-5(a) Tabulated values

Entropy is reported in most tables of properties in exactly the same fashion as internal energy (u) and enthalpy (h) are reported. The subscripts f, g, fg, etc., have the same meaning and the equations for two-phase mixtures

$$s = s_f + xs_{fg} \quad \text{and} \quad s = (1 - x)s_f + xs_g \tag{10-52}$$

apply as well.

10-5(b) Graphical representations

Entropy as a coordinate or a property plot is extremely important and useful. The temperature–entropy (T-s) chart is especially useful and engineers must be as familiar with phase diagrams as these plots. A typical T-s diagram is shown in Fig. 10.2. The two-phase regions appear in the same general

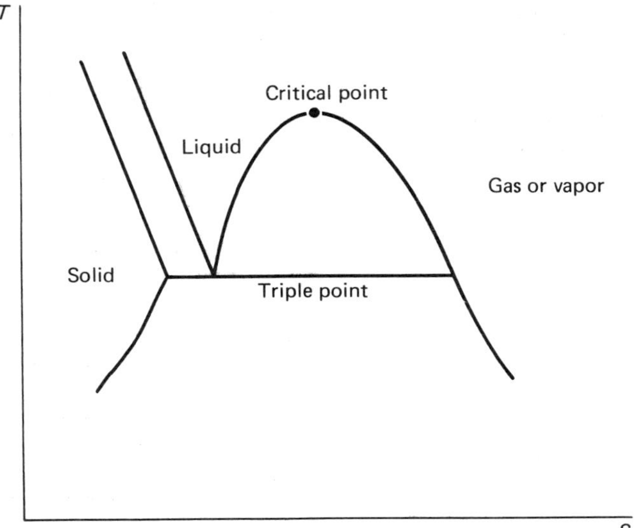

Figure 10.2 Typical T–s phase diagram.

orientation as on a p-v diagram. In Fig. 10.3, the diagram is repeated but with typical lines of constant pressure and constant enthalpy also shown.

Entropy–enthalpy charts are also very useful in some problems in lieu of using tables. They are called *Mollier Charts* and are typically huge (e.g., 40 in. or so on a side!) so that data can be read to nearly as much accuracy as in the tables.

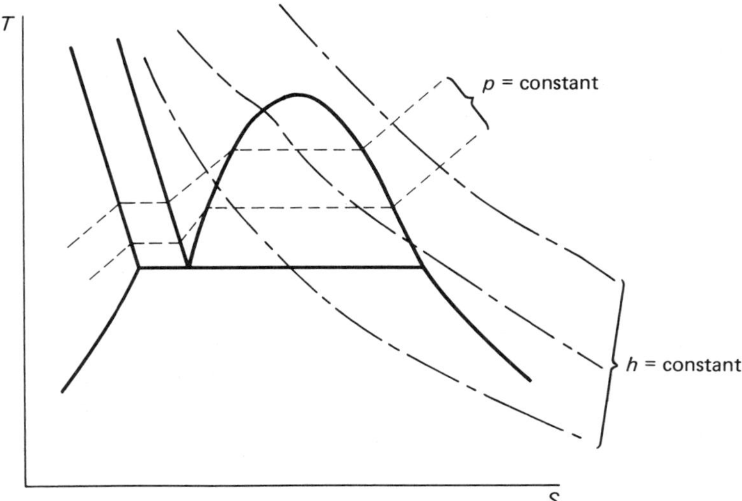

Figure 10.3 Lines of constant p and h on typical T–s phase diagram.

Computerized sources of data give entropy as well as internal energy and enthalpy, usually with p and T as the needed entry information.

EXAMPLE 10.5. Saturated steam vapor at 6 bars absolute enters a reversible adiabatic turbine and is discharged at 0.010 bar absolute. Compute the work per kilogram of steam entering.

SOLUTION: Translation into symbols and diagrams.

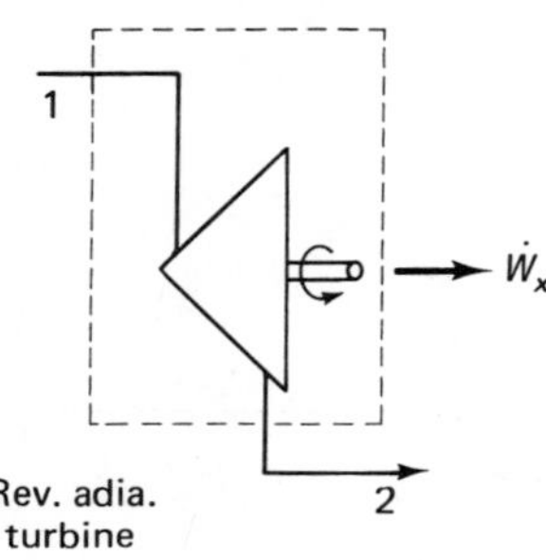

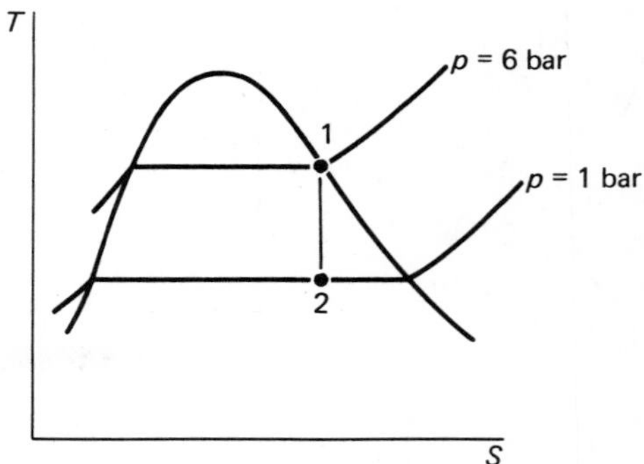

Equation of state: Steam tables.
Process equation: Assume steady flow; reversible adiabatic means constant entropy. Therefore, $s_1 = s_2$ (see process done on sketch).
Application of principles:
Continuity:

$$\dot{m}_1 = \dot{m}_2 = \dot{m}$$

Energy equation:

$$\dot{Q} - \dot{W}_x = \dot{m}\left[h_2 - h_1 + \frac{\mathcal{V}_2^2 - \mathcal{V}_1^2}{2} + g(z_2 - z_1)\right]$$

Neglect K.E. and P.E. terms: $\dot{Q} = 0$ (given).

$$\frac{\dot{W}_x}{\dot{m}} = W_x = h_1 - h_2$$

We can find h_1 from the tables ($h_1 = 2756.8$ j/gm) and if we can find the quality at $2(x)_2$, we can obtain h_2. Since we know that $s_2 = s_1$ and that s_1 can also be found in the tables ($s_1 = 6.76$ j/gm-°K), we can find x_2 as follows:

$$s_2 = s_{f2} + x_2 s_{fg2}$$

$$x_2 = \frac{s_2 - s_{f2}}{s_{fg2}} = \frac{6.76 - 0.1059}{8.8697} = 0.750$$

Then to find h_2,

$$h_2 = h_{f2} + x_2 h_{fg2} = 29.3 + 0.750(2484.9)$$
$$h_2 = 1893 \text{ j/gm}$$

The work per pound is therefore

$$W_x = h_1 - h_2 = 2756.8 - 1893 = 864 \text{ j/gm} = 0.864 \text{ j/kg}$$

This kind of problem should be considered routine!

EXAMPLE 10.6. A typical but idealized refrigeration cycle consists of the following processes:

A—Reversible adiabatic compression of saturated vapor.
B—Constant pressure cooling and condensation to saturated liquid.
C—Constant enthalpy expansion to the initial pressure.
D—Constant pressure evaporation.

Show this process on a typical *T-s* diagram.

SOLUTION:

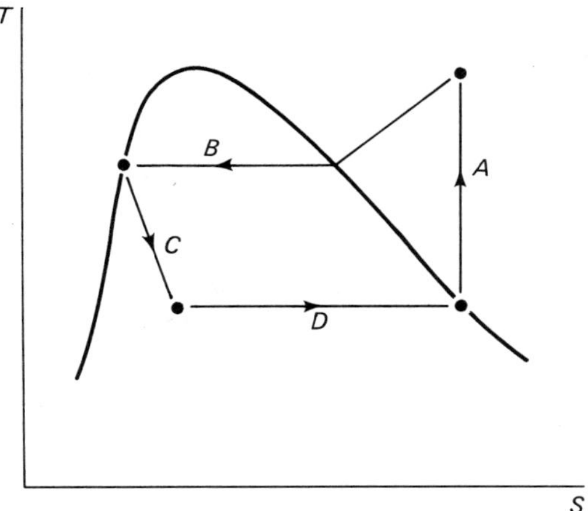

10-5(c) Ideal gases

For ideal gases with constant specific heats, extensive use is made of the $T\,ds$ equations to compute entropy changes. Starting with

$$T\,ds = dh - v\,dp \tag{10-53}$$

and incorporating $dh = c_p\,dT$ and $v = RT/p$, we find

$$ds = c_p \frac{dT}{T} - R\frac{dp}{p} \tag{10-54}$$

which integrates to

$$s_2 - s_1 = c_p \ln \frac{T_2}{T_1} - R \ln \frac{p_2}{p_1} \tag{10-55}$$

In a similar way starting with $T\,ds = du + p\,dv$, we find

$$ds = c_v \frac{dT}{T} + R \frac{dv}{v} \tag{10-56}$$

$$s_2 - s_1 = c_v \ln \frac{T_2}{T_1} + R \ln \frac{v_2}{v_1} \tag{10-57}$$

Equations (10-54)–(10-57) apply to ***any*** process of an *ideal gas* with *constant specific heats.*

If the specific heats are allowed to vary with temperature, then the expressions

$$ds = c_p \frac{dT}{T} - R \frac{dp}{p} = c_v \frac{dT}{T} + R \frac{dv}{v} \tag{10-58}$$

are still valid but the integration cannot be performed unless the $c_p(T)$ function is known.

In the gas tables, the integral $\int_{T_0}^{T} c_p\, dT/T$ is given as ϕ, where T_0 is a reference temperature. Thus,

$$\int_{T_1}^{T_2} c_p \frac{dT}{T} = \int_{T_0}^{T_2} c_p \frac{dT}{T} - \int_{T_0}^{T_1} c_p \frac{dT}{T} = \phi_2 - \phi_1 \tag{10-59}$$

A temperature–entropy chart for an ideal gas (with lines of constant pressure and constant volume illustrated) is shown in Fig. 10.4. *Students must be able to sketch processes and cycles on T-s charts as well as on p-v plots in order to work practical problems.*

If an ideal gas executes a reversible adiabatic process, then $ds = 0$ and we obtain

$$ds = c_p \frac{dT}{T} - R \frac{dp}{p} = 0 \tag{10-60}$$

Integration without limits gives

$$c_p \ln T - R \ln p = C_1 \tag{10-61}$$

where C_1 is a constant. Further manipulation of this results in

$$Tp^{-R/c_p} = C_2 \tag{10-62}$$

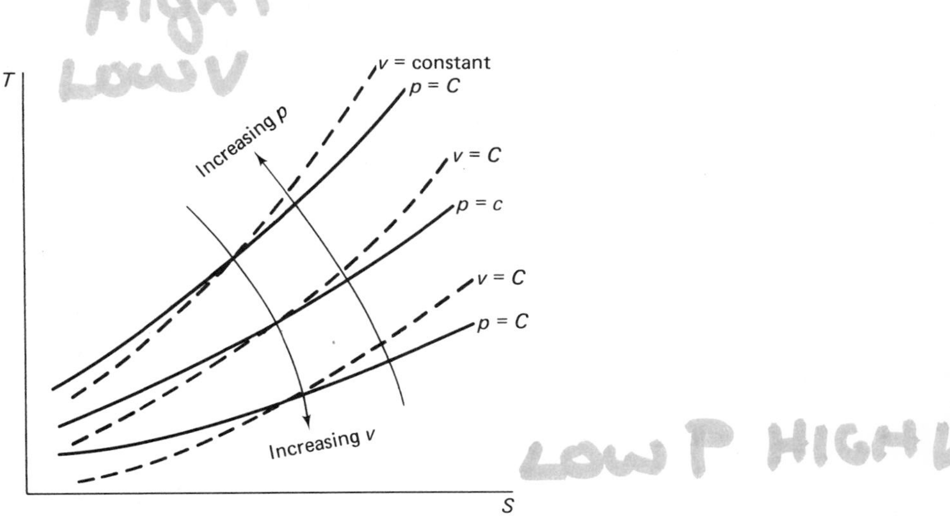

Figure 10.4 *T–s* chart for ideal gas.

For R/c_p we write

$$\frac{R}{c_p} = \frac{c_p - c_v}{c_p} = 1 - \frac{c_v}{c_p} = 1 - \frac{1}{\gamma} = \frac{\gamma - 1}{\gamma} \tag{10-63}$$

Thus, the final result is

$$Tp^{1-\gamma/\gamma} = \text{constant} \tag{10-64}$$

This is the process equation for an ideal gas in a reversible, adiabatic process. The equivalent result with T and v or p and v as independent variables is found by incorporating Eq. (10-64) with $pv = RT$. The results are

$$pv^{\gamma} = \text{constant} \tag{10-65}$$

$$Tv^{\gamma-1} = \text{constant} \tag{10-66}$$

Since we know the range of γ for most substances, we can show constants lines on the p-v plot for ideal gas. For diatomic ideal gases, $\gamma = 1.4$. For monatomic ideal gases, $\gamma = 1.67$. We shall show in Chap. 14 that γ *must* be greater than 1.0 for all substances and it is less than about 2 for almost all. See Fig. 10.5 for how these process lines appear on a p-v plot.

It is interesting, and sometimes very useful, to note that in a reversible process

$$ds = \left(\frac{d'Q}{T}\right)_{\text{reversible}} \quad \text{or} \quad Q = \int T\,ds \quad \text{(Reversible process)} \tag{10-67}$$

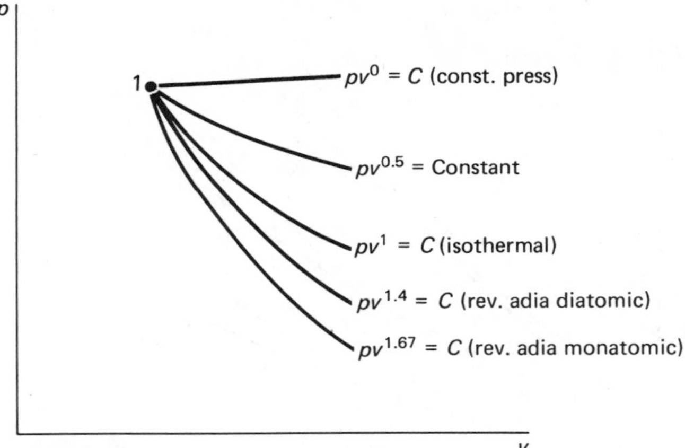

Figure 10.5 Processes of an ideal gas on *pv* plot.

Thus, the heat Q is represented by the area under the process line of a reversible process on a T-s plot. See Fig. 10.6. This is analogous to the work being the area under the process line for a reversible process on a p-v plot. There is, however, nothing analogous to flow work (flow heat?) and so $\int s\, dT$ does *not* have the same implications as $\int v\, dp$ for flow processes.

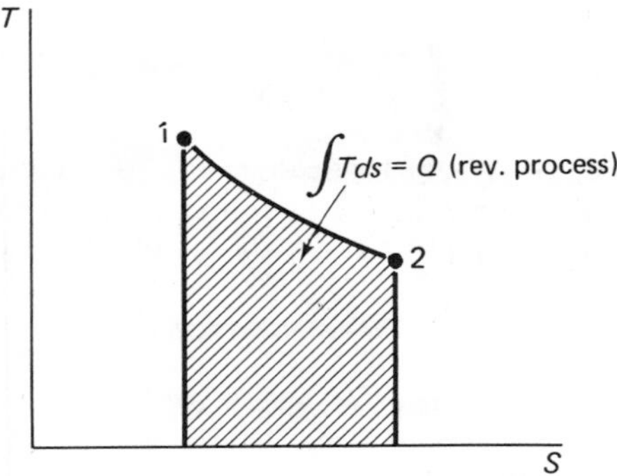

Figure 10.6 *T ds* is heat for reversible processes only.

In the *Gas Tables* by Keenan and Kaye, the specific heats are functions of temperature but only temperatures. To make calculation of entropy changes easier, a function ϕ is plotted as a function of temperature such that

$$\phi = \int_{T_0}^{T} c_p \frac{dT}{T} \tag{10-59a}$$

where T_0 is a constant reference temperature. Thus,

$$\int_{T_1}^{T_2} c_p \frac{dT}{T} = \int_{T_0}^{T_2} c_p \frac{dT}{T} - \int_{T_0}^{T_1} c_p \frac{dT}{T} = \phi_2 - \phi_1 \tag{10-59b}$$

To compute the entropy change of a gas following $pv = RT$, if it is included in the tables, we use the Eq. (10-59a) and (10-59b) as follows:

$$ds = c_p \frac{dT}{T} - R \frac{dp}{p}$$

$$s_2 - s_1 = \int_{T_1}^{T_2} c_p \frac{dT}{T} - R \int_{p_1}^{p_2} \frac{dp}{p} \tag{10-59c}$$

$$s_2 - s_1 = \phi_2 - \phi_1 - R \ln \frac{p_2}{p_1}$$

Also in the gas tables, functions called *relative pressure* (p_r) and *relative volume* (v_r) are defined to assist in making calculations for *constant entropy processes only*. p_r and v_r are defined such that for a given s = constant process 1-2,

$$\left(\frac{p_2}{p_1}\right)_{s=\text{constant}} = \frac{p_{r_2}}{p_{r_1}}$$

$$\left(\frac{v_2}{v_1}\right)_{s=\text{constant}} = \frac{v_{r_2}}{v_{r_1}}$$

p_r and v_r are tabulated as functions of temperature so that if the initial temperature and pressure (or volume) of an isentropic process are known together with the final pressure (or volume), then properties at the final state can be found by entering the tables at p_{r_2} where p_{r_2} is found from

$$p_{r_2} = p_{r_1}\left(\frac{p_2}{p_1}\right)_{s=\text{constant}}$$

EXAMPLE 10.7. Air is compressed from $p_1 = 20$ psia and $T_1 = 80°\text{F}$ to $p_2 = 100$ psia and $T_2 = 380°\text{F}$. Compute the change in entropy of the air.

SOLUTION: Assume air is an ideal gas in this range.

$$T\,ds = dh - v\,dp \quad \text{or} \quad ds = \frac{dh}{T} - \frac{v}{T}dp$$

$$s_2 - s_1 + c_p \ln \frac{T_2}{T_1} - R \ln \frac{p_2}{p_1} = 0.24 \ln \frac{840}{540} - 0.07 \ln \frac{100}{20}$$

$$s_2 - s_1 = 0.24(0.442) - 0.07(1.61) = 0.105 - 0.113 = -0.008 \text{ Btu/°R}$$

EXAMPLE 10.8. An ideal gas executes the following processes in a cycle:

A—Isentropic compression (1-2).
B—Isothermal compression (2-3).
C—Constant volume cooling to p_1 (3-4).
D—Constant pressure heating to state 1 (4-1).

Sketch this cycle on *p-v* and *T-s* diagrams.

SOLUTION:

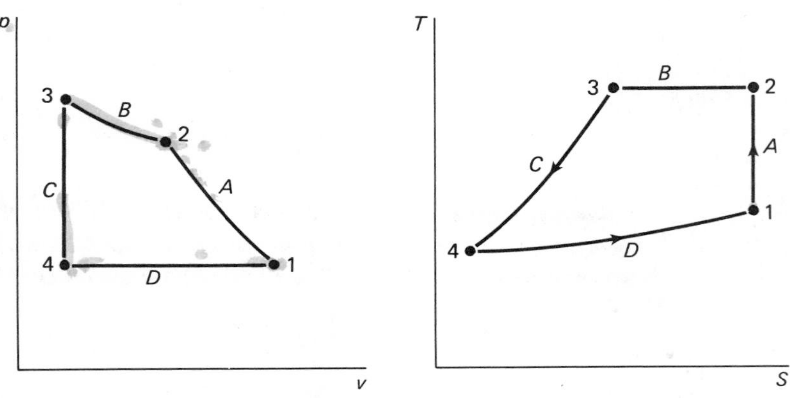

EXAMPLE 10.9. An ideal gas expands in a closed cylinder from 4 bars absolute and 27°C to 2 bars. The process is reversible and adiabatic. Compute the work done per kilogram by the gas if $c_p = 400$ and $c_v = 1000$ j/kg-°K.

SOLUTION:

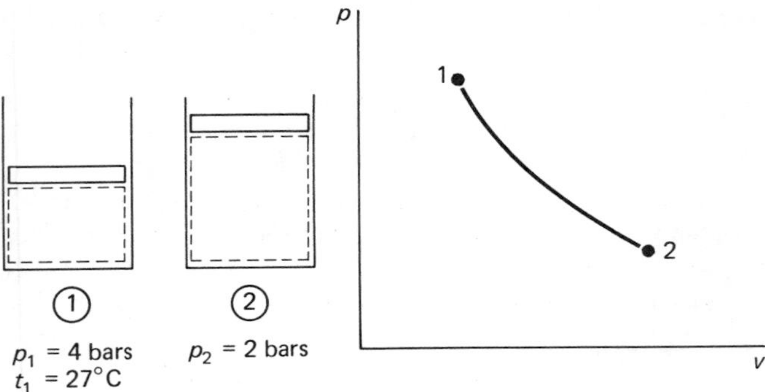

System: Closed as shown.
Equation of state:

$$pv = RT$$

Process equation: Reversible adiabatic ideal gas,

$$pv^{\gamma} = \text{constant}$$

or

$$Tp^{1-\gamma/\gamma} = \text{constant}$$

or

$$Tv^{\gamma-1} = \text{constant}$$

Fundamentals:
Constant of Mass: $M = \text{constant} = 1$ lbm
First law:

$$d'Q - d'W = du \qquad d'Q = 0$$

Second law: Not applicable.
Relations among properties:

$$dh = c_p\, dT$$
$$du = c_v\, dT$$
$$R = c_p - c_v$$

Combining $-d'W = du$ with $du = c_v\, dT$,

$$-d'W = c_v\, dT \quad \text{or} \quad -W_{12} = c_v(T_2 - T_1)$$

To find T_2, we use the process equation $Tp^{1-\gamma/\gamma} = \text{constant}$.

$$T_1 p_1^{1-\gamma/\gamma} = T_2 p_2^{1-\gamma/\gamma}$$

$$T_2 = T_1\left(\frac{p_2}{p_1}\right)^{\gamma-1/\gamma} = 300(2)^{0.286} = 300(1.22)$$

$$T_2 = 366°\text{K}$$

Thus, the work is

$$-W_{12} = 1000(366 - 300) = 66{,}000 \text{ j/kg}$$

Alternate solution:
For the reversible process,

$$W = \int_1^2 p\, dv$$

Combining with the process equation

$$pv^{\gamma} = p_1 v_1^{\gamma} = p_2 v_2^{\gamma} = C_1$$

gives

$$W_{12} = C_1 \int_1^2 v^{-\gamma}\, dv = (v_2^{1-\gamma} - v_1^{1-\gamma}) \frac{C_1}{1-\gamma}$$

Letting $C_1 = p_2 v_2^\gamma$ to multiply $v_2^{1-\gamma}$ and $C_1 = p_1 v_1^\gamma$ to multiply $v_1^{1-\gamma}$ gives

$$W_{12} = (p_2 v_2 - p_1 v_1) \frac{1}{1-\gamma} = \frac{R}{1-\gamma}(T_2 - T_1)$$

But

$$\frac{R}{1-\gamma} = \frac{c_p - c_v}{1 - c_p/c_v} = -\frac{c_p - c_v}{(c_p - c_v)/c_v} = -c_v$$

Or

$$W_{12} = -c_v(T_2 - T_1) = -(u_2 - u_1)$$

which is the same result previously obtained using the first law directly.

Second alternate solution:

Starting with

$$W_{12} = \int_1^2 p\, dv \quad \text{and} \quad pv^\gamma = \text{constant}$$

we derived

$$W_{12} = \frac{1}{1-\gamma}(p_2 v_2 - p_1 v_1)$$

We can put this in terms of the initial conditions and the pressure ratio as follows:

$$p_1 v_1^\gamma = p_2 v_2^\gamma = C_1$$

or

$$v_1 = C_1^{1/\gamma} p_1^{-1/\gamma} \quad \text{and} \quad v_2 = C_1^{1/\gamma} p_2^{-1/\gamma}$$

then

$$W_{12} = \frac{1}{1-\gamma} C_1^{1/\gamma}(p_2^{1-1/\gamma} - p_1^{1-1/\gamma})$$

$$W_{12} = \frac{1}{1-\gamma} C_1^{1/\gamma} p_1^{1-1/\gamma} \left[\left(\frac{p_2}{p_1}\right)^{\gamma - 1/\gamma} - 1\right]$$

$$W_{12} = \frac{1}{1-\gamma} p_1^{1/\gamma} v_1 p_1^{1-1/\gamma} \left[\left(\frac{p_2}{p_1}\right)^{\gamma - 1/\gamma} - 1\right]$$

$$W_{12} = -C_v T_1 \left[\left(\frac{p_2}{p_1}\right)^{\gamma - 1/\gamma} - 1\right]$$

In the numbers of this example, we obtain

$$-W_{12} = +(1000)(300)[(2)^{0.286} - 1]$$
$$= 66{,}000 \text{ j/kg}$$

10-6 The Carnot Cycle

The Carnot cycle is a reversible cycle of very little practical significance but since it is referred to so frequently in discussions of the second law and cycle efficiency, students of engineering should know what it is. See the sketch.

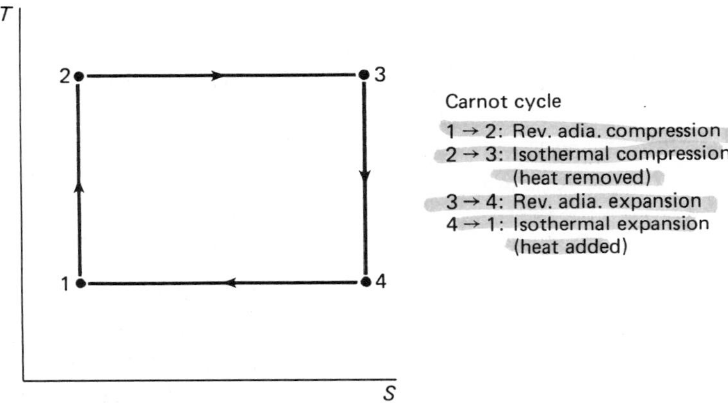

The Carnot cycle is convenient in theoretical discussions not only because it is reversible but also because all the heat is added at a single temperature (T_2) and all the heat is rejected at a single temperature (T_4). Real cycles are much messier with the working fluid changing in temperature during the heat addition and/or heat rejection. Thus when we write that $\eta_{max} = 1 - T_{ce}/T_{he}$ for a Carnot cycle, there really is a T_{ce} and T_{he} to use in the equation numerically.

On a pressure–volume diagram, the Carnot cycle using an ideal gas is shown in the sketch on p. 220. Since the net work done by a reversible cycle is the area enclosed by the cycle on a p-v plot, the practical difficulty with the Carnot cycle can be guessed. Though it is efficient, it delivers little power in relationship to its size (hence also cost and weight).

When the term *Carnot efficiency* is used, it means the efficiency of a Carnot engine operating between the highest and lowest temperatures available to the system, receiving all heat at T_h and rejecting all heat at T_c. Since it is a reversible cycle, its efficiency is the best possible.

$$\eta_{\text{Carnot}} = \eta_{\max} = 1 - \frac{T_c}{T_h}$$

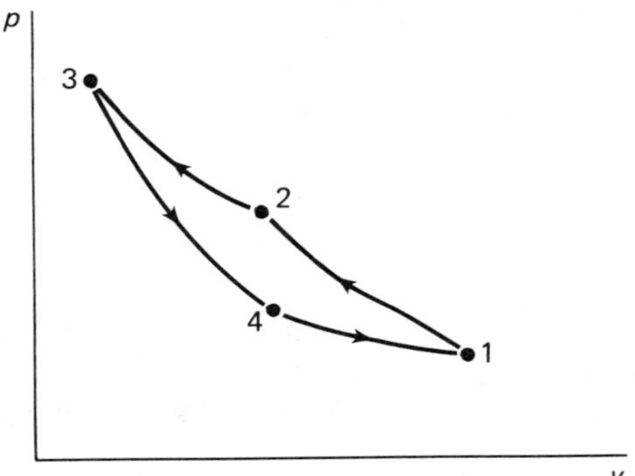

10-7 Making assumptions in applying the second law

In earlier chapters we have already discussed the assumptions commonly made regarding equations of state, heat, work, and the use of the first law and the energy equation. All of these continue to be needed in solving problems but several more can now be added. They are as follows:

1. Reversible processes are best. This is not so much an assumption as it is a statement of the second law. In many problems, however, it is desirable to find the maximum or minimum value of something (such as work in an expansion or compression process) and the student should recognize that the limiting best case is the reversible one. This will usually lead to a very useful process equation for the problem.
2. Though a process may be irreversible in total, certain subsystems may execute internally reversible processes, the criterion being whether or not all properties of the subsystem are uniform throughout at all times during the process. Thus, the work done by a gas expanding in a cylinder can be computed from the reversible equation $W = \int p\, dv$ if the properties of the gas are uniform throughout the cylinder at all times. This condition is met as long as the motion of the piston does not approach the speed of sound in the gas. It is also met by that gas *that remains inside* a tank during a blowdown, unless of course the blowdown is so rapid that significant pressure gradients are present *inside* the tank.
3. Well-designed devices such as turbines, compressors, pumps, and nozzles usually have a high efficiency; that is, they operate fairly close to reversibility. They also normally are reasonably close to adiabatic on a per pound of fluid basis. As a first approximation, therefore, assuming them to be constant entropy is reasonable if actual data are not available. Bear in mind, of course, that what you obtain is the *best* possible performance using this assumption. Actual efficiencies for large, well-designed equipment of this type are in the range 0.80 to 0.90.

4. In contrast to the devices listed in item 3, equipment such as orifices, valves, and capillary tubes are highly irreversible and hence inefficient. Even if they are adiabatic, they are definitely not close to isentropic.

It is now possible to work some quite complex applied problems in which equations of state, process equations, conservation of mass, the first law, the second law, and relations among properties (especially the $T\,ds$ equations) must all be used. Two examples follow.

EXAMPLE 10.10. To illustrate one early technique used to liquefy gases, the following experiment is performed. A tank with volume 0.5 ft³ is to be filled with dry saturated steam at 100 psia and then allowed to blowdown through a valve to atmospheric pressure (14.696 psia). How much liquid water, if any, would you expect to find in the bottom of the tank at the end of the blowdown process?

SOLUTION: Translation into schematic drawings and symbols:

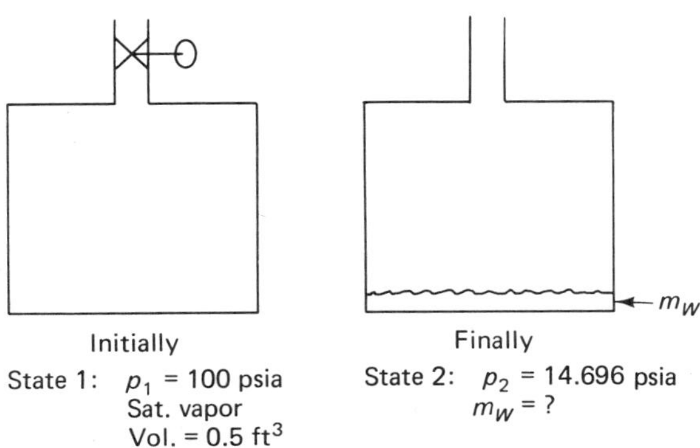

Equations of state: The steam tables will be used for the properties of H_2O.
Process assumptions and equations: The overall process will undoubtedly take place quite rapidly. The H_2O that remains in the tank can thus be assumed to be adiabatic during its expansion. At the same time, as long as sonic velocities are not approached inside the tank itself, the properties inside the tank will be essentially uniform at all times and the process may be assumed to be reversible as well as adiabatic. (Note that the H_2O that leaves the tank during the blowdown executes neither a reversible nor an adiabatic process. It is not reversible because of the nonuniform pressure that it has while passing through the valve and it is not adiabatic, even though rapid, because of the mixing that takes place with the air outside the tank.)
System: Because we are interested in the state of H_2O inside the tank after the expansion, this strongly suggests defining a control mass to be *the* H_2O *that ends up in the tank*. Call this m_2.

This decision should now be incorporated into the schematic diagram.

Application of principles:

Conservation of mass: Included in control mass assumption.

First law:

$$Q_{12} - W_{12} = m_2(u_2 - u_1) \qquad Q_{12} = 0$$
$$- W_{12} = m_2(u_2 - u_1)$$

Note that from the steam tables we can find u_1 but all the other terms are unknown at this time. (W_{12} represents the work done by the H_2O which remains in the tank on the H_2O which leaves.)

Second law: Reversible process

$$s_2 - s_1 = \int_1^2 \frac{d'Q}{T} = 0$$

$$s_1 = s_2$$

This gives us a process equation ($s =$ constant) and enables us to find state 2 completely because we have two independent properties, p_2 and s_2 ($= s_1$, which can be found from tables for saturated vapor). We go no further with the method because with state 2 known (particularly the specific volume V_2) we can solve the problem.

We first find the quality at 2:

$$x_2 = \frac{s_2 - s_{f2}}{s_{g2} - s_{f2}} = \frac{s_1 - s_{f2}}{s_{g2} - s_{f2}} = \frac{1.6034 - 0.3121}{1.7567 - 0.3121} = 0.894$$

Using this quality, we find v_2:

$$x_2 = \frac{v_2 - v_{f2}}{v_{g2} - v_{f2}} \qquad v_2 = v_{f2} + x_2(v_{g2} - v_{f2})$$

$$v_2 = 0.01672 + 0.894(26.8 - 0.01672)$$

$$v_2 = 0.01672 + 23.95 = 23.97 \text{ ft}^3/\text{lbm}$$

This enables us to find m_2 since $V = m_2 v_2$, $m_2 = V/v_2$.

$$m_2 = \frac{0.5}{23.57} = 0.212 \text{ lbm}$$

We want the mass of liquid. The quality is defined as

$$x_2 = \frac{m_{f2}}{m_2} \qquad m_{f2} = x_2 m_2$$

$$m_{f2} = (0.894)(0.0212) = 0.0190 \text{ lbm}$$

EXAMPLE 10.11. Air flows steadily and adiabatically in a constant area duct in which a stationary shock (discontinuity) is observed. The direction of flow is unknown to the observer but the temperature, pressure, and velocity to the left of the shock are 30°F, 10 psia, and 2174 ft/sec, respectively. The velocity to the right of the shock is 815 ft/sec. Determine the direction of flow.

SOLUTION: Translation into schematic diagrams and symbols:

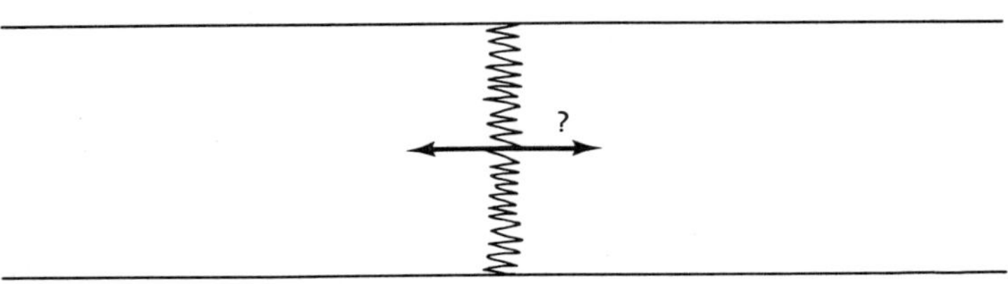

To the left: $T_l = 490°R$
$P_l = 10$ psia
$\mathscr{V}_l = 2174$ ft/sec

To the right: $\mathscr{V}_r = 815$ ft/sec

Control mass or control volume: Steady flow problems are almost always best worked by the control volume approach. Define a control volume that includes the shock. Include it in the schematic sketch as shown.

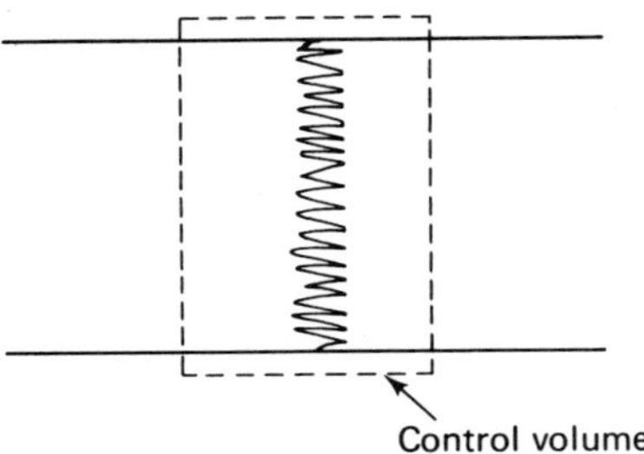

Equation of state: The critical pressure and temperature for air are 37.2 atm and 142°R, respectively. Thus, at pressures of the order of 1 atm and temperatures of the order of 500°R, air will behave very much like an ideal gas. Hence assume $pv = RT$. Applying this to each equilibrium state, we have

$$p_l v_l = RT_l \quad \text{and} \quad p_r v_r = RT_r$$

Application of principles:

Conservation of mass (continuity):

$\dot{m}_l = \dot{m}_r$ in steady flow regardless of direction.

$$\frac{a_l \mathscr{V}_l}{v_l} = \frac{a_r \mathscr{V}_r}{v_r}$$

Using the equation of state above for v_l,

$$\frac{\mathscr{V}_r}{v_r} = \frac{\mathscr{V}_l p_l}{RT_l}$$

Everything is known except v_r.

$$v_r = \frac{\mathscr{V}_r}{\mathscr{V}_l} \frac{RT_l}{P_l}$$

We won't solve this for numbers unless the need arises.

First law:

$$\dot{m}_l\left(h_l + \frac{\mathscr{V}_l^2}{2} + gZ_l\right) = \dot{m}_r\left(h_r + \frac{\mathscr{V}_r^2}{2} + gZ_r\right)$$

in steady flow regardless of direction.

Potential energy changes are zero and we can assume that, for the temperature changes involved, $h_l - h_r = c_p(T_l - T_r)$. Solving for T_r gives

$$T_r = T_l - \frac{\mathscr{V}_r^2 - \mathscr{V}_l^2}{2c_p}$$

Again, we shall simply note this result and only solve for numbers if and when the numbers are needed.

Second law: The process is adiabatic but not reversible. Hence, in the direction of flow, the entropy will increase.

In flow direction:

$$ds > \frac{dQ}{T} \qquad Q = 0$$

$$ds > 0$$

This indicated that if we can solve for the entropy on the left and right (or even determine the sign of the entropy change), we can determine the direction of flow. We note that we have been given the pressure and temperature on the left and can, if we wish, compute v_l, T_r, v_r, and p_r (using $p_r v_r = RT_r$).

Relations among properties:

T ds equations:

$$T\,ds = du + p\,dv = dh - v\,dp$$

Solving for the *ds* because this is what we need,

$$ds = \frac{du}{T} + \frac{p}{T}\,dv$$

For the ideal gas, we use

$$ds = c_v\frac{dT}{T} + R\frac{dv}{v}$$

Integrating from left to right with c_v as a constant,

$$s_r - s_l = c_v \ln\frac{T_r}{T_l} + R \ln\frac{v_r}{v_l}$$

We can now evaluate $(s_r - s_l)$ (and obtain its sign!) if we compute v_l, T_r, and v_r.

$$v_r = \frac{RT_l}{p_l} = \frac{(53.35)(490)}{(10)(144)} = 18.1 \text{ ft}^3/\text{lbm}$$

$$T_r = T_l - \frac{V_r^2 - V_l^2}{2g_0 Jc_p} = 490 - \frac{(815)^2 - (2174)^2}{(2)(32.2)(778)(0.24)} = 828°\text{R} = 368°\text{F}$$

$$v_r = v_l \frac{V_r}{V_l} = 18.1 \frac{815}{2174} = 6.78 \text{ ft}^3/\text{lbm}$$

Now we can solve for entropy:

$$s_r - s_l = c_v \ln \frac{T_r}{T_l} + R \ln \frac{v_r}{v_l} = (0.171) \ln \frac{826}{490} + (0.069) \ln \frac{6.78}{18.1}$$

$$s_r - s_l = +0.0215 \text{ Btu/lbm-°R}$$

Therefore, since $s_r > s_l$, the flow is from left to right.
Students should note that p_r can also be computed:

$$p_r = \frac{RT_r}{v_r} = \frac{(53.35)(826)}{(6.78)} = 6500 \text{ lbf/ft}^3 = 45.1 \text{ psia}$$

and that the flow is from lower to higher pressure.

10-8 Summary

Since $\oint (d'Q/T)_{\text{reversible}}$ is always zero, $(d'Q/T)_{\text{reversible}}$ is a property. That property is called *entropy* and given the symbol ds:

$$ds \equiv \left(\frac{d'Q}{T}\right)_{\text{reversible}} \tag{10-68}$$

In real processes,

$$ds > \frac{d'Q}{T} \tag{10-69}$$

and therefore, in *adiabatic*, real processes,

$$ds > 0 \tag{10-70}$$

Applying the results above to open systems concludes that

$$\oint_A \frac{d'Q}{T} < \frac{\partial}{\partial \theta} \int_V \rho s \, dV + \oint \rho s \vec{V} \cdot \vec{n} \, dA \tag{10-71}$$

The so-called $T\,ds$ equations are extremely general and useful relations among properties that are derived from the first and second laws. They are

$$T\,ds = du + p\,dv \tag{10-72}$$

$$T\,ds = dh - v\,dp \tag{10-73}$$

Tables of properties usually list values for entropy s along with enthalpy h and internal energy u.

When ideal gases are considered in reversible adiabatic processes (that is, constant entropy or isentropic), the process equations are

$$pv^{\gamma} = \text{constant} \tag{10-74}$$

or

$$Tp^{1-\gamma/\gamma} = \text{constant} \tag{10-75}$$

or

$$Tv^{\gamma-1} = \text{constant} \tag{10-76}$$

For suggested reading, students will find a very useful collection of examples in the back part of the *Gas Tables* by Keenan and Kaye. While at the library, look through several recent issues of *Science* magazine for their section of "Energy News." This magazine has reliable, up-to-date information on the changing nature of the "energy crisis."

Problems

10-1 A substance with a specific heat c_p given by $a + bt$ is heated internally reversibly from t_1 to t_2. Compute the change in entropy per pound.

10-2 On a T-s diagram, show the liquid–vapor two-phase region, lines of constant p, and lines of constant v.

10-3 In Example 10.1, compare the heat rejected from the engine to the room air without using the results of part 3. Now check the answer for W in part 3 by using $Q_h - Q_c = W$.

10-4 An engine operates between two constant pressure steam cylinders acting as a source and sink, respectively. The source cylinder is initially at 180 psia and 640°F. The sink cylinder is initially at 180 psia and 420°F. The engine executes an integral number of cycles. What is the maximum work that can be obtained from this system if both cylinders are adiabatic except for their heat exchanges with the engine? Each cylinder contains 1 lbm of steam.

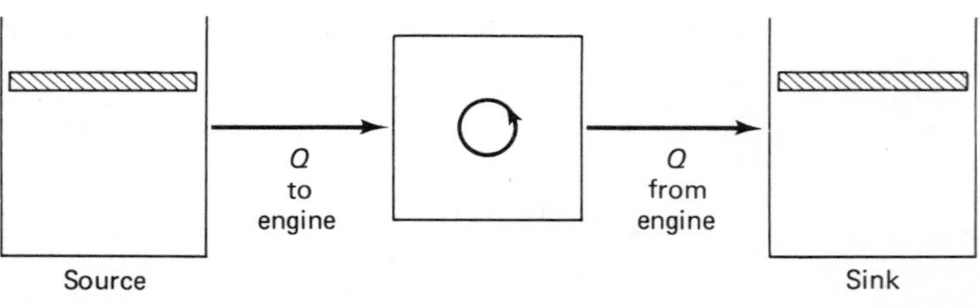

10-5 A tank of volume 1 m³ contains H_2O as saturated vapor at 10 bars absolute. A valve is opened allowing the pressure in the tank to fall to 1.5 bars absolute. How many grams of liquid would you expect to find in the tank?

10-6 Steam is flowing steadily through a horizontal insulated pipeline in which there is friction and an area change. The direction of flow and the rate of flow are unknown and are to be determined. Pressure and temperature readings at a point where the area is 3 in.² are 80 psia and 360°F, respectively. At another section, the area is 2 in.², the pressure is 77 psia, and the temperature is 350°F. *Note:* For students with only abbreviated steam tables, the following data are given:

At 80 psia and 360°F,
$h = 1209.7$ Btu/lbm
$v = 5.888$ ft³/lbm
$s = 1.6541$ Btu/lbm-°F

At 77 psia and 350°F,
$h = 1204.9$ Btu/lbm
$v = 6.039$ ft³/lbm
$s = 1.6523$ Btu/lbm-°F

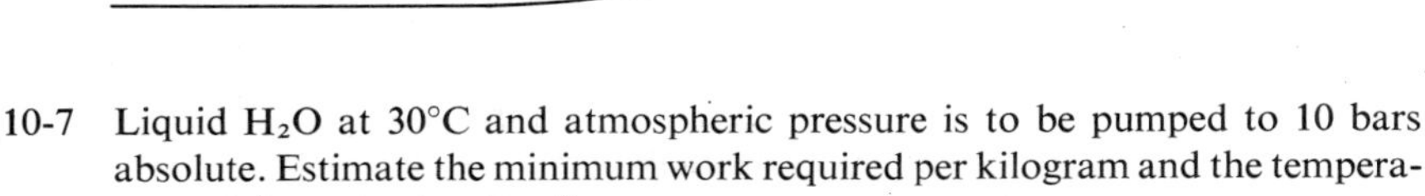

10-7 Liquid H_2O at 30°C and atmospheric pressure is to be pumped to 10 bars absolute. Estimate the minimum work required per kilogram and the temperature of the water leaving the pump.

10-8 Steam flows adiabatically through the pipe section shown ($\text{area}_1 < \text{area}_2$). Which direction is it flowing?

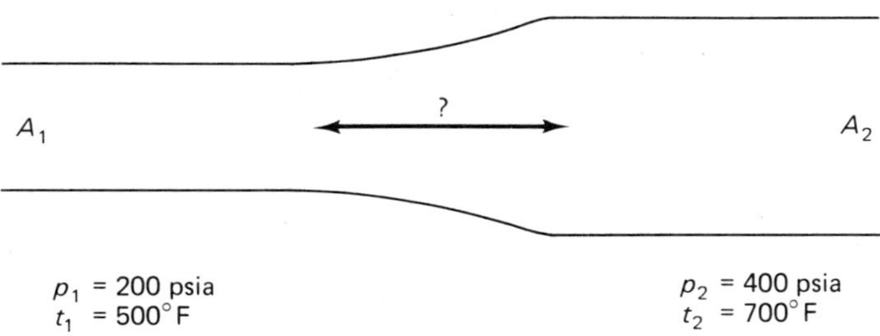

10-9 An evacuated bottle sits in a room where the pressure is atmospheric (15 psia) and the temperature is 80°F. Suppose the cork is pulled suddenly. What would be the temperature in the bottle when the pressure there becomes atmospheric?

10-10 A bottle contains air at 27°C and 0.5 bar absolute in a room at 27°C and 1 bar atmospheric pressure. The cork is pulled. What will the temperature be in the bottle when the pressure is equalized?

10-11 A mortar has the dimensions and other characteristics shown in the sketch below.

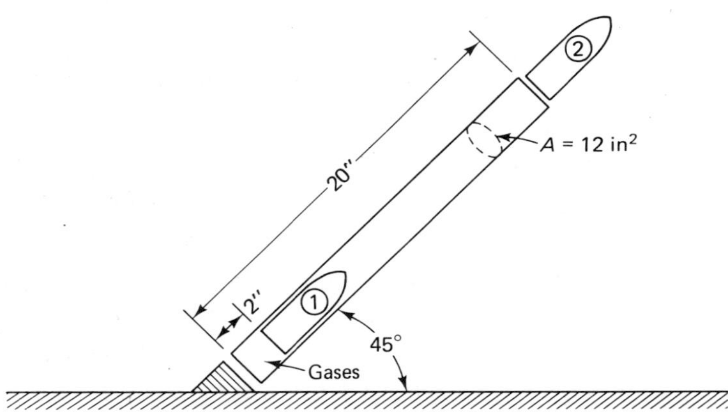

The shell weighs 0.5 lbf. Immediately following firing, the gases behind the shell reach a pressure of 475 psia and a temperature of 1240°F. The gas has a constant of $R = 77.8$ ft-lbf/lbm-°R and $c_p = 0.30$ Btu/lbm-°F. Find the velocity of the shell at position 2 just as it leaves the barrel.

10-12 A tank of compressed air at 30 psia is in thermal equilibrium with its environment of 15 psia and 80°F. A valve is opened and the air in the tank is allowed to escape. When the pressures are equalized, what will be the temperature in the tank?

10-13 One pound of air is initially confined in container *A* at a pressure of 100 psia and a temperature of 70°F. The valve is opened and the air is allowed into container *B*, which is initially evacuated. When the pressure in *A* reaches 60 psia, the valve is closed. The volume of *A* equals the volume of *B*. Both are insulated. The problem is to find the final states and amounts of air in each container.

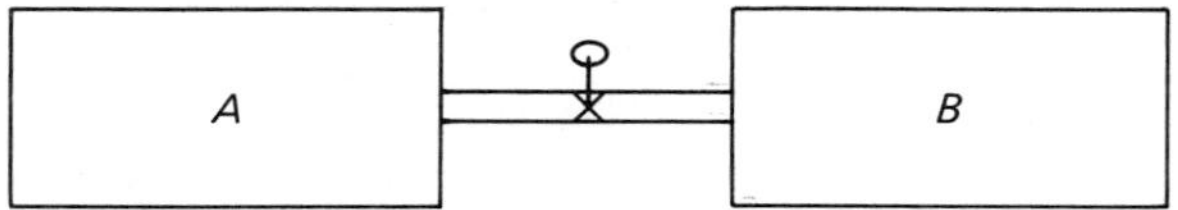

Note: For air, $c_p = 0.24$ Btu/lbm-°F, $c_v = 0.17$ Btu/lbm-°F, and $R = 53.3$ ft-lbf/°R-lbm.

10-14 In Prob. 10-13, air initially confined under pressure in *A* is allowed to expand into an evacuated space until the pressure in *A* is 60 psia. The sketches following give the pertinent data. (Note that the volume of *A* equals the volume of *B*.) Find the entropy change of the air.

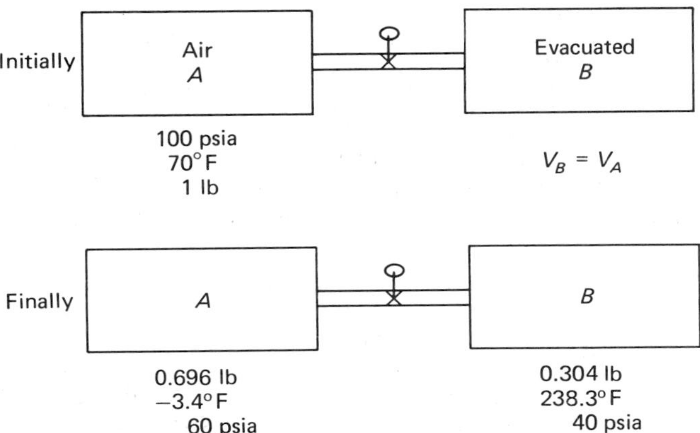

10-15 A steady stream of air at atmospheric pressure and temperature is to be compressed to a higher pressure but returned to or kept at atmospheric temperature. Two processes are proposed:

(a) Adiabatic compression to the final pressure followed by constant pressure cooling back to atmospheric temperature.

(b) Isothermal compression to final pressure.

The problem is to compare the heat removal and work required in the two processes using *T-s* and *p-v* diagrams.

10-16 Saturated steam vapor enters a compressor at 1 bar absolute and leaves at 1.5 bars absolute and 150°C in steady flow. The heat transferred from the compressor is 10^6 j/hr. Steam flow rate is 40 kg/hr. Compute the power required to drive the compressor.

10-17 Steam enters the nozzle of a turbine with a low velocity at a pressure of 400 psia and $t = 600$°F. It leaves at 300 psia and a velocity of 1540 ft/sec. The rate of steam flow is 3000 lbm/hr. Compute the quality and temperature of the steam leaving the nozzle and the exit area of the nozzle. State assumptions.

10-18 One pound of air is confined in a cylinder at a pressure of 30 psia and a temperature of 40°F. Compute the work *and* heat if the air expands to 15 psia

(a) Isothermally.

(b) Adiabatically.

State your assumptions.

(c) Show both processes on the same *p-v* and *T-s* diagrams.

10-19 A steady stream of 10 lbm/min of air at 60°F is to be raised in temperature to 160°F. To do the job, 1 lbm/min of steam at 20 psia and 350°F is available. A counterflow heat exchanger (the two streams exchanging heat flow in opposite directions) is to be used.

(Counter flow heat exchanger)

Compute the net change in entropy (steam *and* air) in the process per minute. For air, $c_p = 0.24$ Btu/lbm-°R, $c_v = 0.17$ Btu/lbm-°R, and $R = 53.35$ ft-lbf/lbm-°R.

10-20 Air enters an adiabatic compressor at 0.9 bar absolute and 20°C. It is to be compressed to 4 bars absolute.
(a) Compute the minimum work required per kilogram in steady flow.
(b) Show the work as an area on a p-v plot.

10-21 Air is compressed in a reversible isothermal steady flow process from 15 psia and 100°F to 100 psia.
(a) Compute the work per pound, Δs, and heat transfer.
(b) Show the work and heat on p-v and T-s plots as appropriate.
(c) How can Δs be negative?

10-22 Air in a closed cylinder is compressed in a reversible isothermal process from 15 psia and 100°F to 100 psia.
(a) Compute the work per pound, Δs, and heat transfer.
(b) Show the work and heat on p-v and T-s plots as appropriate.

10-23 Show by integration that for an ideal gas with constant specific heats in a reversible adiabatic process
(a) $-\int_1^2 p\ dv = c_v(t_2 - t_1)$
(b) $\int_1^2 v\ dp = c_p(t_2 - t_1)$
Show also that these results can be obtained from the first law for a control system and control volume.

11

Degradation and work potential

11-1 Introduction

11-1(a) The concept of work potential

We have already described the work potential (symbol A) as the maximum useful work that can be obtained from a system. And we also showed that the work potential of a quantity Q at a constant temperature T_h is given by

$$A_{Q,T_h} = Q\left(1 - \frac{T_0}{T_h}\right) \tag{11-1}$$

where T_0 is the temperature of the available environmental sink, usually a lake, river, ocean, or the atmosphere. We now wish to extend the concept to more general systems and situations.

We recognize the fact that all systems (open or closed) exist in and processes occur in an environment that in most cases can be considered to be infinite in extent and constant in properties. A system whose potential-type properties (e.g., pressure, temperature, and electrical or magnetic potentials) are different from those of its environment can interact with the environment and produce work until the system reaches a state where such potential differences do not exist. For any system, this state is called the *dead state* because the system can do nothing more. It is dead as far as work is concerned. In its dead state, a system will be at the pressure and temperature of its environment and will also not have any gravity potential energy, kinetic energy, or electrical or magnetic potentials relative to or different from its environment.

The maximum possible work output that can be produced by a system in a change from a given state to its dead state is what we have called, appropriately enough, the *work potential.* The term *availability* is also used.

Usually by the *maximum possible* work is meant only the *useful* portion of the total maximum work done. When a closed system expands, it does work against the atmosphere. Such work is not even potentially useful to man so is excluded. We shall see later in the chapter how this is taken into account in computations of work potential.

It should be noted carefully by students that, for a given environment, work potential is a property of systems. We shall prove this subsequently but it should be clear at this point that the *maximum* work that can be produced by a system is not a function of the process. The actual work, of course, will be a strong function of the process but the maximum is the maximum regardless of how it is obtained. Hence work potential is a property.

11-1(b) The concept of degradation

To obtain a measure of the performance of *cyclic* engines, we used an efficiency defined as the ratio of net useful energy output (work) to the costly energy input (heat). For refrigerators, we defined a coefficient of performance as the ratio of the useful energy taken from the source (heat) to the costly energy input (work). Measuring the performance of a *process* is not quite so straightforward. There may be no energy input but only a change in state of some system. Therefore, *efficiency* in the same sense that it is used for cycles cannot be used to evaluate processes.

We recognize, however, that work is the critical item. In work *producing* processes, we wish to obtain as much work as possible from the state change taking place. In work *utilizing* processes, we wish to use as little work as possible in accomplishing the given change of state. This suggests a process effectiveness for given end states for work producing processes as

$$\text{Process effectiveness} = \frac{W_{\substack{\text{actual}\\ \text{output}}}}{W_{\substack{\text{maximum}\\ \text{possible}}}} \tag{11-2}$$

and for work utilizing processes as

$$\text{Process effectiveness} = \frac{W_{\substack{\text{minimum}\\ \text{required}}}}{W_{\substack{\text{actual}\\ \text{input}}}} \tag{11-3}$$

Unfortunately, and for unfathomable reasons, we do not customarily rate process effectiveness as a ratio but rather as a *difference* between the actual and ideal. We call this difference the *degradation.* The terms *irreversibility* or *dissipation* are also used. We use the symbol D here, defined as follows for work producing processes:

$$D = W_{\substack{\text{maximum}\\ \text{possible}}} - W_{\substack{\text{actual}\\ \text{output}}} \tag{11-4}$$

For work utilizing processes,

$$D = W_{\substack{\text{actual} \\ \text{input}}} - W_{\substack{\text{minimum} \\ \text{required}}} \tag{11-5}$$

Both $W_{\substack{\text{maximum} \\ \text{possible}}}$ and $W_{\substack{\text{minimum} \\ \text{required}}}$ refer to the same end states as in the actual process. Note that D is always positive by definition and is zero only for an ideal process that produces the maximum or uses the minimum work possible. The larger D, the poorer the process in either case. We have already shown, in the preceding section, that the maximum work is produced or the minimum is required in a reversible process.

We have already seen how this works in the case of heat supplied to heat engines operating between T_h and T_0. The maximum possible work obtainable from Q is

$$W_{\substack{\text{maximum} \\ \text{possible}}} = Q\left(1 - \frac{T_0}{T_h}\right) \tag{11-6}$$

If W_{actual} is actually obtained, then the degradation is

$$D = W_{\substack{\text{maximum} \\ \text{possible}}} - W_{\text{actual}} = Q\left(1 - \frac{T_0}{T_h}\right) - W_{\text{actual}} \tag{11-7}$$

Students should review Sec. 9-2.

11-2 Work producing processes in closed systems

11-2(a) Degradation in a closed system

We have shown how the degradation D is a measure of the performance of a process. It's an inverse measure since the smaller it is, the better the process. We now must develop expressions for the computation of D.

We assume a closed system in an environment where properties (e.g., p_0 and T_0) are constant. For any process of the system, the first law is

$$Q_{12} - W_{\substack{12 \\ \text{actual}}} = E_2 - E_1 \tag{11-8}$$

and the second law gives

$$\int_1^2 \frac{d'Q}{T} \leq S_2 - S_1 \qquad \begin{array}{l} < \text{irreversible} \\ = \text{reversible} \end{array} \tag{11-9}$$

Before these equations can be combined, they must have the same units. To accomplish this, we need to multiply Eq. (11-9) by a temperature. We shall choose T_0 arbitrarily at this time but we shall defend this choice later. Multiplying Eq. (11-9) by T_0 and adding Eq. (11-9) to (11-8) gives, with a little rearranging,

$$W_{12\,\text{actual}} \le E_1 - E_2 + Q_{12} + T_0(S_2 - S_1) - T_0 \int_1^2 \frac{d'Q}{T} \tag{11-10}$$

Introducing degradation D for the work producing process,

$$D_{12} = W_{12\,\text{maximum possible}} - W_{12\,\text{actual}} \quad \text{or} \quad W_{12\,\text{actual}} = W_{12\,\text{maximum possible}} - D_{12} \tag{11-11}$$

Combining Eq. (11-10) and (11-11) to eliminate $W_{12\,\text{actual}}$,

$$W_{12\,\text{maximum possible}} \le E_1 - E_2 + Q_{12} + T_0(S_2 - S_1) - T_0 \int_1^2 \frac{d'Q}{T} + D_{12} \tag{11-12}$$

Clearly the equality sign in Eq. (11-12) holds when $D = 0$ so

$$W_{12\,\text{maximum possible}} = E_1 - E_2 + Q_{12} + T_0(S_2 - S_1) - T_0 \int_1^2 \frac{d'Q}{T} \tag{11-13}$$

Substituting Eq. (11-13) into (11-11),

$$D_{12} = E_1 - E_2 + Q_{12} + T_0(S_2 - S_1) - T_0 \int_1^2 \frac{d'Q}{T} - W_{12\,\text{actual}} \tag{11-14}$$

But from the first law, we have

$$E_1 - E_2 + Q_{12} - W_{12\,\text{actual}} = 0$$

so Eq. (11-14) becomes

$$D_{12} = T_0(S_2 - S_1) - T_0 \int_1^2 \frac{d'Q}{T} \tag{11-15}$$

To compute the degradation, either Eq. (11-14) or (11-15) may be used. Note that in the special but quite common case when the system boundaries are at T_0 these reduce to

$$D_{12} = E_1 - E_2 + T_0(S_2 - S_1) - W_{12\,\text{actual}} \tag{11-16}$$

$$D_{12} = T_0(S_2 - S_1) - Q_{12} \tag{11-17}$$

11-2(b) Degradation and entropy

We have been referring to degradation of energy as loss of work potential ever since Chap. 2. Now we can relate that discussion to our new property from the second law, entropy. Equation (11-15) [or Eq. (11-17)] shows exactly what that relationship is. It is more instructive, however, to consider an adiabatic system so that $Q_{12} = 0$. Then

$$D_{12} = T_0(S_2 - S_1) \qquad \text{Adiabatic} \tag{11-18}$$

or in general

$$D = T_0 \, \Delta S \qquad \text{Adiabatic} \tag{11-19}$$

We know from the principle of increase of entropy that ΔS will always be positive for an adiabatic system. Thus, D too is always positive, as expected.

But there is more than a qualitative relationship between degradation and entropy increase. They are directly proportional. Thus, the entropy increase is simply a *measure* of the degradation.

Note that this relates entropy to work, more particularly to lost work potential (degradation). This is tremendously important. The principle of increase of entropy is simply another statement of the principle of energy degradation. Students should understand therefore that entropy is the property derived from the second law that is used to measure energy degradation.

The "adiabatic" restriction is not a severe problem in most practical cases. Usually a sufficiently large system can be selected so that $Q = 0$ without negating whatever aspects are to be considered. If this can't be done, then one of the other equations [Eq. (11-15) to (11-17)] can be used for degradation. The significance of the relationship between degradation and energy remains valid in any case. *Entropy increase* is proportional to *energy degradation!* $D = T_0 \, \Delta s$ in adiabatic systems.

11-2(c) Work potential of closed systems

When a closed system expands, part of the total work accomplished is done in pushing back the atmosphere. As noted earlier, this work is not useful and is designated as $W_{\text{nonuseful}}$. Since the atmospheric pressure is constant, the amount of this nonuseful work is given by

$$W_{\text{nonuseful}} = \int_{V_1}^{V_2} p \, dv = P_0(V_2 - V_1) \tag{11-20}$$

where P_0 is the pressure of the environment (*not* of the system) and V is the volume of the system.

The maximum possible work that a closed system can do in some change from state 1 to state 2 can now be divided into useful and nonuseful.

$$\underset{\substack{\text{maximum}\\\text{possible}}}{W_{12}} = \underset{\substack{\text{useful}\\\text{maximum}\\\text{possible}}}{W_{12}} + P_0(V_2 - V_1) \tag{11-21}$$

or

$$\underset{\substack{\text{useful}\\\text{maximum}\\\text{possible}}}{W_{12}} = \underset{\substack{\text{maximum}\\\text{possible}}}{W_{12}} - P_0(V_2 - V_1) \tag{11-22}$$

Now the work potential is defined as the maximum possible *useful* work that can be obtained from a system in a change from its existing state (1) to the dead state (∞). That is,

$$A_1 = \underset{\substack{\text{useful}\\ \text{maximum}\\ \text{possible}}}{W_{1\infty}} = \underset{\substack{\text{maximum}\\ \text{possible}}}{W_{1\infty}} - P_0(V_\infty - V_1) \tag{11-23}$$

From Eq. (11-13) we have

$$\underset{\substack{\text{maximum}\\ \text{possible}}}{W_{12}} = E_1 - E_2 + Q_{12} + T_0(S_2 - S_1) - T_0 \int_1^2 \frac{d'Q}{T} \tag{11-24}$$

and hence, letting state 2 become the dead state denoted by ∞, we obtain for A_1

$$A_1 = E_1 - E_\infty - T_0(S_1 - S_\infty) + P_0(V_1 - V_\infty) \tag{11-25}$$

$$A_1 = (E_1 + P_0 V_1 - T_0 S_1) - (E_\infty + P_0 V_\infty - T_0 S_\infty) \tag{11-26}$$

Readers should bear in mind that the subscript 0 indicates the environment. The subscript ∞ indicates the properties *of the system* at the pressure and temperature *of the environment*. It is common practice to define an availability function, ϕ, for the closed system

$$\phi_1 = E_1 + P_0 V_1 - T_0 S_1 \tag{11-27}$$

Then

$$\phi_\infty = E_\infty + P_0 V_\infty - T_0 S_\infty \tag{11-28}$$

So

$$A_1 = \phi_1 - \phi_\infty \tag{11-29}$$

Going back to the expression for degradation, we can write

$$D_{12} = (\phi_1 - \phi_2) - \underset{\substack{\text{actual}\\ \text{useful}\\ \text{output}}}{W_{12}} \tag{11-30}$$

or

$$D_{12} = A_1 - A_2 - \underset{\substack{\text{actual}\\ \text{useful}\\ \text{output}}}{W_{12}} \tag{11-31}$$

where $(A_1 - A_2)$ is the *loss* of work potential in the process 1-2. A is a property and this loss is independent of the process but, of course, D_{12} and W_{12} are *not* properties.

The relationship between degradation and work potential defined by Eq. (11-31) should be carefully noted. In a reversible process, $D = 0$ and the work produced is equal to the *change* in work potential. For irreversible processes between the same end states, less work will be obtained than the work potential change and D will be greater than zero. Do not forget that work potential is a property.

The use of work potential is very helpful in determining the maximum work obtainable from a system not only in reaching its dead state but between any specific states. This is often of practical value in comparing or evaluating energy storage or conversion systems. For example, is compressed air a good way to store energy?

EXAMPLE 11.1. One unit mass of air is confined at p_1 and $T_1 = T_0$ in the left half of the adiabatic container shown. The right half is initially evacuated. The partitions are quickly removed. Compute the degradation in the resulting process.

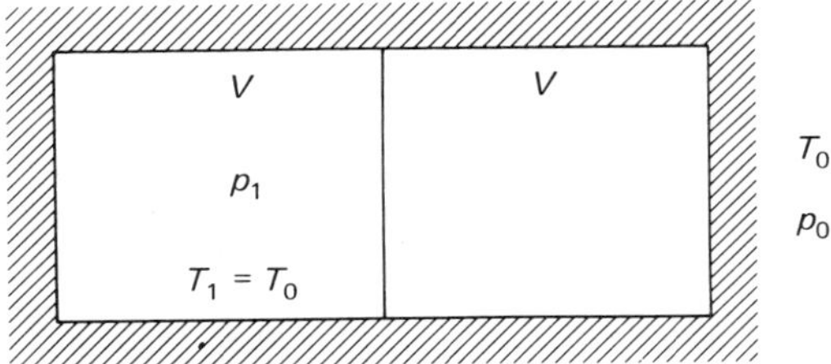

SOLUTION: From our previous work we know that the final state of the air in the whole container is $T_2 = T_0$ and $p_2 = p_1/2$. To compute the degradation, we can use Eq. (11-17):

$$D_{12} = T_0(S_2 - S_1) - Q_{12} \qquad Q_{12} = 0 \quad \text{(given)}$$

$$D_{12} = T_0\left[c_p \ln \frac{T_2}{T_1} - R \ln \frac{p_2}{p_1}\right]$$

$$D_{12} = -RT_0 \ln \frac{1}{2} = RT_0 \ln 2$$

Note that Eq. (11-16) would give the same result since $E_1 = E_2$ and $W_{12\,\text{actual}} = 0$.

In addition to using the equations above to compute D, an alternate method that is sometimes useful when $W_{12\,\text{actual}}$ is known is simply to devise a reversible process between the same ends in order to find $W_{12\,\text{maximum possible}}$ since $W_{12\,\text{reversible}} = W_{12\,\text{maximum possible}}$. In concocting the reversible process, *any* reversible process will do. The only tricky aspect of this procedure is that *everything* involved must end up at the *same* point in the concocted reversible process as in the real process—*except* the environment that may be used as source or sink for heat. All heat must be transferred via reversible engines and the work input or output of these engines taken into account. We illustrate this by reworking Example 11.1 using this alternate technique.

Alternate Solution to Example 11.1

We must find a completely reversible process that changes the state of a closed system of air from $T_1 = T_0$ and p_1 to $T_2 = T_0$ and $p_2 = p_1/2$. We concoct a reversible isothermal expansion of the air from 1 to 2. Since the air is at T_0, this is no problem and the final pressure will then be $p_2 = p_1/2$. The work in this concocted process will be

$$W_{12} = \int_1^2 p\,dv = RT_0 \int_1^2 \frac{dv}{v} = RT_0 \ln \frac{v_2}{v_1} = RT_0 \ln 2$$

EXAMPLE 11.2. Compute the work potential per pound of a tank of compressed air at 90 psia and 80°F. Atmospheric conditions are 15 psia and 80°F.

SOLUTION:

$$A_1 = \phi_1 - \phi_0$$

$$A_1 = (E_1 + p_0 V_1 - T_0 S_1) - (E_\infty + p_0 V_\infty - T_0 S_\infty)$$

Neglect K.E. and P.E. terms:

$$A_1 = (U_1 + p_0 V_1 - T_0 S_1) - (U_\infty + p_0 V_\infty - T_0 S_\infty)$$

$$A_1 = 0 + p_0(V_1 - V_\infty) - T_0(S_1 - S_\infty)$$

$$A_1 = p_0 RT_0\left(\frac{1}{p_1} - \frac{1}{p_0}\right) + T_0 R \ln \frac{p_1}{p_0}$$

$$A_1 = RT_0\left(\frac{p_0}{p_1} - 1\right) + RT_0 \ln 6$$

$$A_1 = (53.35)(540)\left(\frac{15}{90} - 1 + 1.79\right)$$

$$A_1 = 27{,}600 \text{ ft-lbf/lbm}$$

Just as degradation can be computed either by equation or by finding the work in a concocted reversible process, so can work potential. The concocted process must get all parts of the system to the same final state, except sources or sinks. And all heat must be transferred via reversible engines and the work input or output of these engines taken into account. To illustrate, we solve Example 11.2 again.

Alternate Solution to Example 11.2

To obtain the work potential of the compressed air, we find how much work can be obtained from it as it changes from its initial state to the dead state. We imagine it expanding reversibly and isothermally.

$$W_{\text{total}} = \int p\,dV = RT_0 \ln \frac{V_2}{V_1} = RT_0 \ln 6$$

A portion of this work, however, is nonuseful work done against the atmosphere.

$$W_{\text{nonuseful}} = p_0(V_2 - V_1) = RT_0 p_0\left(\frac{1}{p_0} - \frac{1}{p_1}\right)$$

$$W_{\text{nonuseful}} = RT_0\left(1 - \frac{p_0}{p_1}\right)$$

It should be clear that the results are the same as in the first solution.

11-3 Work utilizing processes in closed systems

Equations (11-4) to (11-31) apply to this case except that

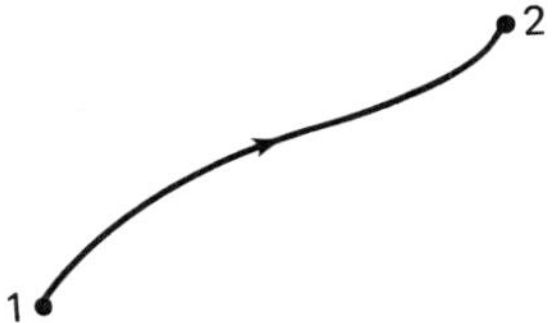

$$\underset{\text{work producing}}{W_{12}} = -\underset{\text{work utilizing}}{W_{12}} \tag{11-32}$$

Thus, the expression for reversible work is now not a maximum but a minimum.

$$\underset{\substack{\text{minimum}\\\text{requirement}\\\text{1-2}}}{W} = \int_1^2 p\,dV + \frac{1}{2}m(\mathscr{V}_2^2 - \mathscr{V}_1^2) + mg(h_2 - h_1) \tag{11-33}$$

For the degradation, which is always positive, we write

$$D = \underset{\substack{\text{actual}\\\text{input}}}{W} - \underset{\substack{\text{minimum}\\\text{requirement}}}{W} \tag{11-34}$$

With this as a starting point, a development parallel to the work producing case leads to

$$D_{12} = \underset{\substack{\text{actual}\\\text{input}}}{W_{12}} + E_1 - E_2 + Q_{12} + T_0(S_2 - S_1) - T_0\int_1^2 \frac{d'Q}{T} \tag{11-35}$$

and

$$D_{12} = T_0(S_2 - S_1) - T_0\int_1^2 \frac{d'Q}{T} \tag{11-36}$$

Again in the common case where the system boundary can be chosen to be at T_0, these become

$$D_{12} = W_{\substack{12\\\text{actual}\\\text{input}}} + E_1 - E_2 + T_0(S_2 - S_1) \tag{11-37}$$

and

$$D_{12} = T_0(S_2 - S_1) - Q_{12} \tag{11-38}$$

Work potential is a property and thus it is unaffected by the nature of any processes involved. Its *change* will be negative in a work producing process and positive in a work utilizing one but its value is a function of state only. Thus, the equations developed in Sec. 11-2(c) are unchanged.

11-4 Work producing processes in open systems

11-4(a) Degradation in open systems

We shall now derive the expressions for degradation in an open system producing work. The general first and second laws are

$$\oint_A d'Q - \oint_A d'W = \frac{d}{d\theta}\int_V \rho e\, dV + \oint_A \left(h + \frac{\mathscr{V}^2}{2} + gZ\right)\rho\vec{\mathscr{V}} \cdot \vec{n}\, dA \tag{11-39}$$

$$\oint_A \frac{d'Q}{T} \leq \frac{d}{d\theta}\int_V \rho s\, dV + \oint_A s\rho\vec{\mathscr{V}} \cdot \vec{n}\, dA \tag{11-40}$$

Multiplying the second law equation by T_0 and combining as before gives

$$\oint_A d'W_{\substack{x\\\text{actual}}} \leq -\frac{d}{d\theta}\int_V (e - T_0 s)\rho\, dV - \oint_A \left(h + \frac{\mathscr{V}^2}{2} + gZ - T_0 s\right)\rho\vec{\mathscr{V}} \cdot \vec{n}\, dA + \oint_A d'\dot{Q} - T_0\oint_A \frac{d'\dot{Q}}{T} \tag{11-41}$$

Introducing degradation D,

$$\dot{D} = -\frac{d}{d\theta}\int_V (e - T_0 s)\rho\, dV - \oint \left(h + \frac{\mathscr{V}^2}{2} + gZ - T_0 s\right)\rho\vec{\mathscr{V}} \cdot \vec{n}\, dA + \oint_A d'\dot{Q} - T_0\oint_A \frac{d'\dot{Q}}{T} - \oint_A d'W_{\substack{x\\\text{actual}}} \tag{11-42}$$

Subtracting Eq. (11-39) results in

$$\dot{D} = \frac{d}{d\theta}\int_V T_0 s\rho\, dV + \oint_A T_0 s\rho\vec{\mathscr{V}} \cdot \vec{n}\, dA - T_0\oint_A \frac{d'\dot{Q}}{T} \tag{11-43}$$

In the case of steady state, these reduce to

$$\dot{D} = -\oint_A \left(h + \frac{\mathscr{V}^2}{2} + gZ - T_0 s\right)\rho\vec{\mathscr{V}} \cdot \vec{n}\, dA + \oint_A d'\dot{Q} - T_0 \int_1^2 \frac{d'Q}{T} - \oint_A d'\dot{W}_{x_{\text{actual}}} \tag{11-44}$$

and

$$\dot{D} = \oint_A T_0 s\rho\vec{\mathscr{V}} \cdot \vec{n}\, dA - T_0 \oint_A \frac{d'\dot{Q}}{T} \tag{11-45}$$

In many cases, the open system boundary is, or can be defined so it is, at the temperature of the environment, T_0. Then the results for $\dot{D}$ in Eq. (11-44) reduce to

$$\dot{D} = -\oint_A \left(h + \frac{\mathscr{V}^2}{2} + gZ - T_0 s\right)\rho\vec{\mathscr{V}} \cdot \vec{n}\, dA - \oint_A d'\dot{W}_{x_{\text{actual}}} \tag{11-46}$$

and

$$\dot{D} = \oint_A T_0 s\rho\vec{\mathscr{V}} \cdot \vec{n}\, dA - \oint_A d'\dot{Q} \tag{11-47}$$

Equations (11-44) and (11-46) may be used to compute the degradation of processes in which no work is done simply letting $d'W_{x_{\text{actual}}} = 0$.

11-4(b) Work potential in open systems

Referring to Eq. (11-42) and remembering that

$$\dot{D} = \dot{W}_{\substack{\text{maximum}\\\text{possible}}} - \dot{W}_{\text{actual}} \tag{11-48}$$

it follows that

$$\dot{W}_{\substack{\text{maximum}\\\text{possible}}} = -\frac{d}{d\theta}\int_V (e - T_0 s)\rho\, dV - \oint_A \left(h + \frac{\mathscr{V}^2}{2} + gZ - T_0 s\right)\rho\vec{\mathscr{V}} \cdot \vec{n}\, dA + \oint_A d'\dot{Q} - T_0 \oint_A \frac{d'\dot{Q}}{T} \tag{11-49}$$

Again let us take the system large enough to have the boundaries at T_0 so this reduces to

$$\dot{W}_{\substack{\text{maximum}\\\text{possible}}} = -\frac{d}{d\theta}\int_V (e - T_0 s)\rho\, dV + \oint_A \left(h + \frac{\mathscr{V}^2}{2} + gZ - T_0 s\right)\rho\vec{\mathscr{V}} \cdot \vec{n}\, dA \tag{11-50}$$

To find the availability of the material flowing into the control volume, we must assume that everything flowing out is at the dead state.

For convenience, we define a work potential function for open systems:

$$b = h + \frac{\mathscr{V}^2}{2} + gZ - T_0 s \tag{11-51}$$

Including this definition, Eq. (11.51) becomes

$$\dot{W}_{\substack{\text{maximum}\\ \text{possible}}} = -\frac{d}{d\theta}\int_V (b - pv)\rho\, dV + \oint_A b\rho\vec{\mathscr{V}}\cdot\vec{n}\, dA \tag{11-52}$$

Dividing the inflow and outflow in separate integrals and denoting the outflow as being at the dead state gives the work potential of the material entering:

$$\dot{A}_1 = -\frac{d}{d\theta}\int_V (b - pv)\rho\, dV - \dot{m}_{\text{total}} b_\infty + \sum_{\text{in}} \dot{m}_{\text{in}} b_{\text{in}} \tag{11-53}$$

If the flow is steady,

$$\dot{A}_1 = +\sum_{\text{in}} \dot{m}_{\text{in}} b_{\text{in}} - \dot{m}_{\text{total}} b \tag{11-54}$$

For a single inlet and single outlet situation,

$$\dot{A}_1 = \dot{m}(b_1 - b_\infty) \quad \text{or} \quad a_1 = b_1 - b_\infty \tag{11-55}$$

The degradation for a single stream in flowing in steady state through an open system from state 1 to state 2 is

$$\dot{D} = \dot{m}(b_1 - b_2) - \dot{W}_{\substack{x\\ \text{actual}}} \tag{11-56}$$

EXAMPLE 11.3. Compute the degradation associated with the steady state transfer of heat Q from a body at a constant temperature of T_1 to the atmosphere at T_0. Work the problem two ways:

1. By devising an equivalent process.
2. By using the derived equations.

SOLUTION: To solve this problem by finding the work in an equivalent reversible process, we first make a schematic sketch of the actual process and of the equivalent reversible process, noting that *equivalent* means that everything except the environment ends up the same.

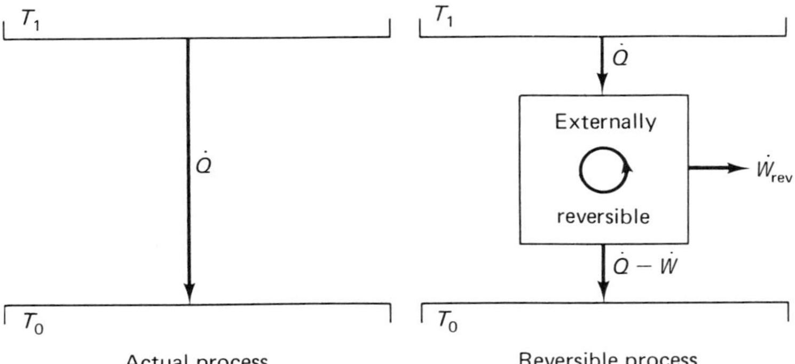

Note that the atmosphere receives $\dot{Q}$ in the real process and only $(\dot{Q} - \dot{W})$ in the equivalent reversible process but that as far as the body at T_1 is concerned, nothing is different. An externally reversible engine is used because we want to obtain the maximum possible work from the process and we have shown that an externally reversible engine has the best efficiency between given reservoirs.

Since $\dot{W}_{\text{actual}}$ in this case is zero, the degradation $\dot{D}$ is just $\dot{D} = \dot{W}_{\text{reversible}} - \dot{W}_{\text{actual}} = \dot{W}_{\text{reversible}}$. To find $\dot{W}_{\text{reversible}}$, we write two expressions for the efficiency of the externally reversible engine.

$$\eta = \frac{\dot{W}_{\text{reversible}}}{\dot{Q}} \quad \text{and} \quad \eta_{\text{reversible}} = 1 - \frac{T_0}{T_1}$$

Combining these gives

$$\dot{W}_{\text{reversible}} = \dot{D} = \dot{Q}\left(1 - \frac{T_0}{T_1}\right)$$

We can also obtain this result by using the derived equations. The easiest way in this case is to use the open system or control volume equation [Eq. (11-45)]

$$\dot{D} = \oint_A T_0 s \rho \vec{\mathscr{V}} \cdot \vec{n}\, dA - T_0 \oint_A \frac{d'\dot{Q}}{T}$$

applied to the region *between* the body at T_1 and the environment.

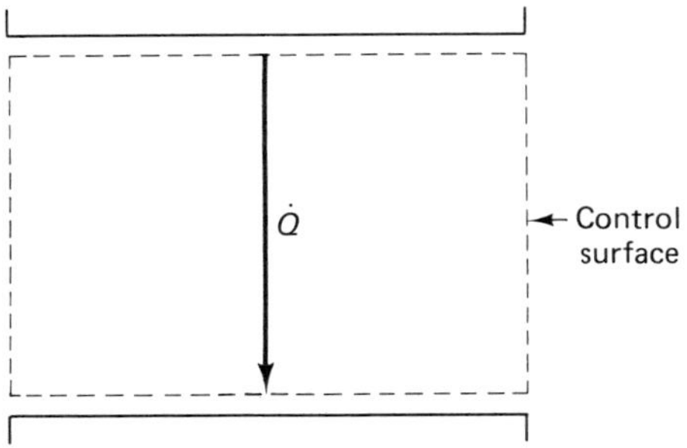

There is no mass flow into or out of the region so

$$\dot{D} = -T_0 \oint \frac{d'\dot{Q}}{T} = -T_0 \frac{\dot{Q}}{T} + T_0 \frac{\dot{Q}}{T_0}$$

$$= \dot{Q}\left(1 - \frac{T_0}{T_1}\right)$$

EXAMPLE 11.4. Compute the degradation associated with the steady state transfer of heat $\dot{Q}$ from a body at a constant temperature T_2 to another body at a constant temperature T_1. Atmospheric temperature is T_0. Work two ways as in Example 11.3.

SOLUTION: To find the work in an equivalent reversible process, we first make a schematic drawing of the real and equivalent situations, making use of the atmosphere at T_0 as the source or sink for any heat required.

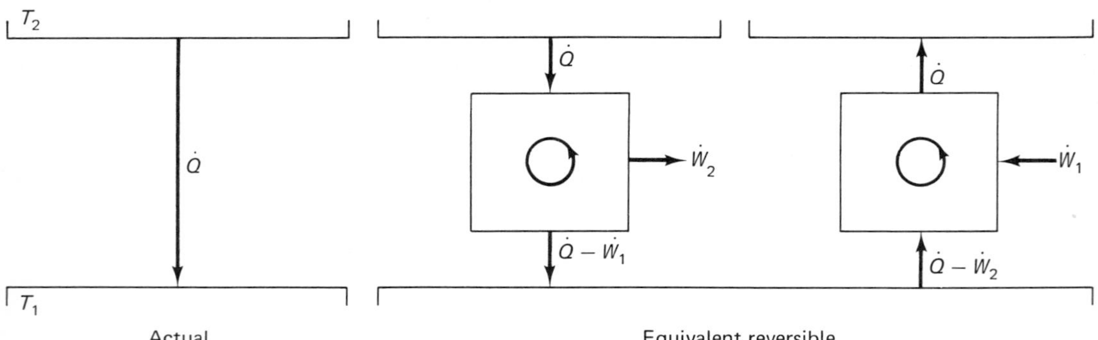

Notice that in the equivalent process the body at T_2 loses $\dot{Q}$ and the body at T_1 gains $\dot{Q}$ but that the atmosphere has net heat equal to $(\dot{W}_2 - \dot{W}_1)$ transferred from it. The engines are assumed to be externally reversible.

The net $\dot{W}_{\text{reversible}}$ for the equivalent process will be $(\dot{W}_2 - \dot{W}_1)$. For $\dot{W}_1$ we find

$$W_1 = \dot{Q}\left(1 - \frac{T_0}{T_1}\right)$$

and, by reversing engine 1, we find

$$\dot{W}_1 = \dot{Q}\left(1 - \frac{T_0}{T_1}\right)$$

Then

$$\dot{W}_{\text{reversible}} = \dot{W}_2 - \dot{W}_1 = Q\left[\frac{T_0}{T_1} - \frac{T_0}{T_2}\right]$$

$$\dot{W}_{\text{reversible}} = \dot{Q}T_0\left(\frac{1}{T_1} - \frac{1}{T_2}\right) = QT_0\left(\frac{T_2 - T_1}{T_1 T_2}\right)$$

and since $\dot{W}_{\text{actual}} = 0$ in the real process,

$$\dot{D} = \dot{W}_{\text{reversible}} = \dot{Q}T_0\left[\frac{T_2 - T_1}{T_2 T_1}\right]$$

It should be carefully noted that the process shown below is *not* equivalent to the actual process shown:

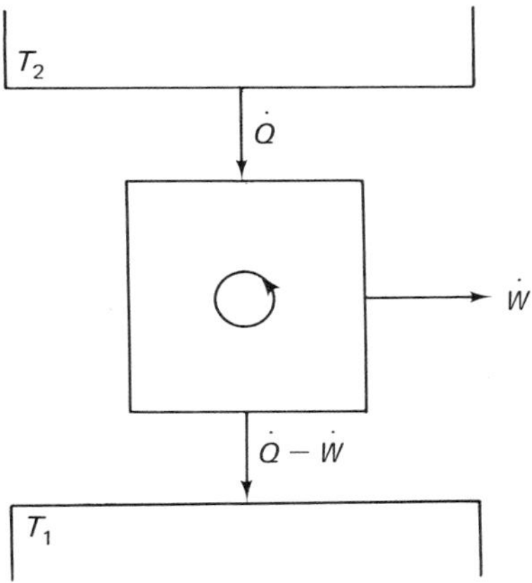

The reason is that the body at T_1 does not now have $\dot{Q}$ transferred to it as in the actual case. Only the atmosphere can have such discrepancies in a proper equivalent process.

It should also be noted that the equations we developed using T_0 as an "arbitrary" multiplier would not give the same results as the equivalent reversible process unless T_0 *had* been included. We incorporate this as justifiable for the general use of T_0.

EXAMPLE 11.5. Many examples of the degradation due to friction could be presented. We shall select a simple one to make the point. Consider a weight w_1 and pulley system that is to do work by lifting another weight w_2. Suppose that the force f required to overcome the friction in the pulley bearings is found to be constant. We shall assume the weights are moving at a constant speed for this analysis. We shall also assume that the pulley is kept at a constant temperature of T_1.

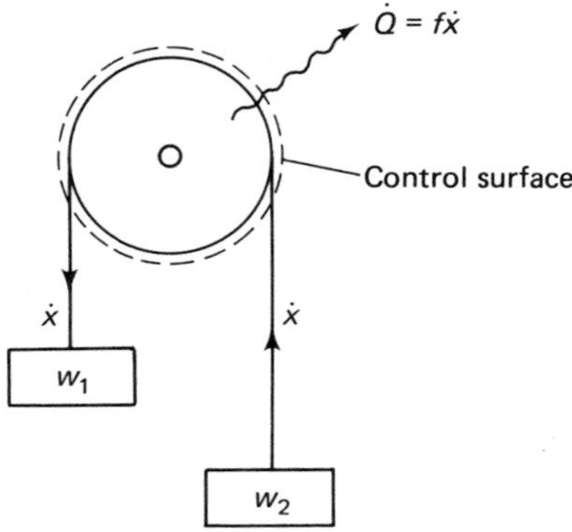

A force balance gives

$$w_1 = w_2 + f$$

The rate of work done against friction in the bearings will be $f\dot{x}$ and to keep the pulley at a constant temperature this must be transferred to the environment as indicated.

Applying Eq. (11-44) to the control volume shown gives

$$\dot{D} = -f\dot{x} + T_0 \frac{f\dot{x}}{T_1} - w_2\dot{x} + w_1\dot{x}$$

where $\dot{Q} = -f\dot{x}$ and $\dot{W}_{\text{actual}} = w_2\dot{x} - w_1\dot{x}$. Combined with the force balance equation, this reduces to

$$\dot{D} = f\dot{x}\frac{T_0}{T_1}$$

In addition to this irreversibility, there is the irreversibility due to the transfer of $\dot{Q} = f\dot{x}$ from T_1 to T_0. This we already know is given by

$$\dot{D} = f\dot{x}\left(1 - \frac{T_0}{T_1}\right)$$

Adding to find the total irreversibility in the process gives

$$\dot{D} = f\dot{x}\frac{T_0}{T_1} + f\dot{x}\left(1 - \frac{T_0}{T_1}\right) = f\dot{x}$$

Another way to obtain this result is simply to note that

$$\dot{W}_{\text{actual}} = w_2\dot{x} \quad \text{and} \quad \dot{W}_{\max} = w_1\dot{x}$$

Thus,

$$\dot{D} = \dot{W}_{\text{max}} - \dot{W}_{\text{actual}} = (w_1 - w_2)\dot{x}$$

Combined with the force balance $(w_1 - w_2 = f)$, this gives

$$\dot{D} = f\dot{x}$$

Students may wish to contemplate that the total degradation in a frictional process is really a combination of the fact that work must be expended to overcome the frictional forces and that the heat generated is usually totally wasted.

The moral is: To reduce degradation, lubricate.

EXAMPLE 11.6. As a final example here of an irreversible process, we consider the steady adiabatic flow of an ideal gas at T_1 through an orifice to a region of significantly lower pressure. We assume pipe sizes so that elevations and kinetic energy changes are zero.

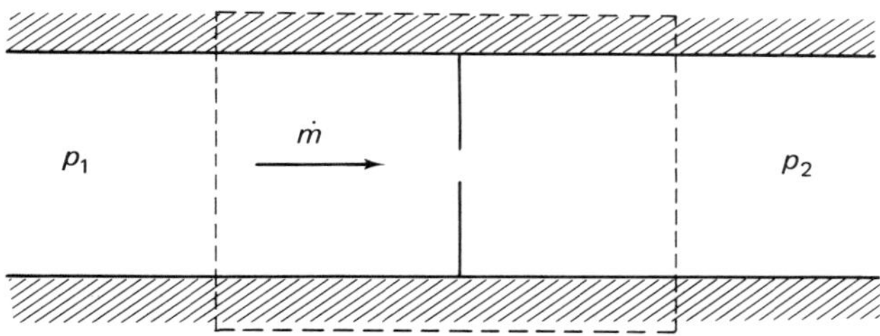

What is the degradation? Work two ways.

SOLUTION: For the control surface shown, we apply Eq. (11-45):

$$\dot{D} = \dot{m}T_0(s_2 - s_1)$$

The first law gives $h_1 = h_2$ and so, for the ideal gas, $T_1 = T_2$ and $U_1 = U_2$. Using the $T\,ds$ equations to evaluate $s_2 - s_1$, we find

$$T\,ds = 0 = -v\,dp = -v\,dp$$

$$ds = -\frac{v}{T_1}\,dp = -R\frac{dp}{p}$$

$$s_2 - s_1 = -R\int_1^2 \frac{dp}{p} = -R\ln\frac{p_2}{p_1} = R\ln\frac{p_1}{p_2}$$

$$\dot{D} = \dot{m}T_0\,R\ln\frac{p_1}{p_2}$$

As an alternate method, an equivalent reversible process might be desired using a reversible, isothermal turbine to expand the gas from p_1 to p_2. The gas will then end up at p_2 and T_1 as before. For steady flow, reversible systems we have shown that

$$W_{\text{reversible}} = -\dot{m} \int v\, dp$$

so in this case with $pv = RT$

$$W_{\text{reversible}} = \dot{m}RT_1 \ln \frac{p_1}{p_2}$$

But, in addition, there is irreversibility due to the heat transfer required to keep the reversible turbine isothermal. Since we take it that $T_1 > T_0$, a reversible heat pump will have to be used as shown:

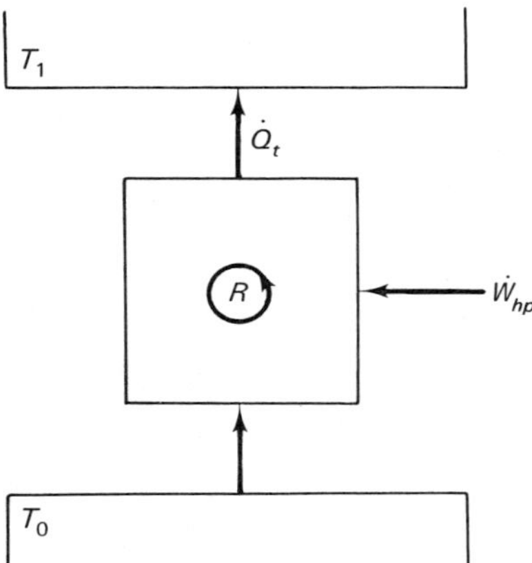

For the turbine,

$$\dot{Q}_t = \dot{W}_t$$

For the heat pump, which should be externally reversible,

$$\dot{W}_{hp} = \dot{Q}_t\left(1 - \frac{T_0}{T_1}\right) = \dot{W}_{\text{reversible}}\left(1 - \frac{T_0}{T_1}\right)$$

$$= \dot{m}RT_1\left(\ln \frac{p_1}{p_2}\right)\left(1 - \frac{T_2}{T_1}\right)$$

The net degradation therefore is

$$\dot{D} = \dot{W}_{\text{reversible}} - \dot{W}_{hp} = \dot{m}RT_0 \ln \frac{p_1}{p_2}$$

which is the same result as using the equation. Again, note that with the process not at T_0 *some* of the degradation can be attributed to heat transfer. But since, in practice, this heat can seldom be used economically, the total figure is realistic for the degradation of unrestrained expansion.

EXAMPLE 11.7. Compute the work potential per pound of a steady stream of high temperature, high pressure air at 30 bars absolute and 327°C if it has a velocity of 30 m/sec. The atmosphere is at 1 bar and 27°C. For air, $c_p = 1000$ j/kg-°K, $c_v = 714$ j/kg-°K.

SOLUTION:

$$a_1 = b_1 - b_\infty$$

$$a_1 = \left(h_1 + \frac{\mathscr{V}^2}{2} + gZ - T_0 s_1\right) - \left(h_\infty + \frac{\mathscr{V}^2}{2} + gZ_0 - T_0 s_\infty\right)$$

Neglecting ΔZ and assuming $\mathscr{V}_0 = 0$,

$$a_1 = (h_1 - h) + \frac{\mathscr{V}^2}{2} - T_0(s_1 - s)$$

Assuming the air is in ideal gas,

$$a_1 = c_p(T_1 - T_0) + \frac{\mathscr{V}^2}{2} - T_0\left[c_p \ln \frac{T_1}{T_0} - R \ln \frac{p_1}{p_0}\right]$$

$$a_1 = (1000)(300) + 450 - 300\left[1000 \ln \frac{600}{300} - 286 \ln \frac{30}{1}\right]$$

$$a_1 = 384{,}270 \text{ j/kg}$$

Students should note that these expressions contain no nonuseful work term. This is because no work is done against the atmosphere since the open system is a fixed region of space. The work of this type that is being done is accounted for in the pv term in h.

11-5 Work utilizing processes in open systems

Proceeding in a manner exactly as in Sec. 11-4 but noting that

$$\dot{W}_{\text{out}} = -\dot{W}_{\text{in}} \tag{11-57}$$

and that

$$\dot{D} = \dot{W}_{\substack{\text{actual}\\\text{in}}} - \dot{W}_{\substack{\text{minimum}\\\text{required}}} \tag{11-58}$$

we obtain the following results.

For the work required to produce a reversible state change in a steady flow single inlet–single outlet system,

Per pound:

$$W_{x \atop \text{reversible}} = \int_1^2 v\, dp - \tfrac{1}{2}(\mathscr{V}_2^2 - \mathscr{V}_1^2) - g(Z_2 - Z_1) \tag{11-59}$$

Per time:

$$\dot{W}_{x \atop \text{reversible}} = \dot{m} \int_1^2 v\, dp - \frac{\dot{m}}{2}(\mathscr{V}_2^2 - \mathscr{V}_1^2) - \dot{m}g(Z_2 - Z_1) \tag{11-60}$$

In case kinetic and potential effects may be neglected, these reduce to

$$W_{x \atop \text{reversible}} = \int_1^2 v\, dp \quad \text{and} \quad \dot{W}_{x \atop \text{reversible}} = \dot{m} \int_1^2 v\, dp \tag{11-61}$$

For the degradation,

$$\dot{D} = \oint_A d'\dot{W}_{x \atop \text{actual input}} - \frac{d}{d\theta}\int_V (e - T_0 s)\rho\, dV + \oint_A \left(h + \frac{\mathscr{V}^2}{2} + gZ - T_0 s\right)\rho \vec{\mathscr{V}} \cdot \vec{n}\, dA + \oint_A d'\dot{Q} - T_0 \oint_A \frac{d'\dot{Q}}{T} \tag{11-62}$$

$$\dot{D} = \frac{d}{d\theta}\int_V T_0 s \rho\, dV + \oint_A T_0 s \rho \vec{\mathscr{V}} \cdot \vec{n}\, dA - T_0 \oint_A \frac{d'Q}{T} \tag{11-63}$$

In case the system boundary is at T_0,

$$\dot{D} = d\dot{W}_{x \atop \text{actual input}} - \frac{d}{d\theta}\int_V (e - T_0 s)\rho\, dV + \oint_A \left(h + \frac{\mathscr{V}^2}{2} + gZ - T_0 s\right)\rho \vec{\mathscr{V}} \cdot \vec{n}\, dA \tag{11-64}$$

For steady flow systems with the boundary at T_0, these become

$$\dot{D} = d\dot{W}_{x \atop \text{actual input}} + \oint_A \left(h + \frac{\mathscr{V}^2}{2} + gZ - T_0 s\right)\rho \vec{\mathscr{V}} \cdot \vec{n}\, dA \tag{11-65}$$

and

$$\dot{D} = \oint T_0 s \rho \vec{\mathscr{V}} \cdot \vec{n}\, dA - \oint_A d'\dot{Q} \tag{11-66}$$

The equations for work potential are unchanged.

EXAMPLE 11.8. Compute the minimum work per pound required in steady flow to change steam at 100 psia and 400°F to saturated liquid at 14.696 psia. Atmospheric temperature is 40°F.

SOLUTION: The *minimum* work required to change the state of a substance is given by its change of work potential. In a steady flow process, this is

$$W_{12\,\substack{\text{minimum}\\\text{required}}} = a_1 - a_2 = b_1 - b_2$$

Neglecting kinetic and potential terms, this is

$$W_{12\,\substack{\text{minimum}\\\text{required}}} = h_1 - h_2 - T_0(s_1 - s_2)$$

From the steam tables we find

$$h_1 = 1227.5 \text{ Btu/lbm} \qquad s_1 = 1.6517 \text{ Btu/lbm-°F}$$

$$h_2 = 180.2 \text{ Btu/lbm} \qquad s_2 = 0.3121 \text{ Btu/lbm-°F}$$

Thus,

$$W_{\substack{\text{minimum}\\\text{required}}} = (1227.4 - 180.2) - 500(1.6517 - 0.3121)$$

$$W_{\substack{\text{minimum}\\\text{required}}} = 1047.2 - 669.8 = 377 \text{ Btu/lbm}$$

EXAMPLE 11.9. Compute the minimum *work* required to raise the temperature of a gram of water from 20 to 70°C in a nonflow process.

SOLUTION:

$$W_{\substack{\text{minimum}\\\text{required}}} = a_1 - a_2 = \phi_1 - \phi_2$$

$$= u_1 - u_2 + p_0(V_1 - V_2) - T_0(s_1 - s_2)$$

Assuming the process is constant pressure at p_0 and remembering $u + pv = h$,

$$W_{\substack{\text{minimum}\\\text{required}}} = h_1 - h_2 - T_0(s_1 - s_2)$$

Estimating h and s as the same as saturated liquid at the same temperature, we find from the tables that

$$W_{\substack{\text{minimum}\\\text{required}}} = (293.0 - 84.0) - 300(0.9549 - 0.2966)$$

$$W_{\substack{\text{minimum}\\\text{required}}} = 11.5 \text{ j/gm}$$

Alternate Solution:

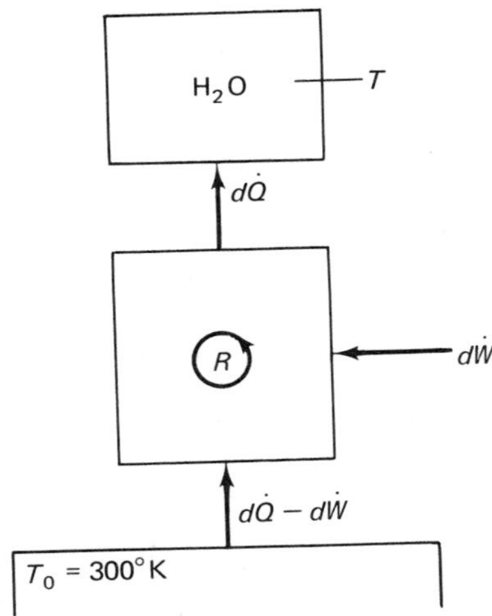

We concoct a reversible heat pump to transfer the heat to the water and find how much work the pump requires to do the job.

The coefficient of performance (COP) of the pump is

$$\text{COP} = \frac{d'Q - d'W}{d'W}$$

Since reversible, it can also be written as

$$\text{COP} = \frac{T_0}{T - T_0}$$

Equating these,

$$\frac{d'Q - d'W}{d'W} = \frac{T_0}{T - T_0}$$

Solving for $d'W$,

$$d'W = d'Q\left(1 - \frac{T_0}{T}\right)$$

As shown in the sketch, $d'Q$ may be related to the temperature rise of the water:

$$d'Q = c_p\, dT$$

Combining,

$$d'W = c_p\left(1 - \frac{T_0}{T}\right) dT$$

Integrating and taking c_p as 4.187 j/gm-°K,

$$W = c_p(T_2 - T_1) - c_p T_0 \ln \frac{T_2}{T_1}$$

$$W = 4.187(50) - 4.187(300) \ln \frac{343}{293}$$

$$W = 11.4 \text{ j/gm}$$

11-6 Helmholtz and Gibbs functions

11-6(a) Helmholtz function

Consider a system at its environment temperature executing an isothermal process from a state 1 to a state 2. The maximum work that can be obtained from this process is given by Eq. (11-13) by taking the system boundaries to be at T_0:

$$\underset{\substack{\text{maximum}\\\text{possible}}}{W_{12}} = (E_1 - E_2) - T(S_1 - S_2) \qquad (T = T_0 = \text{constant}) \qquad (11\text{-}67)$$

In many real situations, potential and kinetic temperature effects are not at all important and in these the situations total energy E equals the internal energy U. Then,

$$\underset{\substack{\text{maximum}\\\text{possible}}}{W_{12}} = (U_1 - TS_1) - (U_2 - TS_2) \qquad (11\text{-}68)$$

Defining $Z = U - TS$ as the Helmholtz function,

$$\underset{\substack{\text{maximum}\\\text{possible}}}{W_{12}} = Z_1 - Z_2 \qquad T = T_0 = \text{constant};\ E = U \qquad (11\text{-}69)$$

That is, the maximum work obtainable in an isothermal process at environment temperature, and where $E = U$, is given by the change in the Helmholtz function. Another way to state the above is that the work potential of a system at environment temperature (neglecting K.E. and P.E., of course) is given by its Helmholtz function, $Z = U - TS$.

In developing this result—that the Helmholtz function change is the maximum work obtainable in an isothermal process—the restriction that the constant temperature used in the equations must be the environmental temperature (T_0) is often not made by some scientists. They make a small error as a result. The effect is to assume that the heat transfer needed to keep the system at constant temperature is transferred irreversibly rather than reversibly through heat engines or heat pumps. Reversible heat transfer in this way is not a practical procedure anyway so the error is in the nature of a small theoretical oversight or maybe even a deliberate omission made in the interest of simplicity or brevity. Thus, it is common practice to say that the Helmholtz function *is* the maximum work obtainable in *any* isothermal process for which $E = U$. This is only precisely true at the environment temperature but it is correct in a practical sense. The Helmholtz function may thus be called the *isothermal* work potential.

The Helmholtz function has another meaning, too. Starting with the definition and then differentiating, we find

$$dZ = dU - T\,ds - S\,dT \tag{11-70}$$

Combining this with $T\,ds = dU + p\,dV$ results in

$$dZ = -p\,dV - S\,dT \tag{11-71}$$

For an isothermal process, $dT = 0$, and so

$$\left(\frac{\partial Z}{\partial V}\right)_T = -p \tag{11-72}$$

With this equation if we know Z as a function of T and V, we can obtain a relation among p, V, and T—the equation of state.

Also from Eq. (11-72), if $dV = 0$, we find

$$\left(\frac{\partial Z}{\partial T}\right)_V = -S \tag{11-73}$$

Again if we have Z as function of T and V, we obtain S as a function of T and V, which makes it possible to determine the specific heat C_v from the equation

$$c_v = T\left(\frac{\partial S}{\partial T}\right)_v \tag{11-74}$$

The Helmholtz function also has implications regarding the equilibrium state of isothermal systems. If we ourselves make the same small error noted above, we can see this quickly. Combining the first law (with $E = U$),

$$d'Q - d'W = dU \tag{11-75}$$

with the second law written as

$$d'Q \leq T\,ds \tag{11-76}$$

we obtain

$$d'W \leq T\,ds - dU \tag{11-77}$$

Combining this with $dZ = dU - T\,ds - S\,dT$, the result can be rearranged to

$$-dZ \geq dW + S\,dT \tag{11-78}$$

For isothermal processes,

$$-dZ \geq dW \qquad T = \text{constant} \tag{11-79}$$

Note that, for any isothermal process in which no work is done, $dZ = 0$. If work is done, however, the Helmholtz function either decreases or remains constant, the latter being the reversible case. Therefore, when the system has done all the work it can do, the Helmholtz function will be a minimum and the system will be at an equilibrium state.

It should be evident that the Helmholtz function is a key property. It is not widely used by energy engineers but it is still very important. The methods of statistical mechanics provide a relatively easy way to compute Z. Thus, its relationship to the equation of state and specific heat is especially useful.

11-6(b) Gibbs function

The Gibbs function ($G = H - TS$) bears the same relationship to work potential in steady flow open systems that the Helmholtz function does to closed isothermal systems. That is, the Gibbs function *is* the work potential in steady flow open systems if kinetic and potential energy effects are neglected. The equations are

$$a_1 = b_1 - b_\infty = (h_1 - T_1 s_1) - (h_\infty - T_0 s_\infty) \tag{11-80}$$

$$a_1 = g_1 - g_\infty \qquad \text{for } T_1 = T_0 \tag{11-81}$$

Its relation to work potential is not, however, the major importance of the Gibbs function. Starting with the definition and differentiating,

$$dG = dH - T\,dS - S\,dT \tag{11-82}$$

Combining this with $T\,ds = dH - V\,dp$ gives

$$dG = V\,dp - S\,dT \tag{11-83}$$

From this, we find

$$\left(\frac{\partial G}{\partial p}\right)_T = V \quad \text{and} \quad \left(\frac{\partial G}{\partial T}\right)_p = -S \tag{11-84}$$

which lead, respectively, to the equation of state and c_p if $G(T, V)$ is known.

Also, starting with Eq. (11-82) and by combining with Eq. (11-77) and the $T\,ds$ equation, one can derive

$$dG \leq p\,dV + V\,dp - S\,dT - d'W \tag{11-85}$$

Now for cases where pressure and temperature are constant, this gives

$$-dG \leq -p\,dV + d'W \tag{11-86}$$

We have already shown that for a reversible closed system $dW = p\,dV$ and that for irreversible expansion processes $p\,dV > dW$. Therefore, Eq. (11-86) shows that $G =$ constant for reversible processes at constant pressure and temperature and that G decreases for irreversible processes of this type. In other words, the Gibbs function reaches a minimum when a system at constant p and T is in its equilibrium state and can do no more work.

Because the Gibbs function is constant in reversible processes at constant pressure and temperature, it is also important in phase change situations because they occur at constant p and T. The Gibbs function is constant in a phase change; that is, g for the liquid is the same as g for the gas in a two-phase liquid–gas mixture. Symbolically, $g_f = g_g$.

11-7 More on the meaning of entropy

You have just seen that the Helmholtz and Gibbs functions are directly related to work potential in special cases. And we have already, in Sec. 11-2, tied entropy intimately to energy degradation. Now let us go to explore some more on the meaning of entropy. This is not intended to be a rigorous discussion but rather it is intended to help you gain further qualitative insight into the meaning of entropy.

We know that heat is *not* a property. Thus we can't talk about the heat *of* a body or the heat contained in a body. (We can talk about hot bodies but that's another matter.) Entropy is a property, however, and it may be thought of as a kind of "energy" if you aren't too fussy about rigor. Entropy is heat, which *is* energy, divided by temperature. If we transfer heat to a system, its entropy increases, so part of our rigorless model is okay. The trouble is that this "energy" we have invented to talk about doesn't have the dimensions of energy. To correct this, we multiply by the absolute temperature as we have done before. TS does have units of energy so we'll use it as our invented "energy." It is sometimes referred to as "bound energy" so we'll use this term too.

Let us now look at the meaning of TS in a reversible isothermal process. The change in TS is

$$d(TS) = T\,dS + S\,dT = T\,dS \qquad (T = \text{constant}) \tag{11-87}$$

From the second law

$$T\,dS = dQ \qquad \text{(reversible)} \tag{11-88}$$

so

$$d(TS) = dQ \qquad \text{(reversible, isothermal)} \tag{11-89}$$

That is, in a reversible isothermal process TS is in fact the amount of heat. This result cannot be generalized to other processes but it does illustrate the appropriateness of thinking about TS as a kind of energy (property) somehow associated especially with heat (not a property).

With this concept, let us look again at the Helmholtz function, $Z = U - TS$. If TS represents the energy associated with "heat," and U is the total energy, then $U - TS$ should be the energy available to do work. This is just what we found the Helmholtz function to be—the work potential in an isothermal process.

The discussion above suggests that one nonrigorous, qualitative way of feeling about entropy is that TS is "energy bound up as heat."

Analogous statements can be made about the TS portion of $G = H - TS$. Enthalpy is analogous to the total energy in flow systems. (We're really rigorless, now!) If TS is *bound* energy, then the total minus the boundup part should be available to do work. That's potential work. And that's just what the Gibbs function is in steady flow open systems: the work potential. So again, thinking of TS as energy bound up as heat, unavailable for work, seems to make sense. It appears that the higher the entropy of a system, the more of its total energy is bound up as heat, unable, by the second law, to be converted to work.

It is also helpful in a qualitative way to write the expression for the work potential of a system in some state 1 as

$$a_1 = \phi_1 - \phi_\infty \tag{11-90}$$

$$a_1 = E_1 - P_0 v_1 - T_0 S_1 - \phi_\infty \tag{11-91}$$

Now ϕ_∞ is a constant so let us drop it for the sake of this qualitative argument. Also, $P_0 v_1$ is usually minor—remember it arose because we did not want to count work done against the atmosphere or environment as useful. Keeping only the major terms, Eq. (11-91) becomes

$$A \simeq E - TS \tag{11-92}$$

or

$$TS \simeq E - A \tag{11-93}$$

Here we see again entropy as almost directly proportional to the difference between energy and work potential. Or, in other words, entropy is an indicator of the difference between the total energy of a system and the *available energy* (to man for his possible use) of the system. The smaller the entropy for a given energy, the greater potential use man can make of that energy. High-grade energy has low entropy. Low-grade energy has lots of TS per unit of energy.

Returning now to somewhat more rigorously written expressions, consider an isolated ($Q = W = 0$) constant volume system that changes from state 1 to state 2 (not its dead state). Its *loss* of work potential will be

$$A_1 - A_2 = E_1 - E_2 + p_0(V_1 - V_2) - T_0(S_1 - S_2) \qquad (11\text{-}94)$$

Since it is isolated, $E_1 = E_2$. Thus,

$$A_1 - A_2 = T_0(S_2 - S_1) \qquad (11\text{-}95)$$

In words, Eq. (11-95) says that for an isolated constant volume system the loss of work potential is directly proportional to the increase in entropy. Since work potential is a little like energy in a bank, once again the entropy increase is seen as bad news, bad but inevitable, however, by the principle of increasing entropy.

Entropy is shown qualitatively to be related to the energy of a system "bound up as heat." The product TS is a kind of unavailable energy. Thus, we find the total energy of a system E approximately equal to $(A - TS)$, where A is the work potential, the *available* energy. When we couple this thought with the principle of increasing entropy and with the importance of work to man, it should give us reason to pause for reflection.

The first law gave us a new property, energy, which is neither good nor bad but there to be used or enjoyed—a little like life. The second law also has given us a new property, entropy, which limits our use of energy and is both bad and inevitable—a little like death. Energy and entropy—life and death. Everyday we use energy and we live a little. At the same time, we generate some entropy and die a little—not just we but our environment too. Philosophers and novelists may take over from here.

For students interested in the social and philosophical implications of the laws of thermodynamics, the following reading is suggested:

Environment, Power, and Society by Howard T. Odum, Wiley, 1971.

Problems

11-1 Consider a steam turbine that has a throttling governor. (That is, the power output of the turbine is controlled by throttling the inlet steam.) The steam in the pipeline flowing to the turbine has a pressure of 400 lbf/in.^{2} and a temperature of 600°F. At a certain load the steam is throttled in an adiabatic process to 300 lbf/in.^{2}. Calculate the availability per pound-mass of steam before and after this process and the irreversibility per pound-mass of steam for this process. Show the initial and final states of the steam on a T-s diagram. The pressures given are absolute.

11-2 A pressure vessel has a volume of 1 m^3 and contains air at 15 bars absolute and 150°C. The air is cooled to 25°C by heat transfer to the surroundings at 25°C. Calculate the availability in the initial and final states and the irreversibility of this process.

11-3(a) 0.79 mole of nitrogen at 14.7 lbf/in.^{2} and 77°F are separated from 0.21 mole of oxygen at 14.7 lbf/in.^{2} and 77°F by a membrane. The membrane ruptures and the gases mix in an adiabatic process to form a uniform mixture. Determine the reversible work and irreversibility for this process.

(b) Determine the minimum work required to separate 1 mole of air (assume composition to be 79% nitrogen and 21% oxygen by volume) at 14.7 lbf/in.^{2} and 77°F into nitrogen and oxygen at 14.7 lbf/in.^{2} and 77°F.

(c) How would you evaluate the performance of an air separation plant regarding work input?

11-4 Two identical blocks of metal have a mass of 10 kg and a specific heat of 400 j/kg-°K. One has an initial temperature of 2000°K and the other an initial temperature of 500°K. The blocks are brought to the same temperature in a reversible process. Determine the final temperature of the blocks and the net work.

11-5 An air preheater is used to cool the products of combustion from a furnace while heating the air to be used for combustion. The rate of flow of products is 100,000 lbm/hr and the products are cooled from 600 to 400°F; for the products at this temperature, $c_p = 0.26$ Btu/lbm-°R. The rate of air flow is 93,000 lbm/hr and the initial air temperature is 100°F; for the air, $c_p = 0.24$ Btu/lbm-°R.

(a) What is the initial and final availability of the products (Btu/hr)?

(b) What is the irreversibility for this process?

(c) Suppose this heat transfer from the products took place reversibly through heat engines. What would be the final temperature of the air? What power would be developed by the heat engines?

11-6 A stream of 1 lb/min of water at 60°F is used to cool a stream of 1 lb/min of water at 100°F to 82°F. Compute the dissipation

(a) If a parallel flow heat exchanger is used.

(b) If a counterflow heat exchanger is used.

Atmospheric temperature is 80°F.

11-7 Compute the minimum work required to raise the temperature of a kilogram of water from 10 to 30°C. Compare this with the heat required. Atmospheric temperature is 30°C.

PART II

Applications

Preface to Part II

The topics in Part II are covered at an introductory level only. The coverage here is not intended to substitute for texts or courses in energy conversion, combustion, air conditioning, etc. Rather, the intention is to illustrate the use of thermodynamic principles in these topics, to provide an introductory survey for students who do not continue further study, and to provide a transition for those who do.

In general, Part II need not be studied in sequence. Prerequisites for the various chapters are listed below:

Chapter	*Prerequisite*
12 Vapor cycles	—
13 Gas cycles	Chapter 12
14 Mixtures of gases	—
15 Combustion	Chapter 14
16 Property relations	—

Thus Part II may be entered at any of the following chapters without regard to order: 12, 14, and 16.

12

Vapor power and refrigeration cycles

12-1 Introduction

Vapor power cycles are so named because the working fluid (usually H_2O in power cycles or a Freon in refrigeration cycles) is condensed and evaporated somewhere in the cycle. Gas cycles are discussed in Chap. 13.

So far in this book we have considered energy conversion systems and cycles in a rather abstract way. The heat engine and heat pump (or refrigerator) have been represented schematically by diagrams such as those in Fig. 12.1. Now we wish to make a more detailed study of what the squares representing the energy conversion devices convey.

In Chap. 9 we noted that no engine can have an efficiency greater than an internally reversible engine operating between the same reservoirs and that all reversible engines have the same efficiency when operating between the same reservoirs. Similar statements can be made about reversible refrigerators.

Let us consider the reversible engine for a moment. Its efficiency is given by

$$\eta_{\text{reversible}} = \frac{W}{Q_h} = \frac{Q_h - Q_c}{Q_h} = \frac{T_{he} - T_{ce}}{T_{he}} = 1 - \frac{T_{ce}}{T_{he}} \tag{12-1}$$

We can see that to improve efficiency we should either increase the temperature at which heat is added (T_{he}) or decrease the temperature at which heat is rejected (T_{ce}). This conclusion applies generally to real cycles as well. Of

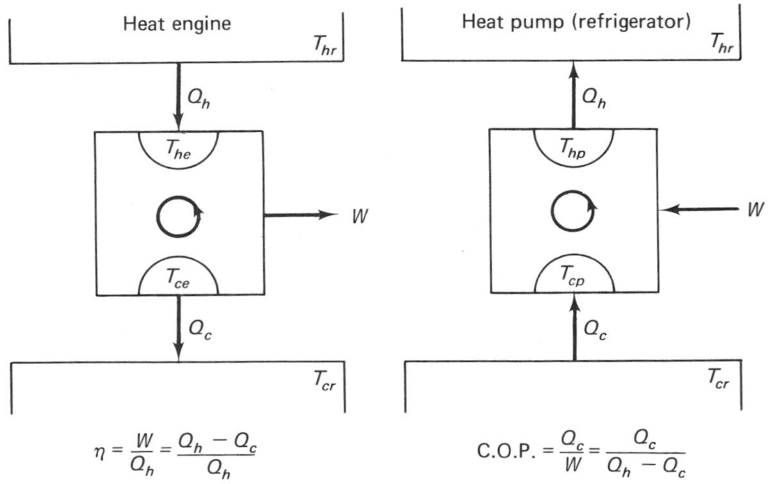

Figure 12.1 Schematics of heat engine and heat pump.

course, in real cycles the heat addition and rejection are usually not done at a single, constant temperature but it is still true that raising the *average* temperature at which heat is supplied, or lowering the average temperature at which it is rejected, will increase the efficiency.

The primary factor that limits the temperature of heat addition is the cost and availability of materials that can withstand the desired temperatures and pressures. Thermodynamic considerations alone require both the temperature and pressure to be as high as possible before the steam enters the turbine but the cost and availability of materials set practical limits at present of about 1200°F (650°C) for steam cycles and 1500°F (800°C) for gas cycles. Nuclear steam cycles must operate at an even lower temperature and hence have a lower efficiency.

The evaluation of an energy cycle cannot be done exclusively on the basis of thermodynamic considerations. In practice other factors are also important. Cost, weight, reliability, pollution, size, and other parameters must all be taken into account in a realistic situation. We can only discuss the thermodynamics here but students must understand that that is only one aspect in design decisions.

Thermodynamic cycle analysis is not basically different from other kinds of engineering analysis. The first steps of setting up the problem are extremely important and if not done well, often lead to later difficulties. In cycle analysis, this is especially true. The first step is to translate the actual system or design idea into a schematic flow diagram showing all equipment and labeling systematically all states and mass flows. Then the analyst must sketch the cycle on a *T-S* plot (and/or a *p-v* plot) again labeling the states corresponding to the schematic. Next, energy balances on all the components of the system must be made. Usually, the efficiency can then be computed with the aid of the needed process equations and equations of state. The most neglected part of this process by beginning students is that which calls for complete, labeled *T-s* and schematic drawings. Before you start any cycle analysis problem, be sure you have the system completely drawn out, all states labeled, and a corresponding *T-s* plot drawn. Note thoroughly how the examples in this chapter are done.

12-2 The basic Rankine cycle

The basic vapor power cycle is called the *Rankine cycle* and is shown schematically in Fig. 12.2. In words, the ideal cycle may be described as shown on the following page.

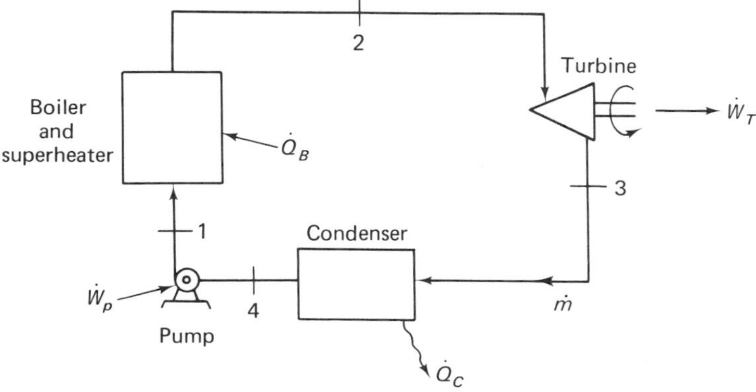

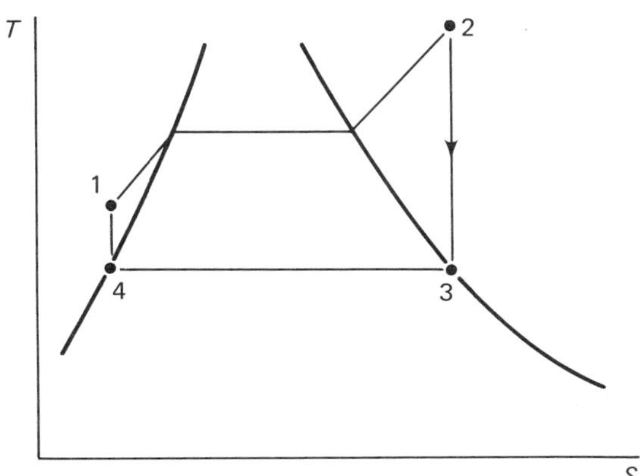

Figure 12.2 The Rankine cycle with superheat.

State change	*Equipment*	*Process*
1–2	Boiler and superheater	Constant high pressure addition of heat changes phase of H_2O from liquid to superheated vapor
2–3	Turbine	Reversible adiabatic expansion to low pressure to nearly saturated vapor producing work
3–4	Condenser	Constant low pressure removal of heat changes phase to saturated liquid
4–1	Pump	Liquid is pumped again to high pressure

The description above and the temperature–entropy diagram of the cycle shown in Fig. 12.2 assume that all the processes are internally reversible. State 3 need not be in the two-phase region but it usually is, though the quality at state 3 is usually not allowed to fall below about 0.90 because the liquid water present tends to erode the turbine blades. The first law written for each of the components, neglecting kinetic and potential energy terms, gives

Boiler: $$\dot{Q}_B = \dot{m}(h_2 - h_1) \tag{12-2a}$$

Turbine: $$\dot{W}_T = \dot{m}(h_2 - h_3) \tag{12-2b}$$

Condenser: $$\dot{Q}_C = \dot{m}(h_3 - h_4) \tag{12-2c}$$

Pump: $$\dot{W}_P = \dot{m}(h_1 - h_4) \tag{12-2d}$$

where $\dot{Q}_B$, $\dot{Q}_C$, $\dot{W}_T$, and $\dot{W}_P$ are all defined positive. The efficiency is therefore given by

$$\eta = \frac{W_{\text{net output}}}{Q_{\text{in}}} = \frac{\dot{W}_T - \dot{W}_P}{\dot{Q}_B} \tag{12-3}$$

$$\eta = \frac{(h_2 - h_3) - (h_1 - h_4)}{(h_2 - h_1)} \tag{12-4}$$

If we note that an overall energy balance gives

$$\dot{W}_T - \dot{W}_P = \dot{Q}_B - \dot{Q}_C \tag{12-5}$$

the efficiency can be written as

$$\eta = \frac{\dot{Q}_B - \dot{Q}_C}{\dot{Q}_B} = 1 - \frac{(h_3 - h_4)}{(h_2 - h_1)} \tag{12-6}$$

A word should be said about the computation of pump work. It is usually small compared to the turbine work in a vapor power cycle because the specific volume of the liquid is small. Except for rough approximations, however, it should be computed. Rather than finding Δh for the pump, which is difficult with the data available in the liquid range, it is best to use the reversible energy equation. Neglecting kinetic and potential terms, this becomes

$$\dot{W}_P = \dot{m} \int_{p_1}^{p_2} v \, dp \tag{12-7}$$

On a per pound basis, and noting that the liquid is highly incompressible, the work can be taken as

$$W_P \simeq v_1(p_2 - p_1) \tag{12-8}$$

v_1 is generally saturated liquid and can easily be looked up in the tables. (Watch out for units though!)

EXAMPLE 12.1. An internally reversible Rankine steam cycle operates at boiler and condenser pressures of 500 and 1 psia, respectively. The temperature of steam entering the turbine is 600°F. Compute the cycle efficiency.

SOLUTION:
Translation:

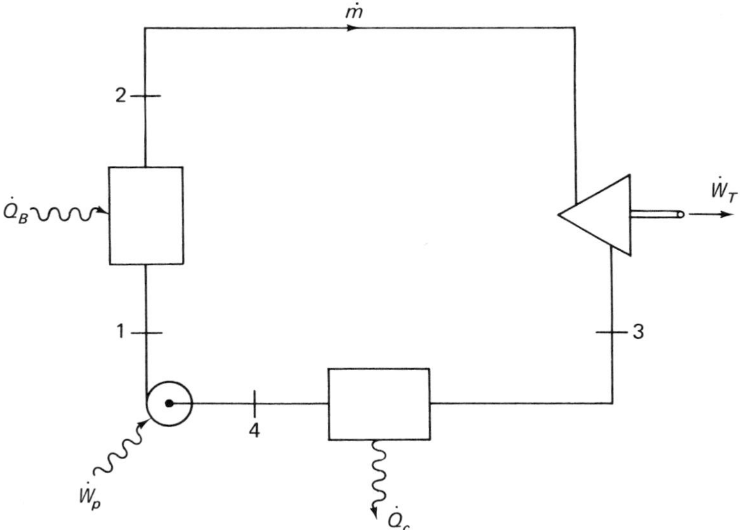

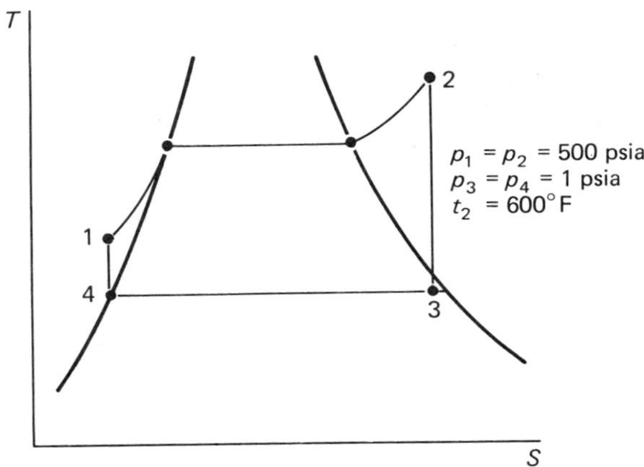

Equation of state: Steam tables.
Efficiency:

$$\eta = \frac{\dot{W}_T - \dot{W}_P}{\dot{Q}_B}$$

To find pump work per pound:

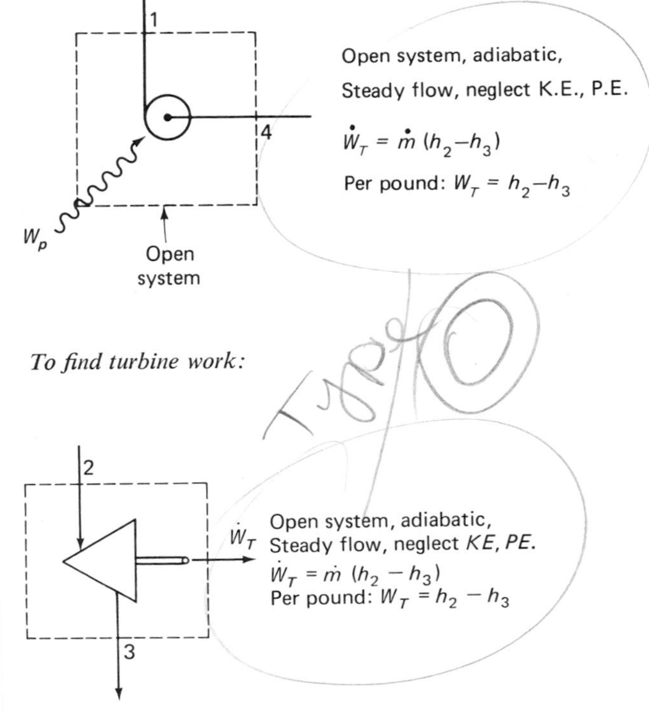

From the tables

$$h_2 = 1298.3 \text{ Btu/lbm}$$
$$S_2 = 1.5585 \text{ Btu/lbm-°R} = S_3$$

At 1 psia:

$$S_f = S_4 = 0.13266$$
$$S_{fg} = 1.8453$$
$$S_g = 1.9779$$
$$S_3 = S_f + x_3 S_{fg}$$
$$x_3 = \frac{S_3 - S_f}{S_{fg}} = \frac{1.5585 - 0.13266}{1.8453} = 0.772$$

Thus,

$$h_3 = h_f + x_3 h_{fg} = 69.74 + 0.722(1036.0)$$
$$h_3 = 871 \text{ Btu/lbm}$$

and

$$W_T = h_2 - h_3 = 1298 - 871 = 428 \text{ Btu/lbm}$$

To find boiler heat addition per pound:

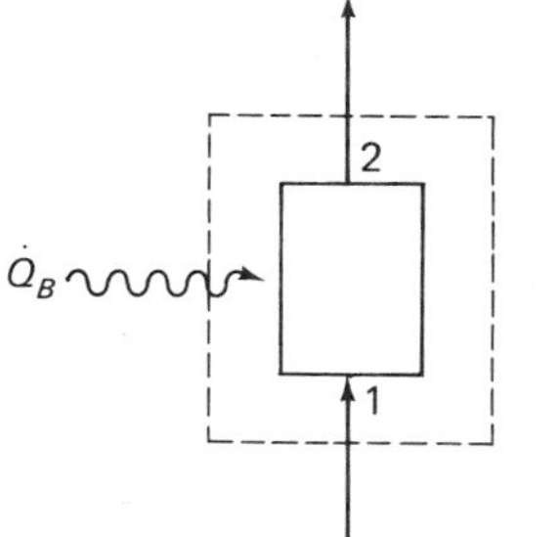

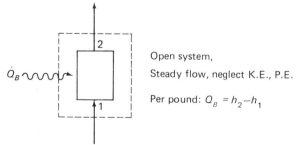

Open system,
Steady flow, neglect K.E., P.E.
Per pound: $Q_B = h_2 - h_1$

$$h_1 = h_4 + W_P = 69.7 + 1.5 = 71.2 \text{ Btu/lbm}$$
$$Q_B = h_2 - h_1 = 1298.3 - 71.2 = 1227.1 \text{ Btu/lbm}$$

Efficiency:

$$\eta = \frac{W_T - W_P}{Q_B} = \frac{428 - 1.5}{1227.1} = 0.347$$

12-3 Reheat and regeneration

It should be noted that in the Rankine cycle all the heat is not added at a constant working fluid temperature but that heat is added at temperatures from T_1 to T_2. As noted, to improve the efficiency we would like to add heat at a higher average temperature. Several modifications of the Rankine cycle help to accomplish this objective. One is called *reheat*. A reheat scheme is shown in Fig. 12.3. The basic idea of reheat is to allow the steam to expand only partially through the first stage of the turbine and then to reheat it again to the maximum temperature before expanding through the rest of the turbine.

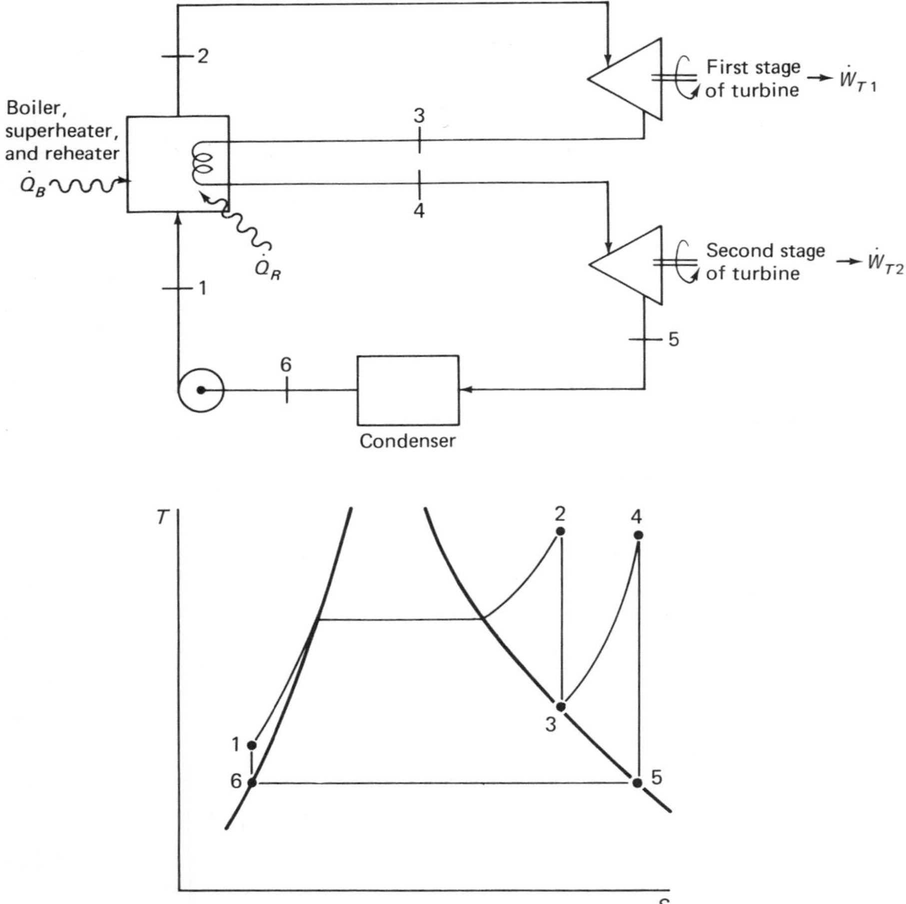

Figure 12.3 The reheat cycle.

This permits a higher percentage of the heat addition to take place at the higher superheated temperatures and hence raises the average temperature at which heat is added. The efficiency of the reheat cycle shown is given by

$$\eta = \frac{\dot{W}_{\text{net output}}}{\dot{Q}_{\text{in}}} = \frac{\dot{W}_{T_1} + \dot{W}_{T_2} - \dot{W}_P}{\dot{Q}_B + \dot{Q}_R} \tag{12-9}$$

$$\eta = \frac{(h_2 - h_3) + (h_4 - h_5) - (h_1 - h_6)}{(h_2 - h_1) + (h_4 - h_3)} \tag{12-10}$$

Or, again using an overall balance,

$$\eta = \frac{\dot{Q}_B + \dot{Q}_R - \dot{Q}_C}{\dot{Q}_B + \dot{Q}_R} = 1 - \frac{\dot{Q}_C}{\dot{Q}_B + \dot{Q}_R} \tag{12-11}$$

$$\eta = 1 - \frac{(h_5 - h_6)}{(h_2 - h_1) + (h_4 - h_3)} \tag{12-12}$$

EXAMPLE 12.2 An internally reversible Rankine cycle with reheat operates between 30 bars absolute and 0.1 bar absolute. Steam enters the first-stage turbine at 300°C. Steam leaves the first-stage turbine at 3.0 bars absolute and is reheated to 300°C. Compute the cycle efficiency.

SOLUTION:

Translation:

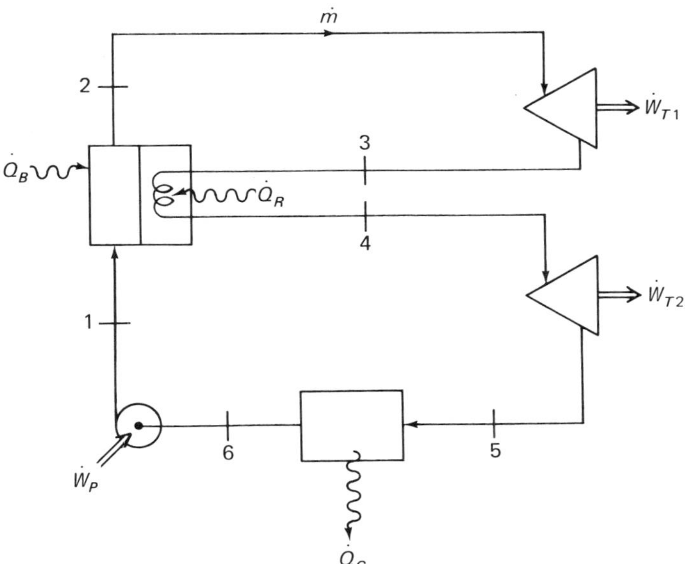

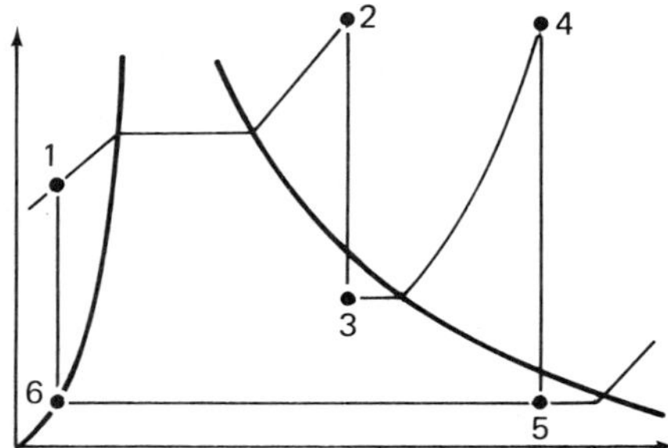

Efficiency:

$$\eta = \frac{\dot{W}_{T_1} + \dot{W}_{T_2} - \dot{W}_P}{\dot{Q}_B + \dot{Q}_R}$$

To find $\dot{W}_{T_1}$*:*
Assume the turbine is adiabatic and neglect kinetic and potential energy changes. Then applying the first law in steady flow to the turbine gives

$$\dot{W}_{T_1} = \dot{m}(h_2 - h_3)$$

From the steam tables,

$$h_2 = 2993.5 \text{ j/gm}$$
$$s_2 = 6.5390 \text{ j/gm-°K}$$

To find h_3, we note that, for the internally reversible, adiabatic process 2-3, $s_2 = s_3 = 6.5390$. At state 3 ($p_3 = 3.0$ bars),

$$s_{f_3} = 1.6718 \qquad h_{f_3} = 561.47 \text{ j/gm}$$
$$s_{fg_3} = 5.3201 \qquad h_{fg_3} = 2163.8$$
$$s_{g_3} = 6.9919 \qquad h_{g_3} = 2725.3$$
$$s_2 = s_3 = s_{f_3} + x_3 s_{fg_3}$$
$$6.5390 = 1.6718 + x_3(5.3201)$$
$$x_3 = 0.915$$

Now for h_3,

$$h_3 = h_{f_3} + x_3 h_{fg_3}$$
$$h_3 = 561.47 + 0.915(2163.8)$$
$$h_3 = 2542 \text{ j/gm}$$

and $\dot{W}_{T_1}$,

$$\frac{\dot{W}_{T_1}}{\dot{m}} = h_2 - h_3$$

$$\frac{\dot{W}_{T1}}{\dot{m}} = 2993.5 - 2542 = 452 \text{ j/gm}$$

To find $\dot{W}_{T_2}$:
Making the same assumptions,

$$\dot{W}_{T_2} = \dot{m}(h_4 - h_5)$$

From the steam tables,

$$h_4 = 3069.3 \text{ j/gm}$$
$$s_4 = 7.7022 \text{ j/gm-°K}$$

To find h_5, we again use the fact that, with the assumptions made, $s_4 = s_5$. At state 5,

$$s_{f_5} = 0.6493 \text{ j/gm-°K} \qquad h_{f_5} = 191.83 \text{ j/gm}$$
$$s_{fg_5} = 7.5009 \qquad h_{fg_5} = 2392.8$$
$$s_{g_5} = 8.1502 \qquad h_{g_5} = 2584.7$$
$$s_4 = s_5 = s_{f_5} + x_5(s_{fg_5})$$
$$7.7022 = 0.6493 + x_5(7.5009)$$
$$x_5 = 0.949$$

$$h_5 = h_{f_5} + x_5(h_{fg_5})$$
$$h_5 = 191.83 + 0.949(2392.8)$$
$$h_5 = 2463 \text{ j/gm}$$

Thus

$$\frac{\dot{W}_{T2}}{\dot{m}} = h_4 - h_5 = 3069.3 - 2463$$

$$\frac{\dot{W}_{T2}}{\dot{m}} = 606 \text{ j/gm}$$

To find $\dot{W}_p$:
Assume

$$\dot{W}_p = \dot{m}v_6(p_1 - p_6)$$

From the steam tables,

$$v_6 = v_{f_6} = 1.0102 \text{ cm}^3/\text{gm}$$

Thus

$$\frac{\dot{W}_p}{\dot{m}} = 1.0102(30 - 0.1)10^5 \times 10^{-6}$$

Units:

$$\text{cm}^3/\text{gm} \times \text{N/m}^2 \times \text{m}^3/\text{cm}^3$$

$$\frac{\dot{W}_p}{\dot{m}} = 3.0 \text{ j/gm}$$

To find $\dot{Q}_B$:

$$\dot{Q}_B = \dot{m}(h_2 - h_1)$$

For h_1,

$$h_1 = h_6 + \frac{\dot{W}_p}{\dot{m}} = 191.8 + 3.0$$

$$h_1 = 194.8 \text{ j/gm}$$

Thus,

$$\frac{\dot{Q}_B}{\dot{m}} = 2993.5 - 194.8 = 2799 \text{ j/gm}$$

To find $\dot{Q}_R$:

$$\dot{Q}_R = \dot{m}(h_4 - h_3)$$

$$\frac{\dot{Q}_R}{\dot{m}} = 3069.3 - 2542 = 527 \text{ j/gm}$$

To find efficiency:

$$\eta = \frac{W_{T_1} + W_{T_2} - W_p}{\dot{Q}_B + \dot{Q}_R}$$

$$\eta = \frac{452 + 606 - 3}{2799 + 527}$$

$$\eta = 0.317$$

Another scheme for improving Rankine cycle efficiency by increasing the average temperature at which heat is added is called *regeneration*. See Fig. 12.4. In this scheme, a portion of the steam ($\dot{m}_e$) is extracted from the main stream after the first-stage turbine. This extracted steam is used later in a feedwater heater to raise the temperature of the main stream from its coolest temperature (t_6) to an intermediate temperature (t_7). Thus heat from this portion is not provided by the boiler and we have reduced the amount of heat added at low temperatures, thereby raising the average temperature of heat addition.

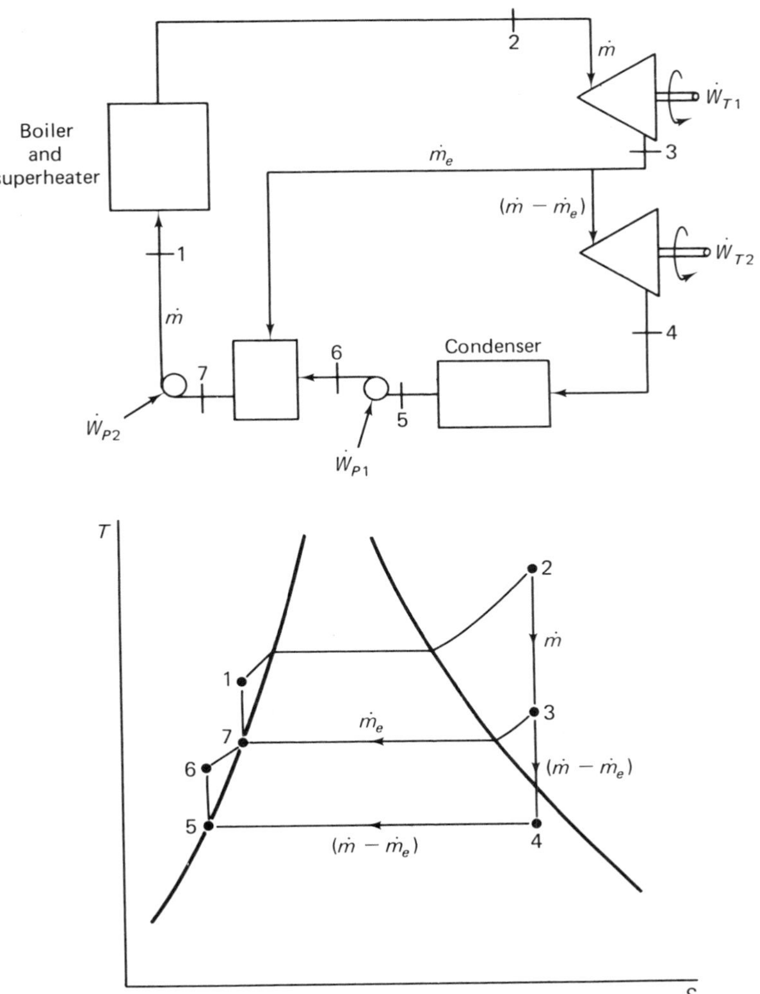

Figure 12.4 The regenerative cycle.

The feedwater heaters in regeneration cycles may be of either the open or the closed type. If open, then they are simply mixing tanks as shown:

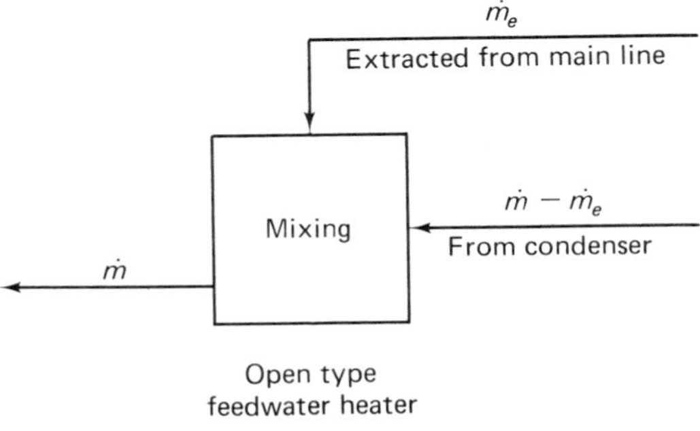

If closed, then they are simply heat exchangers:

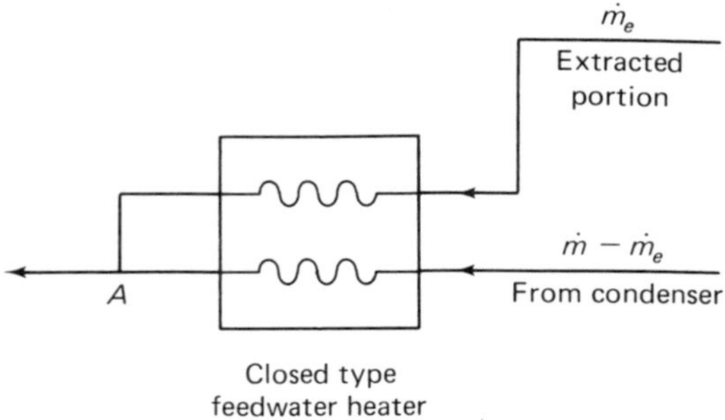

Closed type feedwater heater

Because the heat exchanger is not perfect, the temperatures of the extracted stream and of the main stream are not equal when they are mixed at point *A*. The temperature difference is called the *terminal temperature difference* and it must be known or estimated in order to do an analysis of the closed heater system. Though not shown, there is often a small pump needed to force the extracted steam condensate into the main stream at point *A*.

EXAMPLE 12.3 An internally reversible Rankine cycle with regeneration operates between 30 bars absolute and 0.1 bar absolute. Steam enters the turbine at 300°C. Steam is extracted at 3.0 bars absolute for use in an open feedwater heater. Compute the cycle efficiency.

SOLUTION:

Translation:

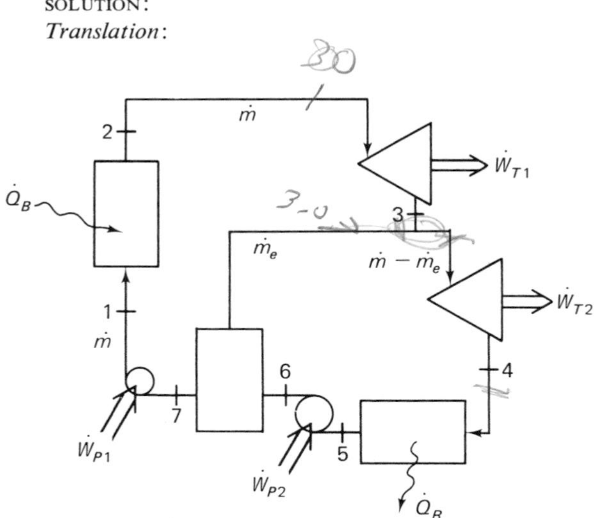

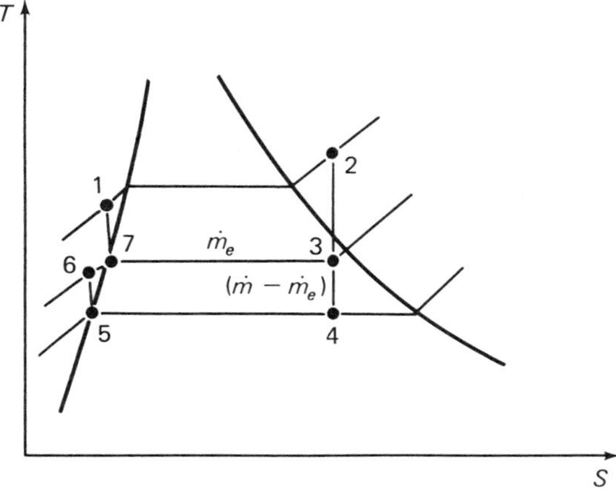

For efficiency:

$$\eta = \frac{\dot{W}_{T1} + \dot{W}_{T2} - \dot{W}_{P1} - \dot{W}_{P2}}{\dot{Q}_B}$$

$$\eta = \frac{\dot{m}(h_2 - h_3) + (\dot{m} - \dot{m}_e)(h_3 - h_4) - \dot{m}(h_1 - h_7) - (\dot{m} - \dot{m}_e)(h_6 - h_5)}{\dot{m}(h_2 - h_1)}$$

$$\eta = \frac{(h_2 - h_3) + y(h_3 - h_4) - (h_1 - h_7) - y(h_6 - h_5)}{(h_2 - h_1)}$$

where

$$y = \frac{\dot{m} - \dot{m}_e}{\dot{m}}$$

To find y:

Apply the steady flow energy equation to the open feedwater heater, assuming K.E. and P.E. terms are negligible:

$$\sum (\dot{m}h)_{\text{in}} = \sum (\dot{m}h)_{\text{out}}$$

$$(\dot{m} - \dot{m}_e)h_6 + \dot{m}_e h_3 = \dot{m}h_7$$

$$\frac{\dot{m}_e}{\dot{m}} = \frac{h_7 - h_6}{h_3 - h_6}$$

From the steam tables,

$$h_7 = h_{f_7} = 561.5 \text{ j/gm}$$

From Example 12.2,

$$h_3 = 2542 \text{ j/gm}$$

To find h_6 and $\dot{W}_{P_2}$:

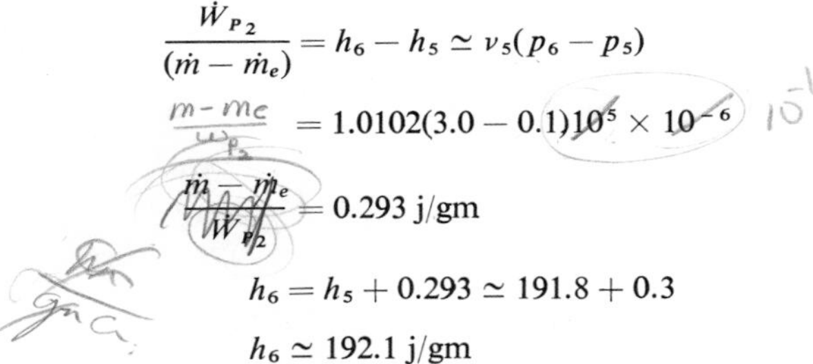

$$\frac{\dot{W}_{P2}}{(\dot{m}-\dot{m}_e)} = h_6 - h_5 \simeq v_5(p_6 - p_5)$$

$$= 1.0102(3.0 - 0.1)10^5 \times 10^{-6}$$

$$\frac{\dot{m}-\dot{m}_e}{\dot{W}_{P2}} = 0.293 \text{ j/gm}$$

$$h_6 = h_5 + 0.293 \simeq 191.8 + 0.3$$

$$h_6 \simeq 192.1 \text{ j/gm}$$

Thus,

$$\frac{\dot{m}_e}{\dot{m}} = \frac{561.5 - 192.1}{2542 - 192.1} = 0.157$$

and

$$y = 1 - \frac{\dot{m}_e}{\dot{m}} = 0.843$$

To find h_1 and $\dot{W}_{P_1}$:

$$\frac{\dot{W}_{P1}}{\dot{m}} = h_1 - h_7 \simeq v_7(p_1 - p_7)$$

$$\frac{\dot{W}_{P1}}{\dot{m}} = (1.0732)(27)(10^{-1}) = 2.9 \text{ j/gm}$$

$$h_1 = h_7 + 2.9 = 561.5 + 2.9 = 546.4 \text{ j/gm}$$

To find h_2:
From the steam tables,

$$h_2 = 2993.5 \text{ j/gm}$$

To find h_4:

$$s_4 = s_2 = s_{f_4} + x_4 s_{fg_4}$$

$$6.5390 = 0.6493 + x_4(7.5009)$$

$$x_4 = 0.785$$

$$h_4 = h_{f_4} + x_4(h_{fg_4})$$

$$h_y = 191.8 + 0.785(2392.8)$$

$$h_4 = 2070 \text{ j/gm}$$

To find efficiency:

$$\eta = \frac{(h_2 - h_3) + y(h_3 - h_4) - (h_1 - h_7) - y(h_6 - h_5)}{(h_2 - h_1)}$$

$$\eta = \frac{(2993.5 - 2542) + 0.843(2542 - 2070) - 2.9 - (0.843)(0.293)}{(2993.5 - 564.4)}$$

$$\eta = 0.348$$

We have shown only the basic reheat and regenerative concepts. In practice, two or more reheat stages and several regenerative extractions may be used. Also, both reheat and regeneration may be used in the same cycle. Thus many combinations and cycle variations are possible.

EXAMPLE 12.4. An internally reversible Rankine cycle with both reheat and regeneration operates between 600°F and 500 psia and 1 psia. The first-stage turbine exhausts at 100 psia where part of the steam is extracted for use in an open feedwater heater. The remainder is reheated to 600°F before entering the second stage. Saturated liquid leaves the feedwater heater. Compute the cycle efficiency.

SOLUTION:
Translation:

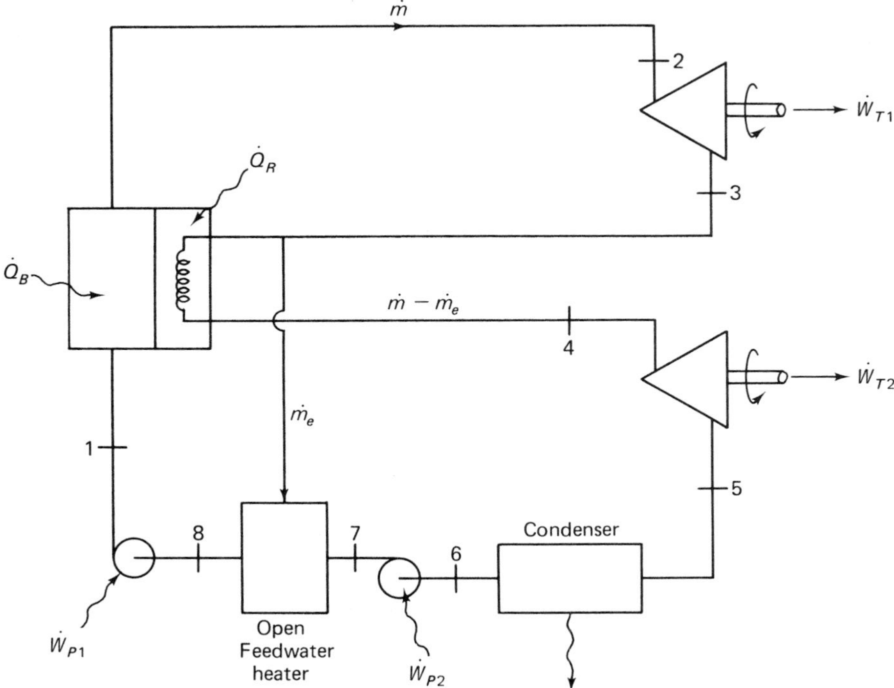

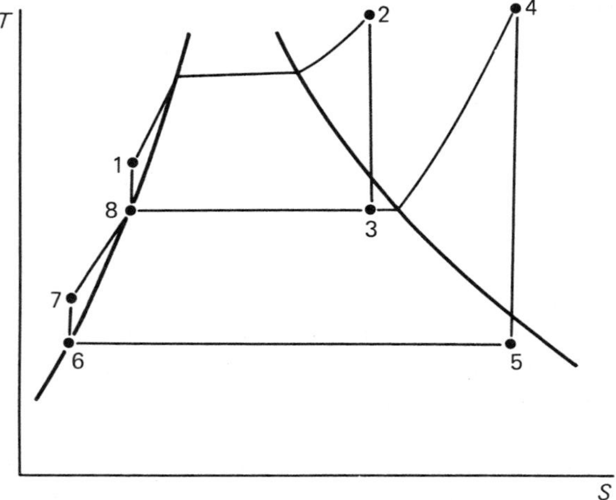

For efficiency:

$$\eta = \frac{\dot{W}_{T_1} + \dot{W}_{T_2} - \dot{W}_{P_1} - \dot{W}_{P_2}}{\dot{Q}_B + \dot{Q}_R}$$

$$\eta = \frac{\dot{m}(h_2 - h_3) + (\dot{m} - \dot{m}_e)(h_4 - h_5) - \dot{m}(h_1 - h_8) - (\dot{m} - \dot{m}_e)(h_7 - h_6)}{\dot{m}(h_2 - h_1) + (\dot{m} - \dot{m}_e)(h_4 - h_3)}$$

To find $(\dot{m} - \dot{m}_e)/\dot{m}$, we write an energy balance on the feedwater heater assuming it is adiabatic and neglecting K.E. and P.E. terms.

$$\dot{m}_e h_3 + (\dot{m} - \dot{m}_e)h_7 = \dot{m}h_8$$

$$\frac{\dot{m}_e}{\dot{m}} h_3 + h_7 - \frac{\dot{m}_e}{\dot{m}} h_7 = h_8$$

$$\frac{\dot{m}_e}{\dot{m}} = \frac{h_8 - h_7}{h_3 - h_7}$$

At state 2:

$$h_2 = 1298.3 \text{ Btu/lbm}$$

$$s_2 = 1.5585 \text{ Btu/lbm-°R}$$

To find h_3:

$$s_2 = s_3 = s_{f3} + x_3 s_{fg3}$$

$$1.5585 = 0.47439 + x_3(1.1290)$$

$$x_3 = 0.960$$

$$h_3 = h_{f3} + x_3 h_{fg3}$$

$$h_3 = 298.6 + 0.960(889.2)$$

$$h_3 = 1152 \text{ Btu/lbm}$$

At state 4:

$$h_4 = 1329.3 \text{ Btu/lbm}$$
$$s_4 = 1.7582 \text{ Btu/lbm-°R}$$

To find h_5:

$$s_4 = s_5 = s_{f5} + x_5 s_{fg5}$$
$$1.7582 = 0.13266 + x_5(1.8453)$$
$$x_5 = 0.881$$
$$h_5 = h_{f5} + x_5 h_{fg5}$$
$$h_5 = 69.7 + 0.881(1036.0)$$
$$h_5 = 982 \text{ Btu/lbm}$$

At state 6:

$$h_6 = h_{f6} = 69.7 \text{ Btu/lbm}$$

For $\dot{W}_{P_2}/\dot{m}_e$:

$$\frac{\dot{W}_{P2}}{\dot{m}} = h_7 - h_6 \simeq v_6(p_7 - p_6)$$
$$\frac{\dot{W}_{P2}}{\dot{m}} = 0.016(99)\frac{144}{778}$$
$$\frac{\dot{W}_{P2}}{\dot{m}} = 0.293 \text{ Btu/lbm}$$

For h_7:

$$h_7 = h_6 + \frac{\dot{W}_{P_2}}{\dot{m}} = 69.7 + 0.3 = 70.0 \text{ Btu/lbm}$$

For $\dot{W}_{P_1}/\dot{m}$:

$$\frac{\dot{W}_{P1}}{\dot{m}} = h_1 - h_8 \simeq v_8(p_1 - p_8)$$
$$\frac{\dot{W}_{P1}}{\dot{m}} = 0.0177(400)\frac{144}{778} = 1.31 \text{ Btu/lbm}$$

For h_1:

$$h_1 = h_8 + \frac{\dot{W}_{P_1}}{\dot{m}} = 298.61 + 1.3 = 299.9 \text{ Btu/lbm}$$

For $\dot{m}_e/\dot{m}$:

$$\frac{\dot{m}_e}{\dot{m}} = \frac{h_8 - h_7}{h_3 - h_7} = \frac{298.6 - 70.0}{1152 - 70.0}$$

$$\frac{\dot{m}_e}{\dot{m}} = 0.211$$

For efficiency:

$$\eta = \frac{(h_2 - h_3) + y(h_4 - h_5) - (h_1 - h_8) - y(h_7 - h_6)}{(h_2 - h_1) + y(h_4 - h_3)}$$

where

$$y = \frac{\dot{m} - \dot{m}_e}{\dot{m}} = 1 - \frac{\dot{m}_e}{\dot{m}} = 1 - 0.211 = 0.789$$

$$\eta = \frac{(1298.3 - 1152) + 0.789(1329.3 - 982) - 1.3 - (0.789)(0.29)}{(1298.3 - 299.9) + 0.789(1329.3 - 1152)}$$

$$\eta = 0.368$$

The student should be aware that not all problems in cycle analysis can be solved directly as in the preceding examples. Many are optimization problems involving trial and error or iterative solutions. For example, in the design of a simple reheat cycle, how does one decide on the intermediate pressure? Within the existing practical constraints, the thermodynamic answer is the pressure that results in the best efficiency. But to determine that pressure is a trial and error process. We must assume a pressure, compute the resulting efficiency, and repeat until the optimum is found. In all trial and error solutions, the most efficient way is to begin by getting the solution bounded; that is, first select a pressure that you are reasonably sure is too high and then one that is too low. Make a plot of the results, η versus intermediate pressure. Next select an intermediate pressure and recompute η; enter the results on your plot; and so on. You will be surprised how quickly you can determine the optimum by this technique.

In Example 12.5, this procedure was followed. The details of each calculation of η are omitted because they are analogous to those of Example 12.2. The results of the trial and error process are shown.

EXAMPLE 12.5. An internally reversible Rankine cycle with reheat is proposed to operate with a boiler pressure of 500 psia, a maximum temperature of 1000°F, and a condenser pressure of 1 psia. Recommend the pressure for the exhaust of the first-stage turbine.

SOLUTION:
Translation:

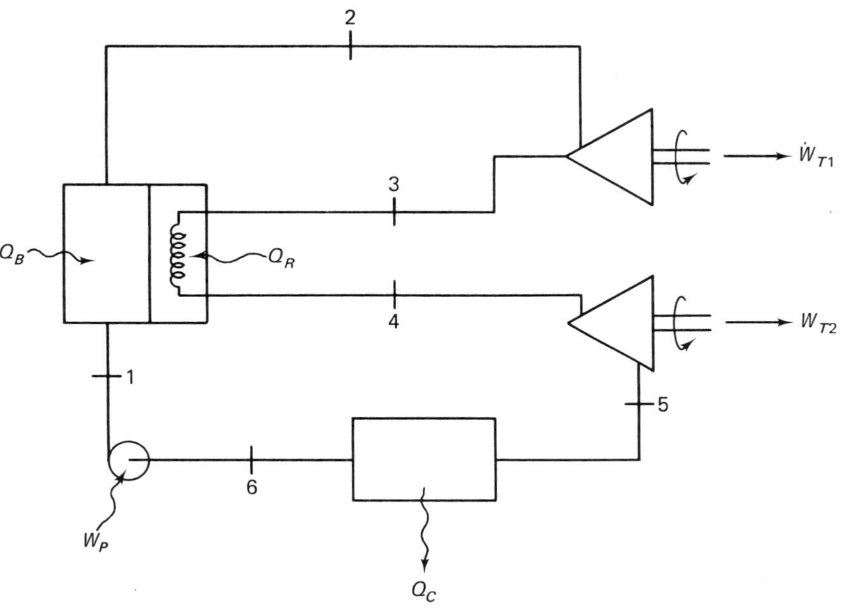

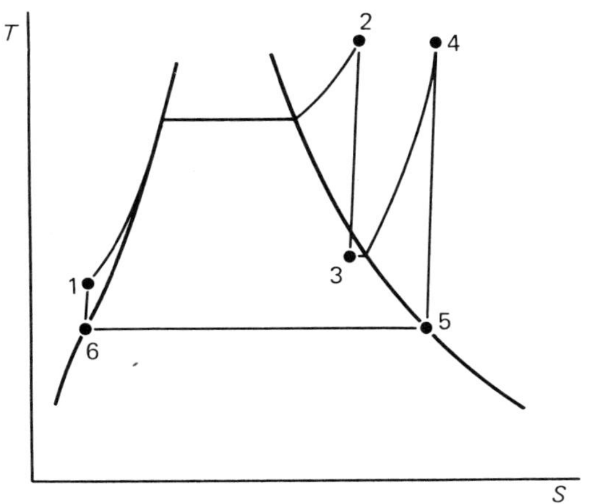

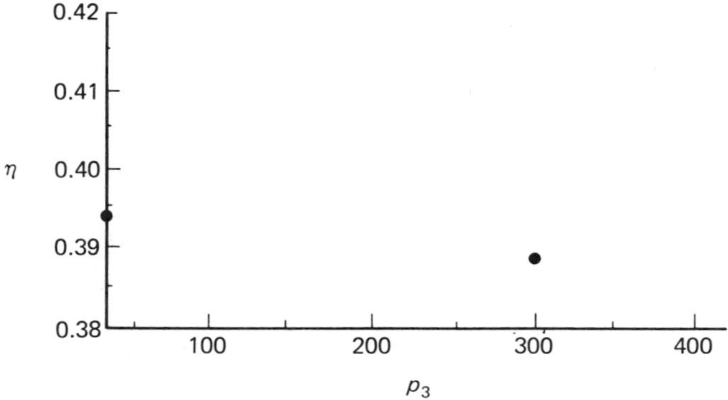

As a first step, assume $p_3 = 300$ psia. Computing efficiency gives $\eta_{300} = 0.389$. Next assume $p_3 = 25$ psia. This gives $\eta = 0.394$. Plotting p_3 versus η:
Next assume $p_3 = 100$ psia. This gives $\eta = 0.40$. Assuming $p_3 = 50$ psia also gives $\eta = 0.40$. Students should enter these points on the plot above and estimate the optimum pressure.

12-4 Deviations from the ideal

The components in the cycles discussed so far in the chapter have been internally reversible. In reality, of course, they are not. We shall discuss only inefficiencies in turbines and pumps here but the student should recognize that piping pressure drops and heat loses from piping and components also cause reduced efficiency.

Turbine efficiency is defined as the ratio of the actual work obtained to the work that would be obtained in a reversible adiabatic turbine with the same inlet conditions exhausting to the same pressure. For an adiabatic but irreversible turbine, the process line is as shown in the sketch:

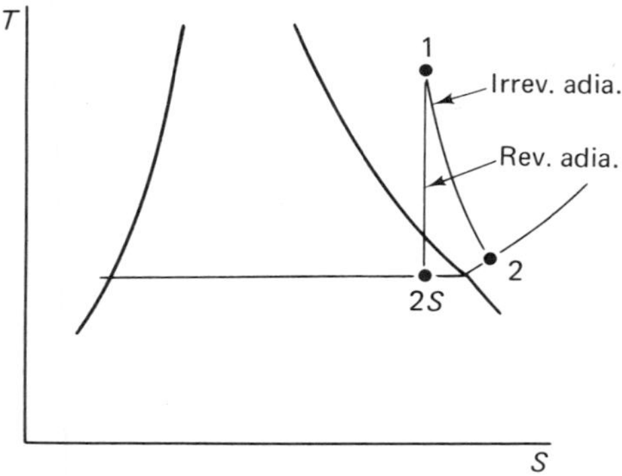

The efficiency is therefore

$$\eta_T = \frac{W_{\text{actual}}}{W_{\substack{\text{reversible}\\\text{adiabatic}}}} = \frac{h_1 - h_2}{h_1 - h_{2S}} \tag{12-13}$$

The subscript ($2S$) is used to denote the end state of the ideal process because it will be a constant entropy (S) process. (Why does the real, but adiabatic, process lead to a state 2 with *greater* entropy?) States 2 and $2S$ have the same pressure.

Real turbine efficiencies are of the order of 50 to 95% and large power plant turbines typically operate in the 85 to 95% range.

Pump efficiency is defined as the ratio of the reversible adiabatic work to the actual work.

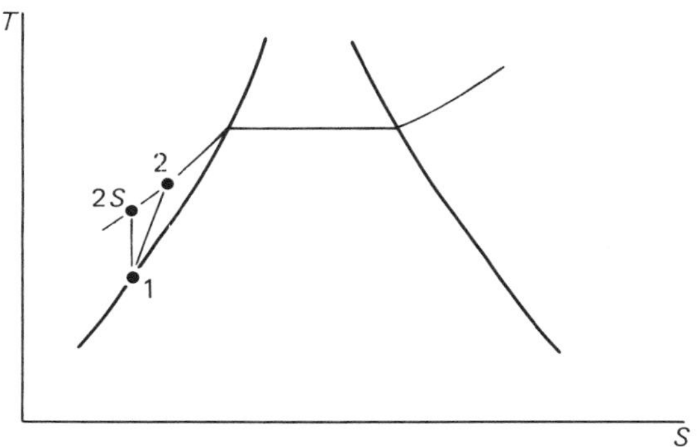

$$\eta_P = \frac{W_{\substack{\text{reversible}\\\text{adiabatic}}}}{W_{\text{actual}}} = \frac{h_{2S} - h_1}{h_2 - h_1} \tag{12-14}$$

With these inefficiencies introduced, an adiabatic Rankine cycle is as shown in Fig. 12.5.

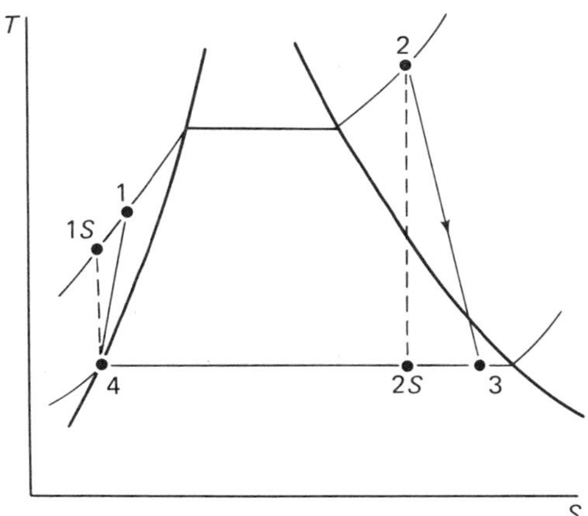

Figure 12.5 Rankine cycle inefficiencies.

EXAMPLE 12.6. Consider the cycle given in Example 12.1, but assume turbine and pump efficiency each to be 0.90. Recompute the cycle efficiency.

SOLUTION:

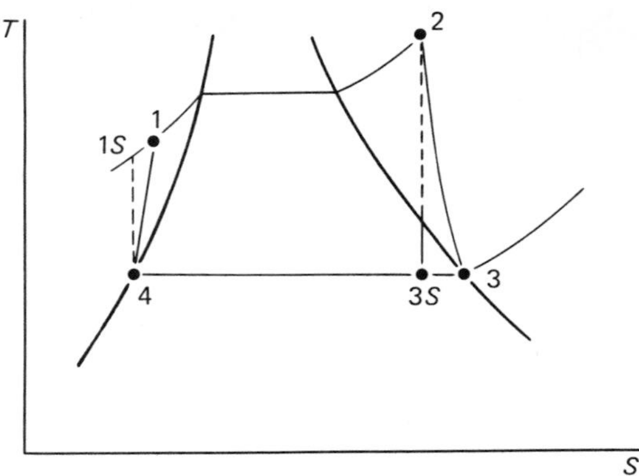

From Example 12.1:

$$W_{\text{turbine}|\text{ideal}} = 428 \text{ Btu/lbm}$$
$$W_{\text{pump}|\text{ideal}} = +1.5 \text{ Btu/lbm}$$
$$h_2 = 1298.3 \text{ Btu/lbm}$$
$$h_4 = 69.74 \text{ Btu/lbm}$$

Actual turbine work:

$$\eta_{\text{turbine}} = \frac{W_{\substack{\text{turbine}\\ \text{actual}}}}{W_{\substack{\text{turbine}\\ \text{ideal}}}}$$

$$\therefore \quad W_{\text{turbine}|\text{actual}} = (0.90)(428 \text{ Btu/lbm}) = 386 \text{ Btu/lbm}$$

Actual pump work:

$$\eta_{\text{pump}} = \frac{W_{\text{pump}|\text{ideal}}}{W_{\text{pump}|\text{actual}}}$$

$$\therefore \quad W_{\text{pump}|\text{actual}} = +\frac{1.5 \text{ Btu/lbm}}{0.90} = +1.66 \text{ Btu/lbm}$$

Actual inlet condition to boiler:

$$-W_{\text{pump}|\text{actual}} = h_1 - h_4$$
$$\therefore \quad h_1 = 1.66 + 69.74 = 71.40 \text{ Btu/lbm}$$

Actual heat added in boiler and superheater:

$$Q_B = h_2 - h_1 = 1298.3 \text{ Btu/lbm} - 71.4 \text{ Btu/lbm} = 1226.9 \text{ Btu/lbm}$$

Efficiency:

$$\eta = \frac{W_{\text{net}}}{Q_{\text{added}}} = \frac{386 - 1.66}{1226.9} = 0.313$$

12-5 Vapor compression refrigeration cycles

The vapor compression refrigeration cycle is used in almost all modern refrigerators and air conditioners. The basic cycle is as shown in Fig. 12.6. The working fluid or refrigerant is usually one of the various Freons but it may also be ammonia.

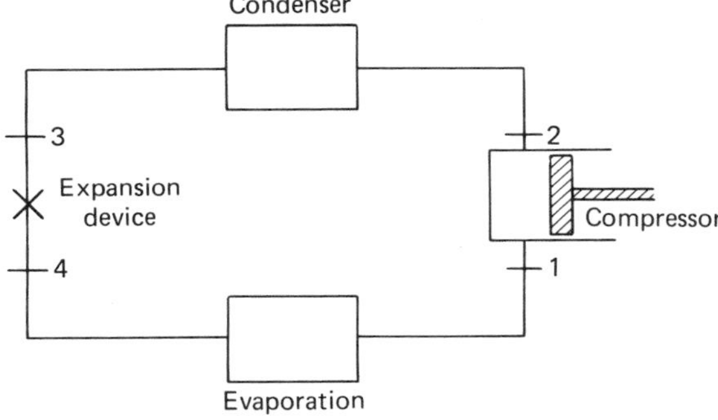

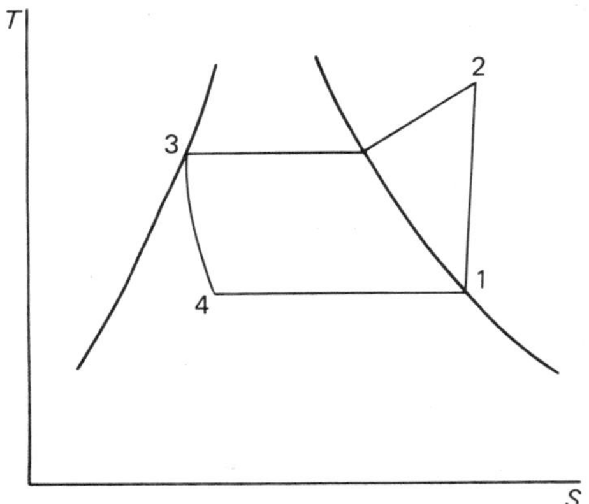

Figure 12.6 Vapour compression refrigeration cycle.

The steps in the basic cycle are the following:

State change	*Equipment*	*Process*
4–1	Evaporator	Heat is transferred from refrigerated space to Freon, bringing to saturated vapor state
1–2	Compressor	Vapor is compressed to high pressure and temperature
2–3	Condenser	High pressure vapor is condensed to saturated liquid by heat removal
3–4	Valve or capillary tube	Saturated liquid expands irreversibly to low evaporator pressure into the two-phase region

The *T-s* diagram in Fig. 12.6 shows an idealized cycle. The processes shown are internally reversible except for the throttling process 3-4. It would be more efficient thermodynamically to replace this highly irreversible process with an expansion engine but in practice such engines are expensive to maintain and the work produced is not readily usable so this is almost never done.

The coefficient of performance of the cycle shown is given by

$$\text{COP} = \frac{\dot{Q}_E}{\dot{W}_C} = \frac{h_1 - h_4}{h_2 - h_1} \tag{12-15}$$

or

$$\text{COP} = \frac{\dot{Q}_E}{\dot{Q}_C - \dot{Q}_E} = \frac{(h_1 - h_4)}{(h_2 - h_3) - (h_1 - h_4)} \tag{12-16}$$

Refrigeration capacities are usually referred to in *tons*. A ton of refrigeration is a rate of heat transfer sufficient to melt 1 ton of ice in 24 hr. Taking the latent heat of fusion to be 144 Btu/lbm,

$$1 \text{ ton of refrigeration} = \frac{(2000)(144)}{(24)} = 12{,}000 \text{ Btu/hr} \tag{12-17}$$

Of course, compressors in refrigeration are neither internally reversible nor ideal. In practice, the actual compression may look something like shown in Fig. 12.7. The heat transfer will usually be from the refrigerant to the surroundings and is not a bad thing because it reduces the heat that must be removed anyway in the condenser.

Other deviations from ideal included piping and heat exchanger pressure drops, heat losses or gains, etc. In addition, the state of the refrigerant leaving the evaporator and condenser is not precisely saturated but includes some superheat and subcooling, respectively. The result is an actual cycle more or less as shown in Fig. 12.8.

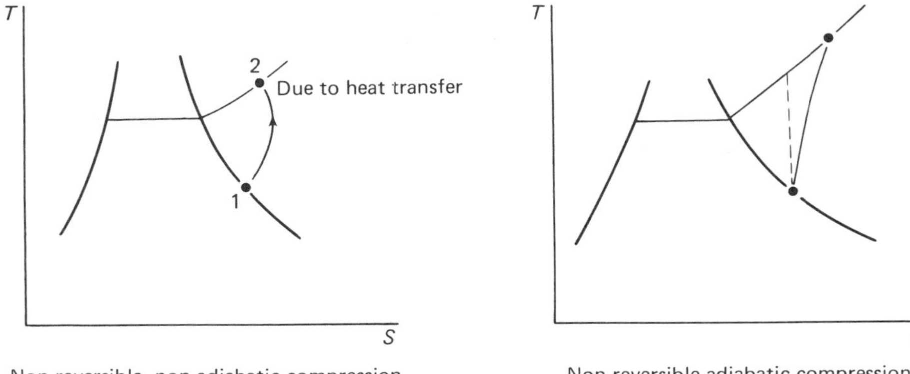

Figure 12.7 Inefficiencies in the vapor compression cycle.

For special applications, the basic vapor compression refrigerator cycle is often modified by use of two or more stages of compression, bypasses, and extractions of various kinds.

Refrigeration systems that use heat rather than work as input are often more economical if an inexpensive source of heat is readily available. Sometimes, for example, unused steam generating capacity is available at low cost in the summer months and can be used to operate an air-conditioning system. An absorption refrigeration system can be used in such cases instead of the vapor compression cycle.

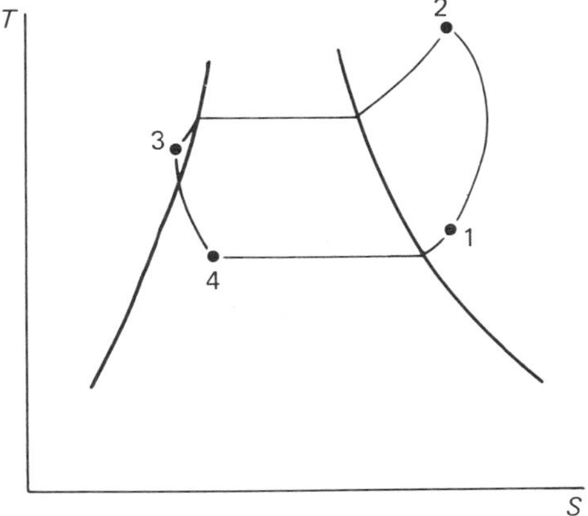

Figure 12.8 The effect of heat transfer on compression process in vapor compression cycle.

EXAMPLE 12.7. An ideal vapor compression refrigeration cycle using Freon-12 operates with an evaporator saturation temperature of 0°F and a condensor pressure of 100 psia. Compute the required compressor compression ratio and the coefficient of performance.

SOLUTION:
Translation:

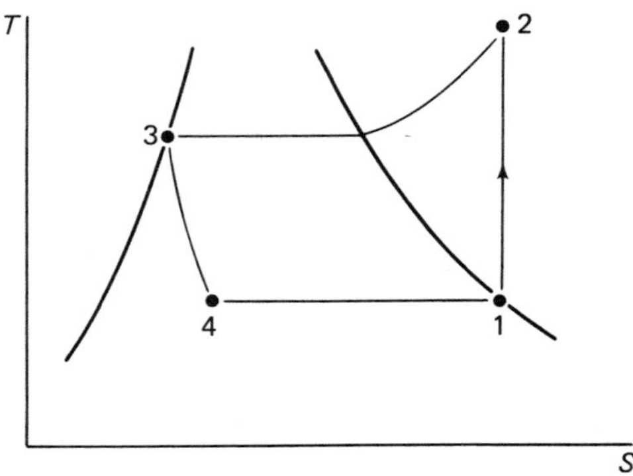

Equation of state:
From tables in Appendix D.
For the ideal cycle, process 1-2 is reversible, adiabatic so $S_1 = S_2$.
From the tables, $S_1 = S_2 = 0.16888$ Btu/lbm-°R.
To find h_2, we know that $p_2 = 100.0$ psia and $S_2 = 0.16888$ Btu/lbm-°R.
From the superheated table by interpolation,

$$h_2 = 88.11 \text{ Btu/lbm}$$

Thus

$$\text{COP} = \frac{77.27 - 26.54}{88.11 - 77.27} = \frac{50.73}{10.84} = 4.68$$

For pressure ratio: Saturation p at 0°F = 23.849 psia.

$$\text{Ratio} = \frac{100.0}{23.85} = 4.19$$

For COP:

$$\text{COP} = \frac{Q_{\text{pumped}}}{W_{\text{net}}} = \frac{h_1 - h_4}{h_2 - h_1}$$

Energy balance on expansion device gives

$$h_4 = h_3$$

Thus

$$\text{COP} = \frac{h_1 - h_3}{h_2 - h_1}$$

From the tables:

$$h_1 = 77.27 \text{ Btu/lbm}$$
$$h_3 = h_4 = 26.541 \text{ Btu/lbm}$$

12-6 Cycles and the second law

In Chap. 2 we introduced the concept of degradation by noting that the maximum possible work obtainable from a quantity of heat Q at a temperature T_1 with an available sink at T_0 is

$$W_{\substack{\text{maximum}\\\text{possible}}} = Q\left(1 - \frac{T_0}{T_1}\right) \tag{12-18}$$

Thus if the actual work obtained is W_{actual}, then the degradation is

$$D = W_{\substack{\text{maximum}\\\text{possible}}} - W_{\text{actual}} \tag{12-19}$$

Such an overall calculation is interesting conceptually but has little practical usefulness. First, most cycles do not have heat input or rejection at a single temperature. In the Rankine cycle in Fig. 12.2, for example, heat is added at temperatures from t_1 to t_2. But, more important, the overall degradation does not tell us which processes in the cycle are causing the most serious degradation. Cycle efficiency is an indicator of the amount of degradation, too, but it doesn't tell us how to improve the cycle. Similarly, total cycle degradation does not tell us where to put our efforts in order to reduce the waste of useful energy. From Eq. (12-18), we did learn that we should attempt to raise the average temperature at which heat is added and to lower the average temperature at which heat is rejected. But to learn more we must compute the degradation in the various processes in the cycle and then put our effort into reducing those that can be reduced at appropriate costs.

Engineers should adopt the habit of looking for such payoffs. Where are the major sources of degradation (for example, high ΔT heat transfers, high ΔT mixing, excessive friction, etc.)? Which of these can be reduced at worthwhile cost? With energy costs increasing, the years ahead will see many new opportunities to reduce degradation that, though acceptable at today's energy prices, will be worth improving in the future. Look for and compute the degradations as shown in Chap. 11. Compute the payoff; that is, how much energy can you save for how many dollars?

12-7 Summary

The basic Rankine superheat cycle consists of heat addition at high pressure to superheated steam, expansion in a turbine to low pressure and high quality, condensing to a liquid, and pumping. To improve the efficiency—that is, to raise the average temperature at which heat is added and/or to lower the average temperature at which heat is rejected—the basic cycle is often modified with reheat or regeneration.

Reheating utilizes either two turbines or two stages of a single turbine. Steam is taken from the exhaust of the first turbine or stage and reheated in a section of the boiler. It is then delivered to the second turbine or stage.

Regeneration involves extracting some steam between stages and using that steam to heat the condensed liquid being supplied to the boiler. Regeneration and reheat in several stages and combinations are often combined.

Internally reversible cycles, involving reversible adiabatic ($s = \text{constant}$) turbine and pump processes, of course do not exist in reality. There are frictional losses and heat transfers. These must be taken into account in design and analysis. The frictional losses in turbines and pumps are usually handled by use of an efficiency defined for a turbine as

$$\eta = \frac{\text{work actually obtained}}{\text{work if reversible adiabatic from actual inlet to outlet pressure}}$$

In a vapor compression refrigeration cycle, the refrigerant receives heat and is boiled in the evaporator; then the vapor is compressed and the resulting high pressure, high temperature gas is cooled and condensed. Finally, the condensed liquid is expanded through a valve or capillary tube to a low pressure, low temperature liquid–vapor mixture ready to receive heat again.

Cycle design and analysis require students familiar with schematic cycle, T-s, and h-s (Mollier) diagrams. It is nearly always essential to sketch the layout of any cycle and draw it on a skeleton T-s or h-s diagram *before* starting an analysis. Good cycle designers and analysts will also be aware of the sources and causes of degradation in a cycle. These are the places where thermodynamic improvements can be made and very often thermodynamic improvements will result in economic and environmental improvements as well.

Problems

12-1 Without looking in the book or at your notes, sketch schematic diagrams and T-s diagrams (labeling all states) for the following cycles:

(a) An internally reversible Rankine cycle.

(b) An internally reversible Rankine cycle with reheat.

(c) An internally reversible Rankine cycle with regeneration and open-type feedwater heater.
(d) Repeat part c using a closed feedwater heater.
(e) An internally reversible Rankine cycle with two stages of reheat.
(f) Repeat all the above with adiabatic turbines that have $\eta < 1.0$.
(g) An internally reversible vapor compression refrigeration cycle.

12-2 The exhaust from the turbine in a Rankine cycle is mostly vapor. We want vapor to supply to the turbine. Why, then, do we go to the trouble of condensing the steam and then reboiling it? Why not just pump it back to the high pressure, add the little heat needed, and send it back to the turbine?

12-3 If reversible pump work is $\int v\,dp$, then the order of magnitude of v will be important in determining how much pump work must be done.
(a) At 1 psia, compare v_g with v_f. Would pump work still be negligible if we pumped the vapor?
(b) Make some comparison at 0.060 bars.

12-4 An internally reversible Rankine cycle with reheat operates between 500 and 1 psia. Steam enters the first-stage turbine at 600°F. Steam leaves the first-stage turbine at 100 psia and is reheated to 600°F. Compute the cycle efficiency.

12-5 An internally reversible Rankine cycle with regeneration operates between 600 psia absolute and 1 psia. Steam enters the turbine at 600°F. Steam is extracted at 100 psia for use in an open feedwater heater. Compute the cycle efficiency.

12-6 An internally reversible Rankine steam cycle operates at boiler and condenser pressures of 30 bars absolute and 0.1 bar absolute, respectively. The temperature of steam entering the turbine is 300°C. Compute the cycle efficiency and the mass of H_2O that must be circulated per kilowatt-hour.

12-7 Repeat Prob. 12-6 but assume turbines have 0.90 efficiency and pump has 0.80 efficiency.

12-8 An internally reversible Rankine cycle with reheat operates between 600 and 2 psia. Steam enters the first-stage turbine at 700°F. The intermediate pressure is 60 psia. The steam is reheated to 500°F. Compute the cycle efficiency and the pounds of steam that must be circulated per kilowatt-hour.

12-9 Repeat Prob. 12-8 but with turbine and pump efficiencies of 0.85.

12-10 An internally reversible Rankine cycle with regeneration operates between 600 and 2 psia. Steam is extracted at 60 psia for use in an open feedwater heater. Compute the cycle efficiency and the circulation of H_2O required per kilowatt-hour. The extraction steam temperature is 700°F.

12-11 Repeat Prob. 12-10 but with turbine and pump efficiencies of 0.92.

12-12 An internally reversible Rankine cycle has two stages of reheat. Steam enters the first turbine at 40 bars absolute and 300°C; the second turbine, at 10 bars absolute and 250°C; the third turbine, at 0.50 bar absolute and 150°C. The condenser pressure is 0.01 bar absolute. Compute the cycle efficiency.

12-13 An internally reversible Rankine cycle with both reheat and regeneration operates between 300°C and 40 bars absolute and 0.01 bar absolute. The first-stage turbine exhausts at 1 bar absolute where part of the steam is extracted for use in an open feedwater heater. The remainder is reheated to 150°C before entering the second stage. Saturated liquid leaves the feedwater heater. Compute the cycle efficiency.

12-14 About how many tons of coal would you estimate are required per year to produce 1 million kw-hr of electricity? For the same output, how much cooling water is needed for the condenser if the temperature of the water is raised 15°F? How large a population will 1 million kw-hr/year service?

12-15 Repeat Example 12.2 but assume turbines have 0.85 efficiency and pump has 0.90 efficiency.

12-16 Repeat Example 12.3 but assume turbines have 0.92 efficiency and pump has 0.80 efficiency.

12-17 Repeat Example 12.4 but assume pumps and turbines have 0.90 efficiency.

12-18 Repeat Example 12.7 but assume the compressor is adiabatic with an efficiency of 0.75.

12-19 Repeat Example 12.5 but assume the turbine efficiencies are 0.80.

12-20 An ideal vapor compression cycle operates on Freon-12 with an evaporation temperature of −10°F and a compressor discharge pressure of 90 psia. Compute the coefficient of performance and the horsepower required per ton of refrigeration.

12-21 Repeat Prob. 12-20 but with an adiabatic compressor efficiency of 0.75.

12-22 At your library, find out how an ammonia absorption refrigeration system works.

12-23 In a refrigeration system, two steady flow streams of Freon-12 are to be simultaneously mixed and heated in an insulated tank. The heating is accomplished by passing air through pipes that pass through the tank. One Freon stream has a flow rate of 100 lbm/min and enters as saturated liquid at 12 psia. The other Freon flow rate is 50 lbm/min and it enters at 12 psia as saturated vapor. The Freon mixture leaves the tank at 5 psia and 20°F. The air enters at 80°F and leaves at 50°F.

(a) Compute the required air flow rate.

(b) Show that the device above satisfies the second law.

For air, $c_p = 0.24$, $c_v = 0.17$ Btu/lbm-°R.

12-24 The shaft of a steam turbine is connected to an air compressor. 100 lb/hr of steam is supplied to the turbine at 200 psi and 500°F. It leaves the turbine at 20 psi. The turbine efficiency is 0.90. Air enters the compressor at 15 psi and 70°F and is pumped isothermally at a steady flow rate to a pressure of 150 psi. What is the maximum flow rate of air that can be delivered? State assumptions.

12-25 A refrigeration cycle using a *flash intercooler* is sketched below. The flash intercooler is simply a liquid–vapor separator tank with no moving parts. The refrigerant is Freon-12. The evaporator pressure is 20 psia, the intercooler pressure is 60 psia, and the condenser pressure is 200 psia. Neglect inefficiences. Saturated vapor leaves the evaporator and saturated liquid leaves the condenser.

Compute the coefficient of performance.

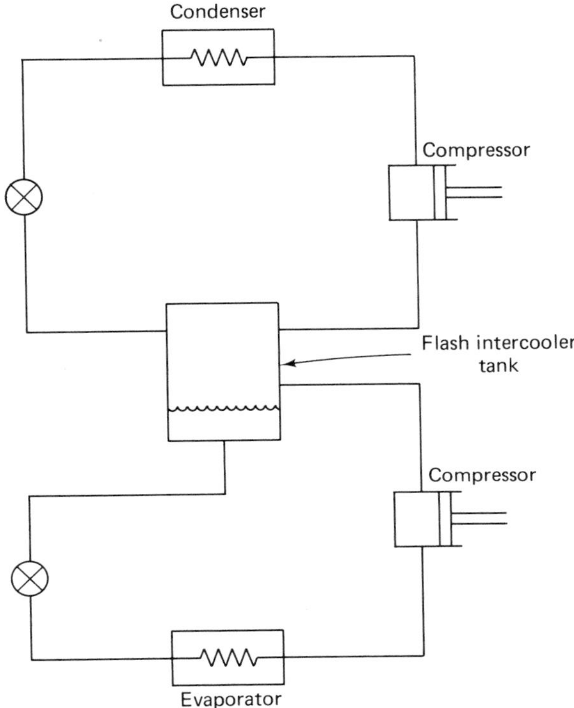

12-26 Steam enters a turbine at 500 psia, 500°F, and a velocity of 600 ft/sec. The process is isentropic and the exit velocity is 200 ft/sec. Calculate the work output of the turbine per pound mass of steam for discharge at 1 psia. What would be the turbine efficiency if the exit condition was 1 psia saturated vapor and the same velocities as above?

12-27 The Mt. Tom power plant in Holyoke, Mass., produces electricity at the rate of 140,000 kw. The heat input to the boiler is 1.32×10^9 Btu/hr. The temperature of the steam entering the turbine is 1000°F and the temperature in the condenser is 100°F. The condenser is cooled by the Connecticut River.

(a) What is the efficiency of this power plant?
(b) If the power plant were reversible, with all heat added and rejected at 1000°F and 100°F, respectively, what fraction of the present coal requirement would be needed for the same power output?
(c) What river flow rate is needed if the temperature rise of the river is to be 10°F?

12-28 A heat pump is simply a vapor compression refrigeration cycle. Heat is pumped from a source, usually the earth or cool well water, and rejected from the condenser to heat the space.

(a) What measure of performance would you recommend for a heat pump?
(b) In terms of your measure, what is best possible performance for a heat pump operating between 40 and 70°F?
(c) What performance would you expect from a standard cycle using Freon-12 operating with the temperatures above?

12-29 Compute the degradation in the expansion device in the refrigeration cycle of Prob. 12-20.

12-30 Compute the degradation in the feedwater heater in Example 12-4.

12-31 Compute the percent error that would result from neglecting pump work in the examples in this chapter.

12-32 Obtain a large Mollier (h-s) chart from your instructor or your library. (One source is the *Steam Tables* by Keenan and Keyes or *Steam Tables* by Keenan, Keyes, Hill, and Moore.) Then rework the following, comparing the results with those obtained when tables are used:
(a) Example 12.1.
(b) Problem 12-4.
(c) Problem 12-5.

12-33 A refrigeration-type unit operates a heat pump using water as the fluid. Evaporator temperature is 40°F. Dry saturated fluid enters the compressor and is compressed to 1 psia. By means of jacket cooling, the vapor leaves the compressor at 260°F. Liquid is subcooled in the condenser to 90°F.
(a) Calculate the number of pounds per minute of fluid that must be circulated in order to provide sufficient heat rejected to meet heating requirements of 10,000 Btu/min in a room.
(b) If the overall compressor efficiency is 75%, what is the horsepower required to drive the compressor?
(c) Compute the coefficient of performance of the heat pump.

12-34 A power plant is proposed that is to combine features of reheat and regeneration cycles. See sketch. Compute the thermal efficiency.

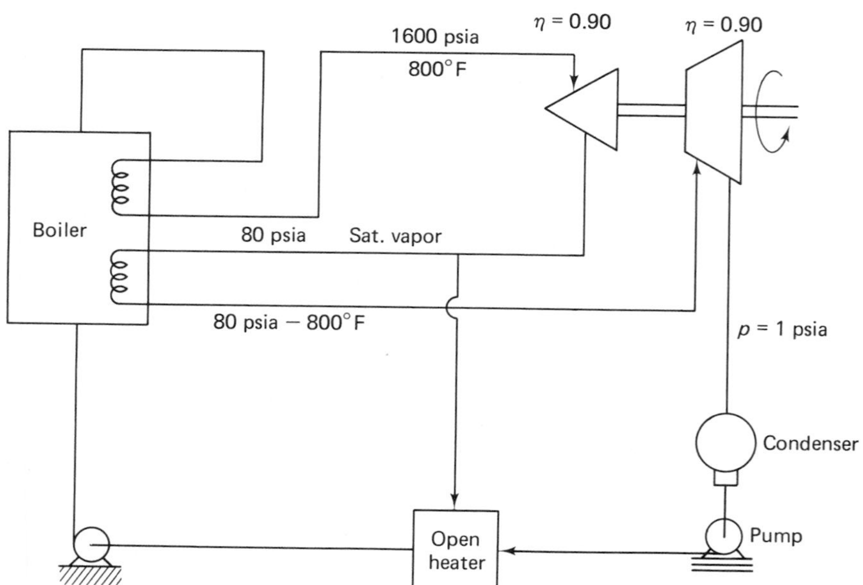

12-35 An ideal vapor compression refrigerating cycle operates between temperature limits of 40 and 100°F. Dry saturated vapor leaves the compressor and saturated liquid enters the expansion valve. For a refrigerating capacity of 5 tons, calculate the required power input if the refrigerant is Freon-12.

12-36 A household electric refrigerator is operated in a kitchen that is closed and thermally insulated. If it is operated continuously for 2 hr, will the average temperature of the air in the kitchen increase, decrease, or remain constant if the refrigerator is kept

(a) Open?

(b) Closed?

13

Gas power engines

13-1 Introduction

Gas engines burn fuel oil or gasoline to produce mechanical power. Because of their use in transportation, gas engines are of tremendous technical and social importance. As noted in Chap. 1, current automobiles using the Otto cycle (to be described in this chapter) are only about 20 to 25% efficient; yet they are major users of increasingly scarce fuel oil reserves. The diesel engines used in trucks, trains, buses, ships, etc., are better, about 30 to 35% efficient. Efficiency (and hence operating cost and conservation of resources) is not the only concern, however. These engines are also a major source of pollution from unburned hydrocarbons (HC), carbon monoxide (CO), and oxides of nitrogen (NO_x).

Gas engines are classified as internal or external combustion engines depending on whether or not the fuel is burned inside or outside the power producing chamber. Thus the Otto and diesel engines are internal combustion engines; the gas turbine and steam engine are external combustion engines. The pollution problems of internal combustion (IC) engines are directly related to the intermittent combustion, which limits the time available to complete combustion, requires cool walls to prevent preignition of new charges, etc. External combustion systems do better because the combustion is hot and continuous and the chamber is designed to do only one thing: burn fuel.

Therefore one avenue being put forward to help with the automotive emission problem is to convert to external combustor systems. Steam cars and gas turbines are the leading alternatives being explored but without much interest as yet on the part of either the manufacturers or the consumers of the typical automobile.

On the other hand, there is great interest in an alternative internal combustion engine called the *Wankel* (after its inventor) *engine*. (Wankel is pronounced *Vahnkle*.) It is a *rotary* internal combustion engine (described later in this chapter).

The working fluid in gas engines does not operate on a cycle in the same sense as the H_2O in, say, a Rankine cycle. Fuel and oxygen enter, are burned, used to produce power, and exhausted. It is an open rather than a closed system. However, an analytical model of open gas power systems is often made using the following assumptions:

1. The working fluid is air throughout and executes a closed cycle.
2. The air is an ideal gas with constant specific heats.
3. The real processes in the cycle are approximated as constant volume, constant pressure, etc., whichever best models the real system.
4. Heat addition is used to replace combustion.
5. Heat rejection at atmospheric pressure is used to complete the open part of the cycle.

This model of the real situation is called an *air-standard cycle*. It neglects the effects of combustion on the properties of the gases and it ignores some of the features of exhaust and intake (since it assumes a closed circulation) but it still allows some simple and useful, if approximate, calculations and comparisons. In this chapter, we shall describe the actual cycles briefly but we shall make analyses based on the air-standard model.

13-2 Gas turbine cycles

13-2 (a) The open system

The gas turbine is a rotating external combustion gas power device.

In the basic gas turbine system, air is compressed and sent to a combustor where fuel is injected and burned. The products of the combustion are then expanded through a turbine and finally exhausted to the atmosphere. See Fig. 13.1. The efficiency, as always, is defined as the net useful output divided by the costly input. The net output is $(\dot{W}_T - \dot{W}_C)$. The costly input in thermodynamic terms is the lower heating value of the fuel. Thus

$$\eta = \frac{\dot{W}_T - \dot{W}_C}{\dot{m}_f(\text{LHV})} \tag{13-1}$$

where $\dot{W}_T$, $\dot{W}_C$, and $\dot{m}_f$ are defined as shown in Fig. 13.1, and LHV stands for the *lower heating value* of the fuel. (For a discussion of lower and higher heating values of fuels, see Chap. 14.)

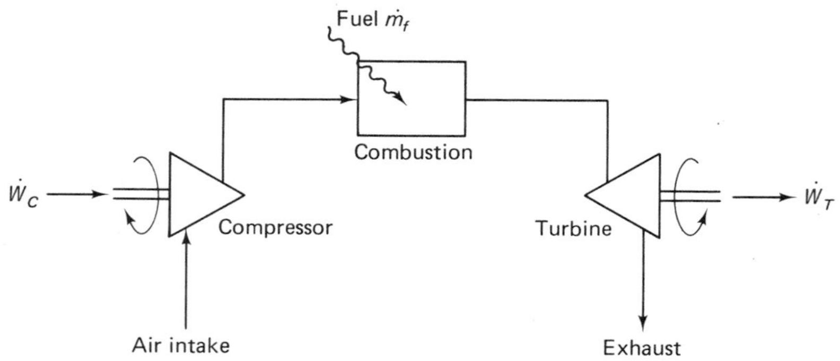

Figure 13.1 Basic gas turbine system.

13-2(b) The air-standard Brayton cycle

The air-standard model of the gas turbine system is called the *Brayton cycle.* It is shown in Fig. 13.2. The efficiency of the Brayton cycle is

$$\eta = \frac{\dot{W}_T - \dot{W}_C}{\dot{Q}_B} = \frac{(h_3 - h_4) - (h_2 - h_1)}{(h_3 - h_2)} = 1 - \frac{(h_4 - h_1)}{(h_3 - h_2)} \tag{13-2}$$

and if the air is assumed to be an ideal gas with constant specific heats, $\Delta h = c_p \, \Delta T$ then this becomes

$$\eta = \frac{(T_3 - T_4) - (T_2 - T_1)}{(T_3 - T_2)} = 1 - \frac{(T_4 - T_1)}{(T_3 - T_2)} \tag{13-3}$$

For internally reversible processes with an ideal gas, the compression and expansion will be isentropic and the process equation will be

$$Tp^{(1-\gamma)/\gamma} = \text{constant} \tag{13-4}$$

so that we can write

$$T_1 p_1^{(1-\gamma)/\gamma} = T_2 p_2^{(1-\gamma)/\gamma} \quad \text{or} \quad \frac{T_2}{T_1} = \left(\frac{p_2}{p_1}\right)^{(\gamma-1)/\gamma} \tag{13-5}$$

and

$$T_3 p_3^{(1-\gamma)/\gamma} = T_4 p_4^{(1-\gamma)/\gamma} \quad \text{or} \quad \frac{T_3}{T_4} = \left(\frac{p_3}{p_4}\right)^{(\gamma-1)/\gamma} \tag{13-6}$$

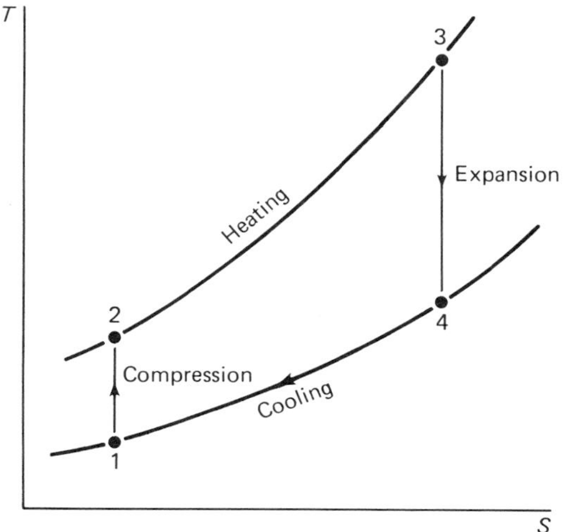

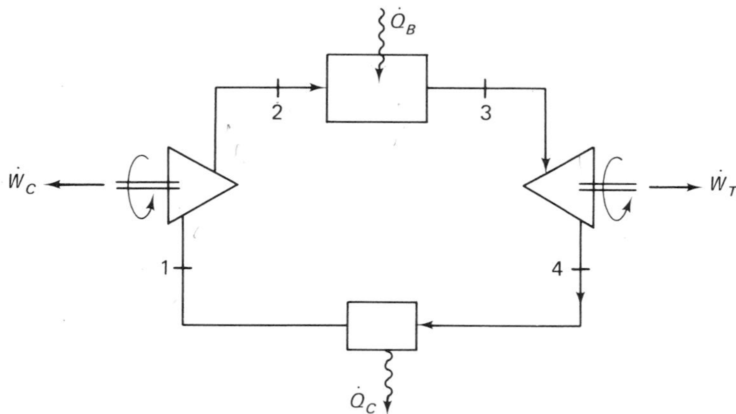

Figure 13.2 Air-standard Brayton cycle.

Since $p_2 = p_3$ and $p_1 = p_4$, we have

$$\frac{T_2}{T_1} = \frac{T_3}{T_4} \quad \text{or} \quad \frac{T_4}{T_1} = \frac{T_3}{T_2} \tag{13-7}$$

Subtracting 1 from both sides gives

$$\frac{T_4 - T_1}{T_1} = \frac{T_3 - T_2}{T_2} \quad \text{or} \quad \frac{T_4 - T_1}{T_3 - T_2} = \frac{T_1}{T_2} \tag{13-8}$$

Incorporating this result into the expression for efficiency gives

$$\eta = 1 - \frac{T_4 - T_1}{T_3 - T_2} = 1 - \frac{T_1}{T_2} = 1 - \left(\frac{p_1}{p_2}\right)^{\gamma - 1/\gamma} \tag{13-9}$$

Thus, when air is assumed to be the working fluid and an ideal gas, these kinds of theoretical analyses can be made. In more realistic cases, tables of properties for the products of combustion are used.

EXAMPLE 13.1 An air-standard internally reversible Brayton cycle operates between 15 and 75 psia. The inlet temperature is 80°F and the maximum temperature is 600°F. Compute the cycle efficiency.

SOLUTION:

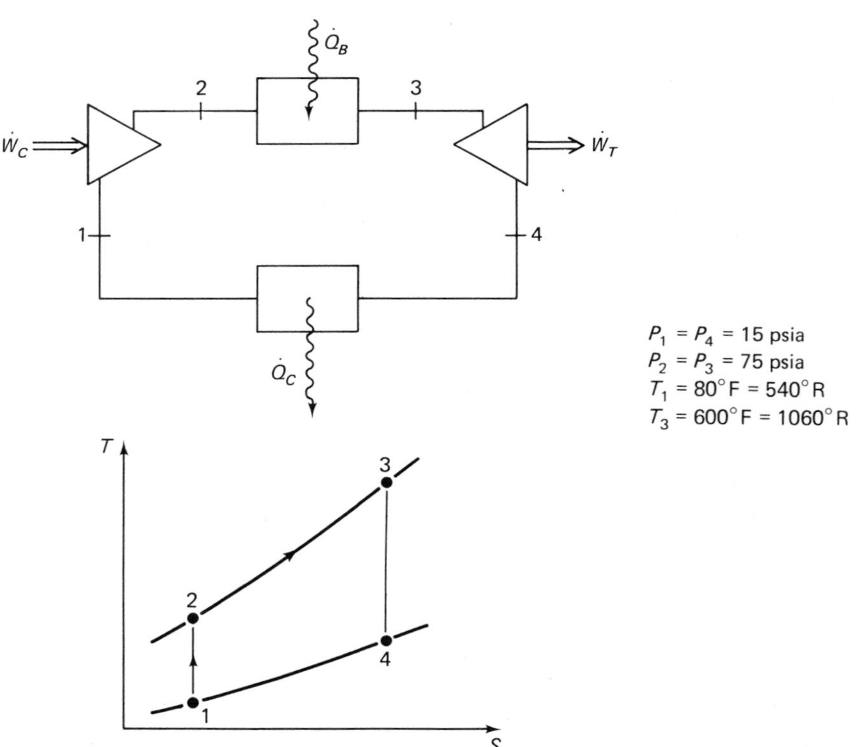

Assumptions implied in solution:

1. Air is ideal gas.
   (a) Equation of state, $PV = mRT$.
   (b) c_p and c_v are constant and not functions of temperature.

2. Consider compressor and turbine adiabatic and since they are also internally reversible, the expansion and compression processes are isentropic.
3. Negligible potential and kinetic energy changes between states.
4. Steady flow, steady state operation of power plant.

Temperature at the end of compression:

$$T_2 = T_1\left(\frac{P_2}{P_1}\right)^{\gamma-1/\gamma} = (540)\left(\frac{75}{15}\right)^{1.4-1/1.4} = 856°\text{R}$$

Temperature at the end of expansion:

$$T_4 = T_3\left(\frac{P_4}{P_3}\right)^{\gamma-1/\gamma} = (1060)\left(\frac{15}{75}\right)^{1.4-1/1.4} = 670°\text{R}$$

Work from turbine per pound:

$$W_{\text{turbine}} = h_3 - h_4 = c_p(T_3 - T_4) = (0.24)(1060 - 670) = 93.6 \text{ Btu/lbm}$$

Work to compressor per pound:

$$W_{\text{compressor}} = h_1 - h_2 = c_p(T_1 - T_2) = (0.24)(540 - 856) = -75.8 \text{ Btu/lbm}$$

Heat added in burner per pound:

$$Q_B = h_3 - h_2 = c_P(T_3 - T_2) = (0.24)(1060 - 856) = 49 \text{ Btu/lbm}$$

Cycle efficiency:

$$\eta = \frac{W_{\text{net}}}{Q_{\text{added}}} = \frac{W_{\text{turbine}} + W_{\text{compressor}}}{Q_B} = \frac{93.6 - 75.8}{49} = 0.363$$

Alternatively, using Equation (13-9),

$$\eta = 1 - \left(\frac{p_1}{p_2}\right)^{(\gamma-1)/\gamma} - 1 - (5)^{-0.286}$$

$$\eta = 1 - \frac{1}{1.585} = 0.369$$

13-2(c) Reheat, regeneration, and intercooling

Gas turbine engines often employ reheat and regeneration and the terms mean much the same as in vapor power cycles. A single stage of reheat is shown in Fig. 13.3.

The air-standard model of the reheat cycle is shown in Fig. 13.3 with its *T-s* plot. The efficiency of the air approximation system is

$$\eta = \frac{\dot{W}_{T_1} + \dot{W}_{T_2} - \dot{W}_C}{\dot{Q}_{B_1} + \dot{Q}_{B_2}} = \frac{(h_3 - h_4) + (h_5 - h_6) - (h_2 - h_1)}{(h_3 - h_2) + (h_5 - h_4)} \tag{13-10}$$

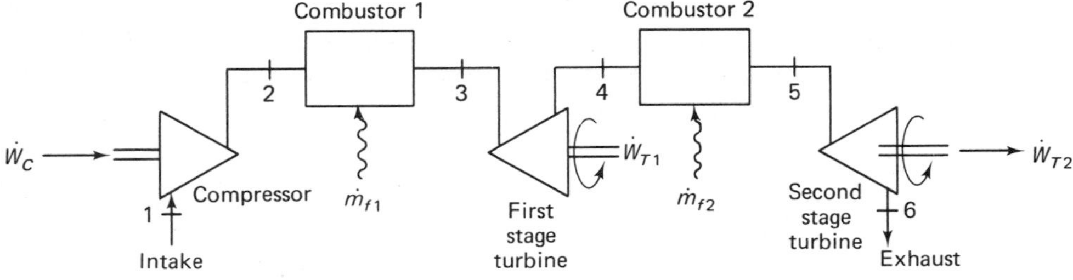

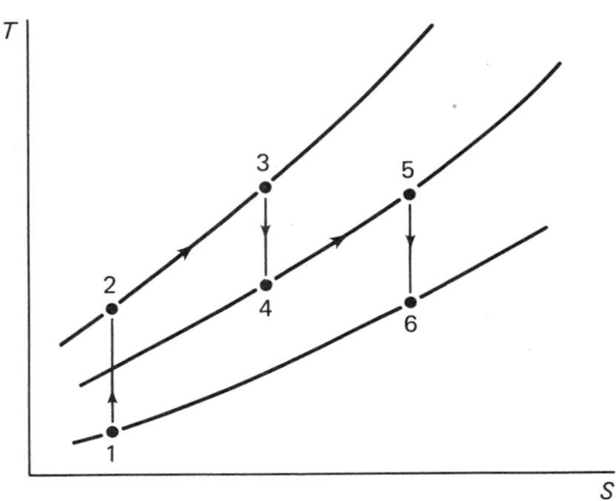

Figure 13.3 The gas reheat cycle.

For constant specific heat, this becomes

$$\eta = \frac{(T_3 - T_4) + (T_5 - T_6) - (T_2 - T_1)}{(T_3 - T_2) + (T_5 - T_4)} \tag{13-11}$$

and again the process equation for reversible adiabatic turbines and compressors handling ideal gases is

$$Tp^{(1-\gamma)/\gamma} = \text{constant} \tag{13-12}$$

Intercooling is analogous to reheat except that it is done in the compression process. The intake air is compressed to an intermediate pressure and then cooled in a heat exchanger before being compressed again. See Fig. 13.4. Remember that $Tp^{(1-\gamma)/\gamma} = \text{constant}$ applies to reversible adiabatic processes of ideal gas and this is applicable for such processes in the model as 5-6, 1-2, and 3-4. The pressures, or at least their ratios, are usually known or specified.

The highest and lowest temperatures are also usually known or specified, the high temperature being limited metallurgically, the low being limited by the available environmental sink temperature. The following example shows how the cycles are analyzed.

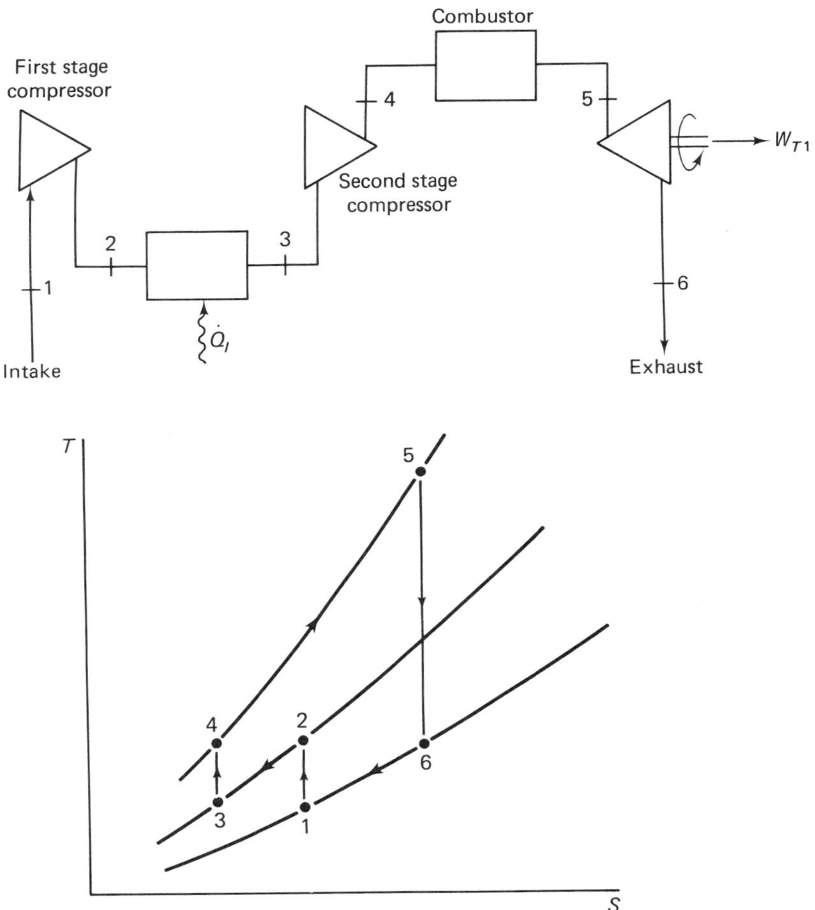

Figure 13.4 The gas intercooling cycle.

EXAMPLE 13.2. An internally reversible air-standard Brayton cycle operates between 1 and 5 bars absolute and between 27 and 327°C. It employs a single stage of intercooling and a single stage of reheat, the intermediate pressure in both cases being 3 bars absolute. Compute the cycle efficiency.

SOLUTION:

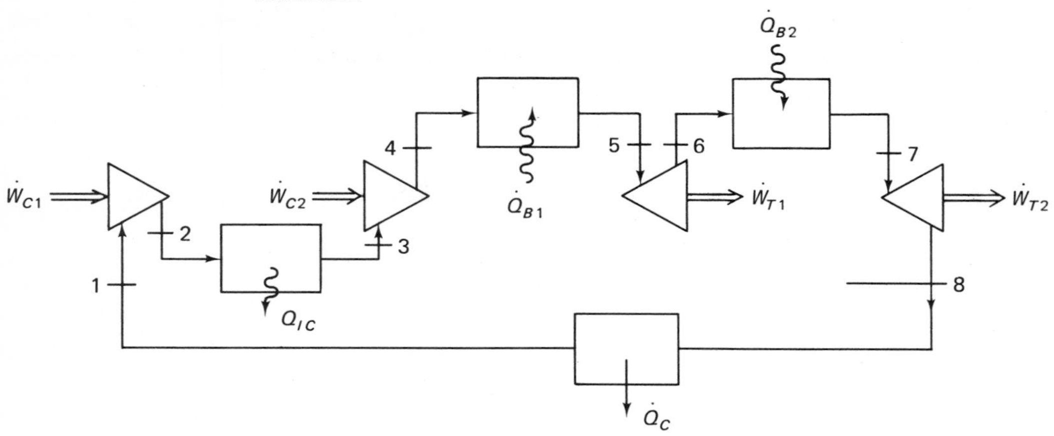

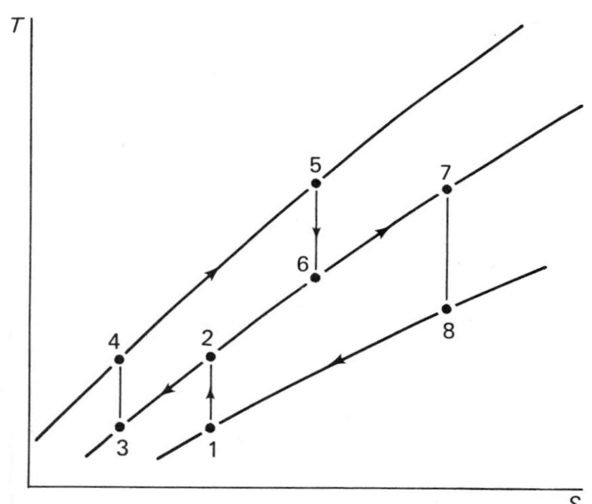

$P_1 = P_8 = 1$ bar
$P_4 = P_5 = 5$ bars
$P_2 = P_3 = P_6 = P_7 = 3$ bars
$T_1 = T_3 = 27°C = 300°K$
$T_5 = T_7 = 600°K$

Assumptions implied in solution:

1. Air is ideal gas.
   (a) Equation of state, $PV = mRT$.
   (b) c_p and c_v are constant and not functions of temperature.
2. Consider both compressors and both turbines adiabatic and since they are also internally reversible, the expansion and compression processes are isentropic.

3. Negligible potential and kinetic energy changes between states.
4. Temperature at end of intercooling is the same as first stage compressor inlet temperature.
5. Temperature at end of reheating is the same as first-stage turbine inlet temperature.
6. Steady flow, steady state operation of power plant.

Temperature at end of first-stage compression:

$$T_2 = T_1\left(\frac{P_2}{P_1}\right)^{(\gamma-1)/\gamma} = (300)\left(\frac{3}{1}\right)^{(1.4-1)/1.4} = 411°\text{K}$$

Temperature at end of second-stage compression:

$$T_4 = T_3\left(\frac{P_4}{P_3}\right)^{(\gamma-1)/\gamma} = (300)\left(\frac{5}{3}\right)^{(1.4-1)/1.4} = 347°\text{K}$$

Temperature at end of first-stage expansion:

$$T_6 = T_5\left(\frac{P_6}{P_5}\right)^{(\gamma-1)/\gamma} = (600)\left(\frac{3}{5}\right)^{(1.4-1)/1.4} = 519°\text{K}$$

Temperature at end of second-stage expansion:

$$T_8 = T_7\left(\frac{P_8}{P_7}\right)^{(1.4-1)/1.4} = (600)\left(\frac{1}{3}\right)^{(1.4-1)/1.4} = 438°\text{K}$$

Work to first-stage compressor:

$$W_{C_1} = h_2 - h_1 = c_p(T_2 - T_1) = (1)(411 - 300) = 111 \text{ j/gm}$$

Work to second-stage compressor:

$$W_{C_2} = h_4 - h_3 = c_p(T_4 - T_3) = (1)(347 - 300) = 47 \text{ j/gm}$$

Work from first-stage turbine:

$$W_{T_1} = h_5 - h_6 = c_p(T_5 - T_6) = (1)(600 - 519) = 81 \text{ j/gm}$$

Work from second-stage turbine:

$$W_{T_2} = h_7 - h_8 = c_p(T_7 - T_8) = (1)(600 - 438) = 162 \text{ j/gm}$$

Heat added in first-burner:

$$Q_{B_1} = h_5 - h_4 = c_p(T_5 - T_4) = (1)(600 - 347) = 253 \text{ j/gm}$$

Heat added in second-burner:

$$Q_{B_2} = h_7 - h_6 = c_p(T_7 - T_6) = (1)(600 - 519) = 81 \text{ j/gm}$$

Cycle efficiency:

$$\eta = \frac{W_{\text{net}}}{Q_{\text{added}}} = \frac{W_{T_1} + W_{T_2} - W_{C_1} - W_{C_2}}{Q_{B_1} + Q_{B_2}}$$

$$\eta = \frac{81 + 162 - 111 - 47}{253 + 81} = 0.254$$

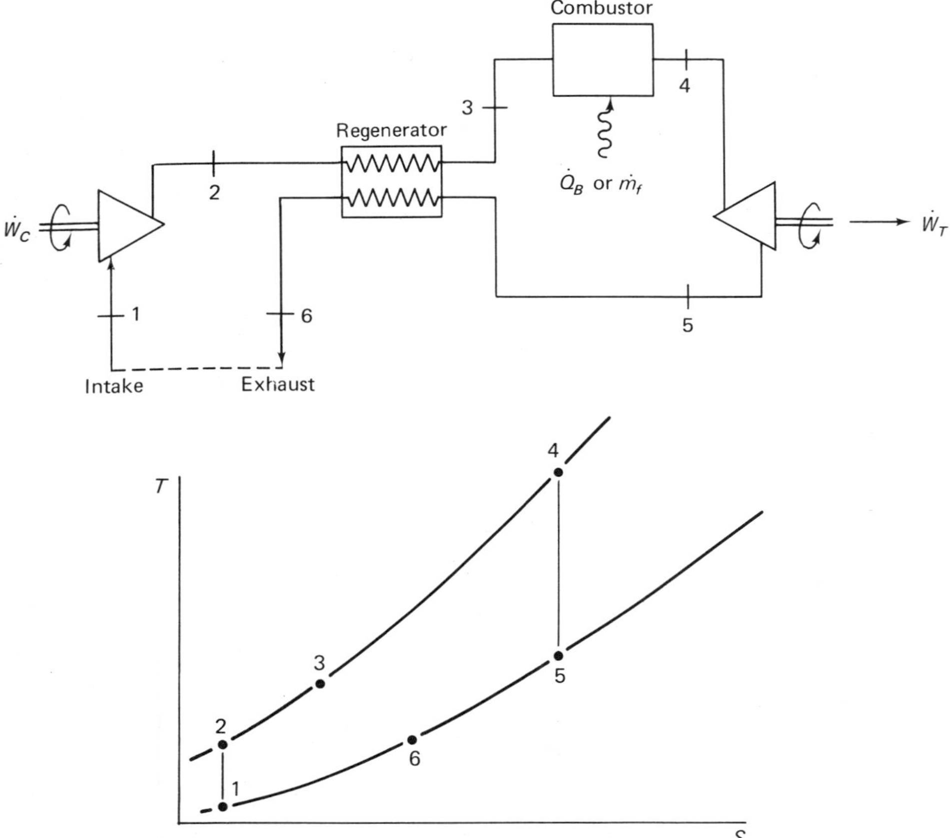

Figure 13.5 The gas regenerative cycle.

Ideal regeneration in the gas cycle is shown schematically and with the air approximation *T-s* diagram in Fig. 13.5. Notice how this cycle makes use of relatively high temperature exhaust gases before they are thrown away by using them to heat the high pressure intake air before combustion, thereby saving fuel. For the air-standard cycle, the efficiency of the ideal regeneration cycle is

$$\eta = \frac{\dot{W}_T - \dot{W}_C}{\dot{Q}_B} = \frac{(h_4 - h_5) - (h_2 - h_1)}{(h_4 - h_3)} \simeq \frac{(T_4 - T_5) - (T_2 - T_1)}{(T_4 - T_3)} \tag{13-13}$$

Also note that if the heat exchanger is adiabatic, the energy equation applied just to it in the ideal case gives

$$(h_3 - h_2) = (h_5 - h_6) \tag{13-14}$$

or

$$(T_3 - T_2) = (T_5 - T_6) \tag{13-15}$$

EXAMPLE 13.3. An internally reversible air-standard Brayton cycle operates between 15 and 75 psia and between 80 and 1000°F maximum. It employs an ideal regenerator. Compute the cycle efficiency.

SOLUTION:

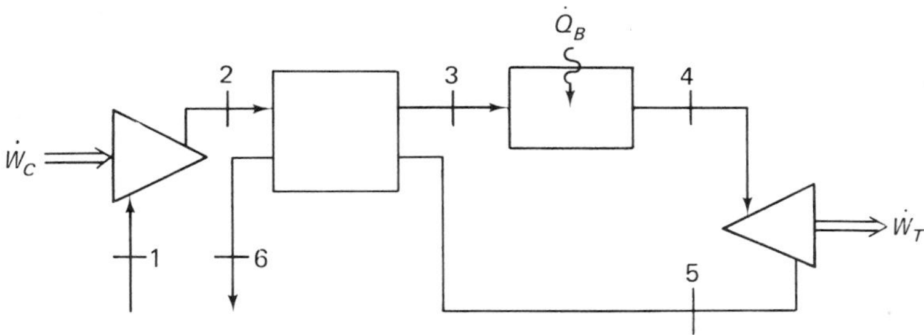

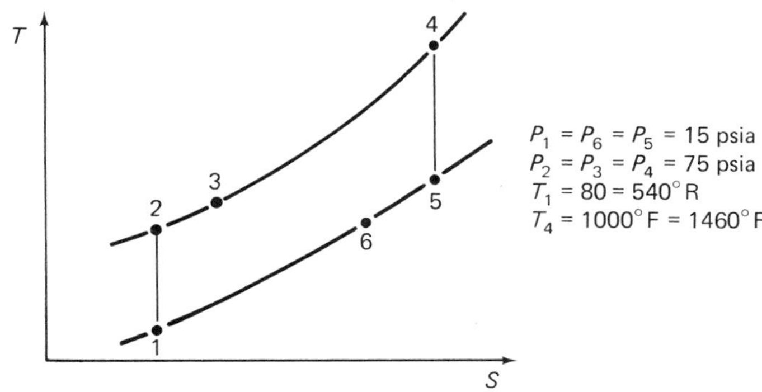

Assumptions implied in solution:

1. Air is ideal gas.
   (a) Equation of state, $PV = mRT$.
   (b) c_p and c_v are constant and not functions of temperature.

2. Consider compressor and turbine adiabatic and since they are also internally reversible, the expansion and compression processes are isentropic.
3. Negligible potential and kinetic energy changes between states.
4. Steady flow, steady state operation of power plant.
5. Ideal regenerator, $T_5 = T_3$.

Temperature at end of compression:

$$T_2 = T_1\left(\frac{P_2}{P_1}\right)^{(\gamma-1)/\gamma} = (540)\left(\frac{75}{15}\right)^{(1.4-1)/1.4} = 856°\text{R}$$

Temperature at end of expansion:

$$T_5 = T_4\left(\frac{P_5}{P_4}\right)^{(\gamma-1)/\gamma} = (1460)\left(\frac{15}{75}\right)^{(1.4-1)/1.4} = 922°\text{R}$$

$$T_3 = T_5 = 922°\text{R}$$

Work of turbine per pound:

$$W_{\text{turbine}} = h_4 - h_5 = c_p(T_4 - T_5) = (0.24)(1460 - 922) = 129 \text{ Btu/lbm}$$

Work to compressor per pound:

$$-W_{\text{compressor}} = h_2 - h_1 = c_p(T_2 - T_1) = (0.24)(856 - 540) = 75.7 \text{ Btu/lbm}$$

Heat added in burner per pound:

$$Q_B = h_4 - h_3 = c_p(T_4 - T_3) = (0.24)(1460 - 922) = 129 \text{ Btu/lbm}$$

Cycle efficiency:

$$\eta = \frac{W_{\text{net}}}{Q_{\text{added}}} = \frac{W_{\text{turbine}} + W_{\text{compressor}}}{Q_{\text{added}}} = \frac{129 - 75.7}{129} = 0.414$$

13-2(d) Deviations from the ideal

The gas turbine cycles described so far have assumed internally reversible processes. In practice, of course, deviations occur because of compressor and turbine inefficiencies, ducting pressure losses, heat exchanger irreversibility, etc. We shall consider only the compressor, turbine, and heat exchanger deviations from ideal here because they are the major deviations. Students should bear in mind, however, that these are not the *only* ones.

Gas compressor and turbine efficiencies are defined the same as liquid pump and steam turbine efficiencies. Refer to Fig. 13.6.

The adiabatic turbine efficiency is given by

$$\eta_T = \frac{W_{\text{actual}}}{W_{\substack{\text{reversible}\\ (\text{same } \Delta P)}}} = \frac{h_1 - h_2}{h_1 - h_{2S}} \tag{13-16}$$

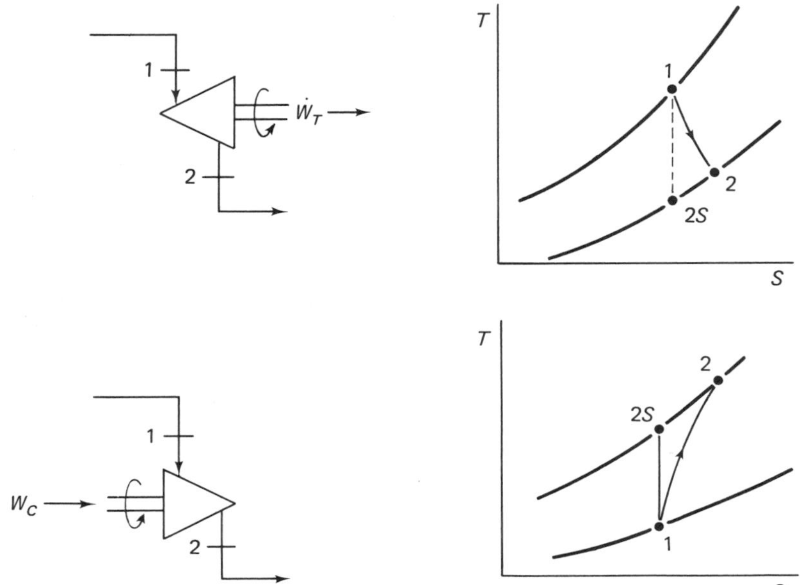

Figure 13.6 Turbine and compressor inefficiencies.

The adiabatic compression efficiency is given by

$$\eta_C = \frac{W_{\substack{\text{reversible}\\ \text{(same } \Delta P)}}}{W_{\text{actual}}} = \frac{h_{2S} - h_1}{h_2 - h_1} \tag{13-17}$$

If the working fluid is an ideal gas with constant specific heats, $\Delta h = c_p \, \Delta T$, then these become

$$\eta_T = \frac{T_1 - T_2}{T_1 - T_{2S}} \qquad \eta_C = \frac{T_{2S} - T_1}{T_2 - T_1} \tag{13-18}$$

Regenerative heat exchangers are always of the counterflow type. See the notation shown in Fig. 13.7. The subscript c denotes the stream that *enters* cold and subscript h the stream that enters hot; i and e stand for inlet and exit, respectively. The energy equation for a well-insulated exchanger gives

$$\dot{m}_c(h_{c_e} - h_{c_i}) = \dot{m}_h(h_{h_i} - h_{h_e}) = \dot{Q}_t \tag{13-19}$$

where $\dot{Q}_t$ stands for the heat transfer rate between the two streams. We define heat exchanger efficiency as

$$\eta_x = \frac{\text{actual heat transferred}}{\substack{\text{maximum possible heat transfer}\\ \text{consistent with both the first and second laws}}} \tag{13-20}$$

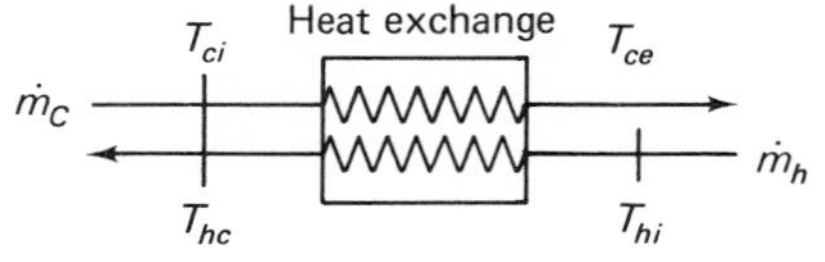

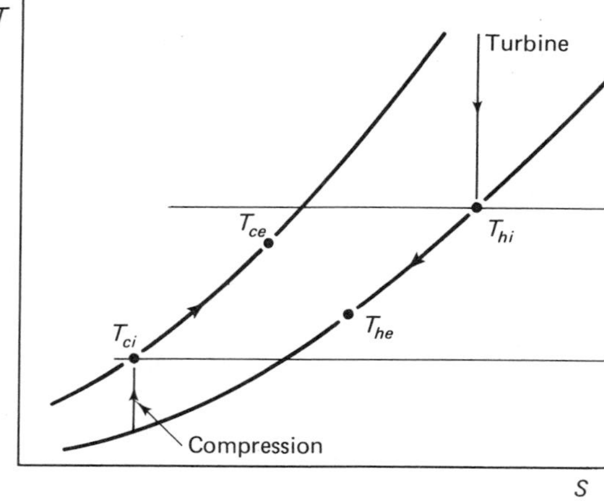

Figure 13.7 Regenerator inefficiency.

If the air standard approximations are made, then $\dot{m}_c - \dot{m}_h = \dot{m}$ and $\Delta h = c_p \, \Delta T$ and $c_{p_h} = c_p$, and the efficiency becomes

$$\eta_x = \frac{\dot{Q}_t}{\dot{m}c_p(T_{c_e} - T_{c_i})} = \frac{\dot{Q}_t}{\dot{m}c_p(T_{h_i} - T_{h_e})} \tag{13-21}$$

Under these conditions if the exchanger is ideal (infinite area), we will obtain $T_{c_i} = T_{h_e}$ and $T_{h_i} = T_{c_e}$. But, of course, they are not ideal or of infinite area (and would cost more than we would save even if they were). Efficiencies in practice run anywhere from 40 to 90%. The effect of regenerator inefficiency in the Brayton cycle is shown in the *T-S* diagram in Fig. 13.7. A perfect exchanger could raise the entering air stream to as high as T_{h_i} and cool the exhaust air to T_{c_i}. Since this is not possible, more heat must be added in the "burner" and more energy is discarded in the exhaust.

EXAMPLE 13.4. Repeat Example 13.1 assuming the compressor and turbine each have 95% efficiency.

SOLUTION:

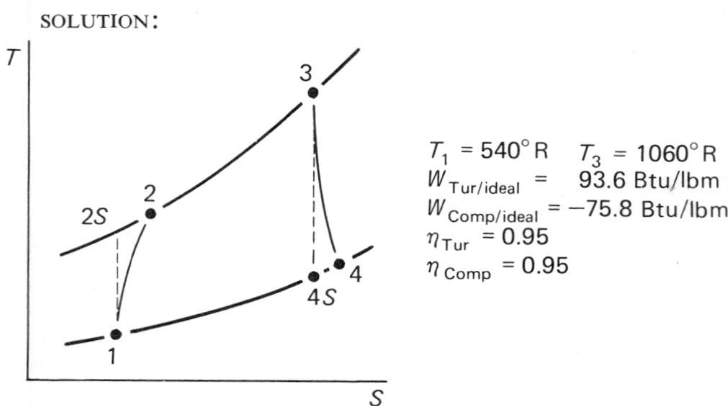

Actual work of turbine:

$$W_{\text{turbine/actual}} = \eta_{\text{turbine}} W_{\text{turbine/ideal}} = (0.90)(93.6) = 89 \text{ Btu/lbm}$$

Actual work to compressor:

$$W_{\text{compressor/actual}} = \frac{W_{\text{compressor/ideal}}}{h_{\text{compressor}}} = \frac{-75.8}{0.95} = -80 \text{ Btu/lb}$$

$$-W_{\text{compressor/actual}} = h_2 - h_3 = c_p(T_2 - T_3)$$

$$T_2 = T_1 - \frac{W_{\text{compressor}}}{c_p} = 540 + (80)/(0.24)$$

$$T_2 = 873°\text{R}$$

Heat added in burner:

$$Q_B = h_3 - h_2 = c_p(T_3 - T_2) = (0.24)(1060 - 873) = 45 \text{ Btu/lbm}$$

Cycle efficiency:

$$\eta = \frac{W_{\text{net}}}{Q_{\text{added}}} = \frac{89 - 80}{45} = 0.20$$

EXAMPLE 13.5. Repeat Example 13.3 assuming the turbine and compressor are internally reversible but that the regenerator is 80% efficient.

SOLUTION:

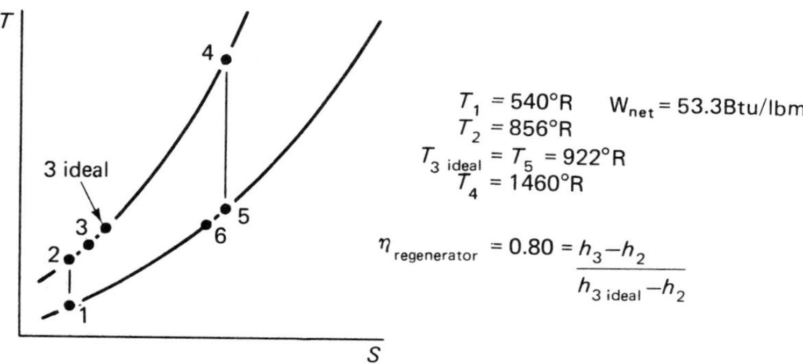

Actual temperature at end of regeneration:

$$\eta_{\text{regenerator}} = \frac{h_3 - h_2}{h_{3\text{ ideal}} - h_2} = \frac{c_p(T_3 - T_2)}{c_p(T_{3\text{ ideal}} - T_2)}$$

$$T_3 = (0.80)(922 - 856) + 856 = 909°\text{R}$$

Actual heat added in burner:

$$Q_B = h_4 - h_3 = c_p(T_4 - T_3) = (0.24)(1460 - 908) = 132.5 \text{ Btu/lbm}$$

Cycle efficiency:

$$\eta = \frac{W_{\text{net}}}{Q_{\text{added}}} = \frac{53.3}{132.5} = 0.402$$

Reheat, regeneration, and intercooling can all be incorporated into one engine system. As with vapor cycles, students must be able to translate actual systems, real or proposed, into schematic sketches and (for first analyses) air approximation T-S diagrams. Then you should be able to apply the energy equation to the components, the process equations (such as $T^{1-\gamma/\gamma} = C$) when appropriate, and the definition of efficiency to evaluate the system.

13-3 Jet engines

The jet engine system most commonly used in aircraft propulsion is best described as a turbojet. See Fig. 1.3, p. 16. The products of combustion are passed through a small turbine just large enough to drive the compressor and then expanded through a nozzle to a high velocity to provide the thrust. See Fig. 13.8. The turbine work is made just sufficient to drive the compressor. The remaining energy is then converted to kinetic energy in the nozzle.

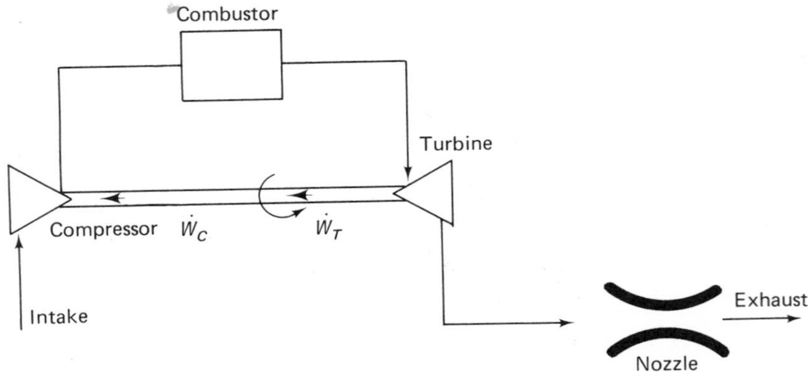

Figure 13.8 Basic jet engine system.

To obtain the velocity of the gas at the nozzle exit, we apply the steady flow energy equation to the nozzle. Assuming it is adiabatic, on a per pound basis we find

$$\frac{V_1^2 - V_2^2}{2g} = h_2 - h_1 \tag{13-22}$$

Since $V_2^2 \gg V_1^2$,

$$V_2^2 \simeq 2g(h_2 - h_1) \tag{13-23}$$

For an ideal gas,

$$\Delta h = c_p \, \Delta T \tag{13-24}$$

so

$$V_2^2 = 2gc_p(T_2 - T_1) \tag{13-25}$$

If, but only if, the nozzle is also reversible,

$$T_1 p_1^{(1-\gamma)/\gamma} = T_2 p_2^{(1-\gamma)/\gamma} \tag{13-26}$$

And thus

$$V_2^2 = 2gc_p T_1 \left[\left(\frac{p_1}{p_2}\right)^{(1-\gamma)/\gamma} - 1\right] \tag{13-27}$$

13-4 Internal combustion engines

13-4(a) The Otto engine

Most engineering students are familiar with the open Otto engine because it is the one commonly used in automobiles. (Otto and "auto" have no other connection. Otto was the name of the man who proposed the system in 1876.) It is a spark ignition rapid combustion system; it is also a major air pollutor. The cycle begins with the piston forward as shown in Fig. 13.9. The exhaust valve closes and the intake valve opens. As the piston moves back, the air–fuel mixture is drawn in; then the valve closes and the piston moves forward again, compressing the mixture. With the piston forward, the spark ignites the mixture and the piston moves back on its power stroke; then the exhaust valve opens and the products are swept out as the piston moves forward again and the cycle repeats.

Analysis of the realistic, open Otto engine is beyond the scope of this book and is properly left to a course in internal combustion engines. We can, however, study the corresponding air-standard cycle. As before, the assumption is that the working fluid is air, an ideal gas with constant specific heats, and that it executes a closed cycle. The air-standard Otto cycle, by analogy to the real cycle, consists of the following processes:

$1 \rightarrow 2$: Constant pressure intake.
$2 \rightarrow 3$: Adiabatic compression.
$3 \rightarrow 4$: Constant volume heat addition.
$4 \rightarrow 5$: Adiabatic expansion.
$5 \rightarrow 2$: Constant volume cooling.
$2 \rightarrow 1$: Constant pressure exhaust.

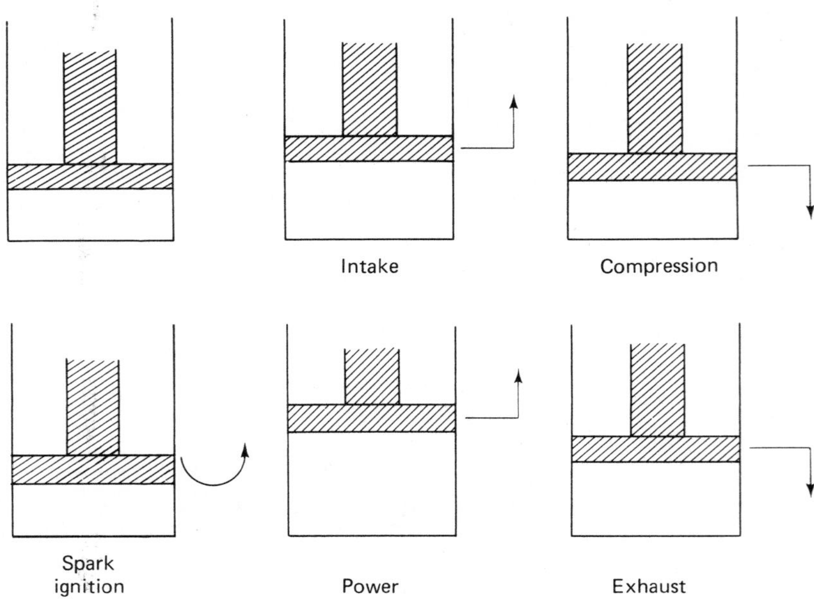

Figure 13.9 The basic Otto engine system.

The first and last processes involve no net work or heat addition and so may be ignored in the analysis of the air-standard cycle. A typical real pressure–cylinder volume diagram is shown in Fig. 13.10. The air-standard cycle approximation is shown in Fig. 13.11.

The efficiency of the air-standard Otto cycle is

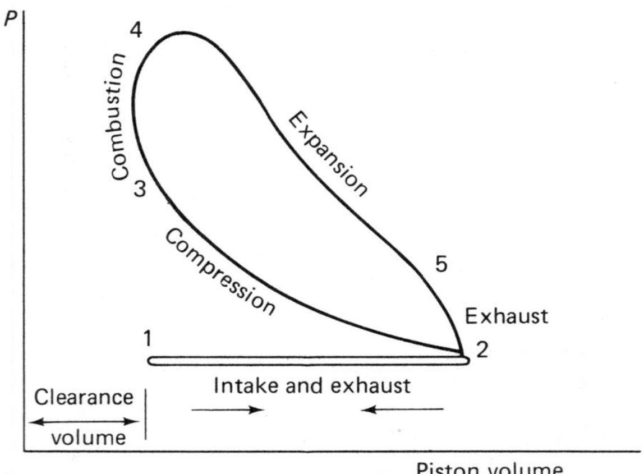

Figure 13.10 A typical pressure-cylinder volume diagram for an Otto engine.

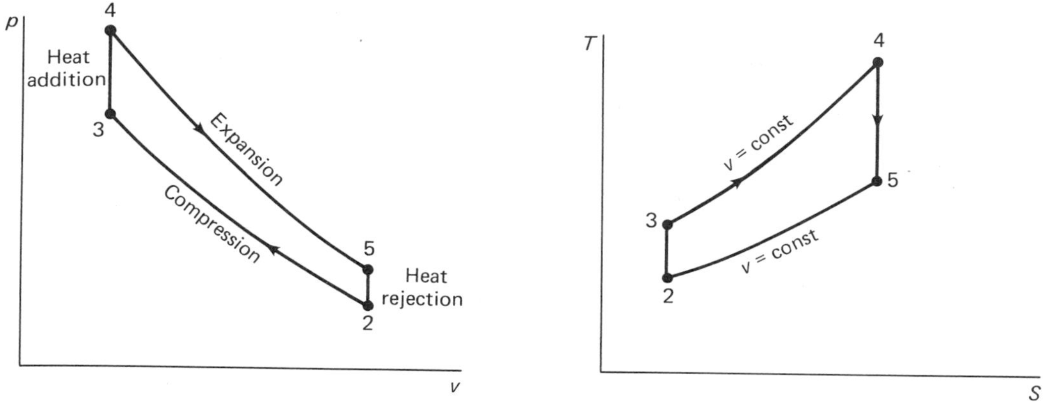

Figure 13.11 Air standard *T–S* and *p–v* diagrams for the Otto cycle.

$$\eta = \frac{W_{\text{net}}}{Q_{\text{in}}} = \frac{Q_{\text{in}} - Q_{\text{out}}}{Q_{\text{in}}} = 1 - \frac{Q_{\text{out}}}{Q_{\text{in}}} \tag{13-28}$$

$$\eta = 1 - \frac{mc_v(T_5 - T_2)}{mc_v(T_4 - T_3)} \tag{13-29}$$

The process equation for the isentropic compression and expansion is $T_v^{\gamma-1} = \text{constant}$. Thus,

$$\frac{T_3}{T_2} = \left(\frac{v_2}{v_3}\right)^{(\gamma-1)} = \left(\frac{v_5}{v_4}\right)^{(\gamma-1)} = \frac{T_4}{T_5} \tag{13-30}$$

These may be combined and manipulated to give

$$\eta = 1 - \frac{T_2}{T_3} = 1 - \left(\frac{v_2}{v_3}\right)^{(1-\gamma)} \tag{13-31}$$

The ratio (v_2/v_3) is called the *compression ratio* and is commonly given the symbol r.

Because the efficiency increases with increasing pressure ratio, automobile engines have been designed for higher and higher ratios. The result has been increased power for the same size engine (though engine sizes have increased too) and accompanying this has been more sophisticated fuels for rapid burning without preignition—*and* more air pollution.

EXAMPLE 13.6. An internally reversible air-standard Otto cycle operates between the temperature limits of 27 and 650°C. Calculate the thermal efficiency for compression ratios of 8 and 12. Also compute the work output per pound of air.

SOLUTION:

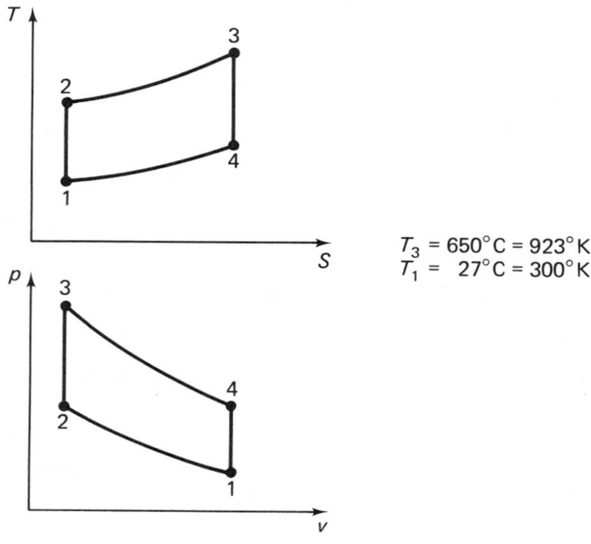

Assumptions implied in solution:

1. A closed cylinder of air executes the cycle.
2. Air is ideal gas.
   (a) Equation of state, $pV = mRT$.
   (b) c_p and c_v are constant and not functions of temperature.

$$\eta = \frac{W_{\text{net}}}{Q_{\text{added}}} = 1 - \frac{Q_{\text{rejected}}}{Q_{\text{added}}}$$

$$\eta = 1 - \frac{(T_4 - T_1)}{(T_3 - T_2)} = 1 - \frac{T_1(T_4/T_1 - 1)}{T_2(T_3/T_2 - 1)}$$

$$\frac{T_2}{T_1} = \left(\frac{V_1}{V_2}\right)^{\gamma-1} = \left(\frac{V_4}{V_3}\right)^{\gamma-1} = \frac{T_3}{T_4}$$

$$\therefore \quad \frac{T_3}{T_2} = \frac{T_4}{T_1}$$

$$\eta = 1 - \frac{T_1}{T_2} = 1 - \frac{1}{(\text{compression ratio})^{(\gamma-1)}}$$

(A) *Compression ratio of 8*

Cycle efficiency:

$$\eta = 1 - \frac{1}{(8)^{1.4-1}} = 0.565$$

Temperature at end of compression:

$$T_2 = T_1(\text{compression ratio})^{\gamma-1} = (300)(8)^{(1.4-1)} = 690°\text{K}$$

Heat added:

$$Q_{\text{added}} = U_3 - U_2 = c_v(T_3 - T_2) = (0.17)(923 - 690) = 39.6 \text{ j/gm}$$

Net work:

$$W_{\text{net}} = \eta Q_{\text{added}} = (0.565)(39.6) = 22.4 \text{ j/gm}$$

(B) *Compression ratio of 12*
Cycle efficiency:

$$\eta = 1 - \frac{1}{(12)^{(1.4-1)}} = 0.630$$

Temperature at end of compression:

$$T_2 = T_1(\text{compression ratio})^{\gamma-1} = (300)(12)^{(1.4-1)} = 810°\text{K}$$

Heat added:

$$Q_{\text{added}} = U_3 - U_2 = c_v(T_3 - T_2) = (0.17)(923 - 810) = 19.2 \text{ j/gm}$$

Net work:

$$W_{\text{net}} = \eta Q_{\text{added}} = (0.63)(19.2) = 12.1 \text{ j/gm}$$

13-4(b) The Diesel engine

The Diesel engine differs from the Otto engine only in the combustion portion of the cycle. Slower burning fuel is injected beginning at the end of the compression stroke through part of the power stroke. A realistic pressure–cylinder volume plot is shown in Fig. 13.12 together with the analogous air-standard cycle.

The Otto cycle is generally more efficient than a diesel cycle with the same compression ratio. On the other hand, the diesel engine can operate at higher pressure ratios and can burn cheaper fuel because the fuel is not injected until combustion is wanted and hence preignition is not a problem.

The efficiency of the air-standard diesel cycle is given by

$$\eta = \frac{W_{\text{net}}}{Q_{\text{in}}} = \frac{Q_{\text{in}} - Q_{\text{out}}}{Q_{\text{in}}} = 1 - \frac{Q_{\text{out}}}{Q_{\text{in}}} \tag{13-32}$$

$$= 1 - \frac{c_v(T_4 - T_1)}{c_p(T_3 - T_2)} \tag{13-33}$$

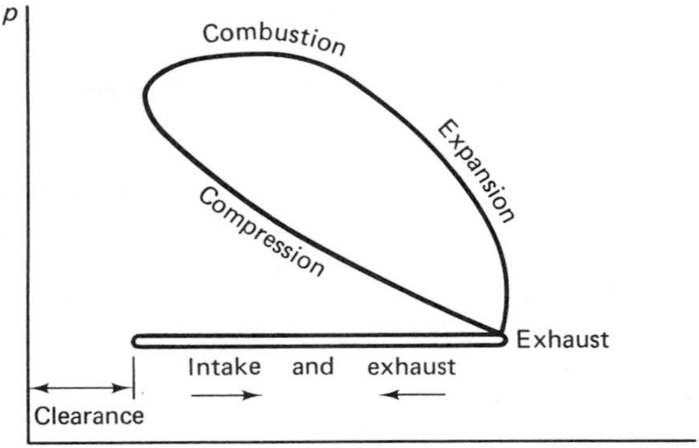

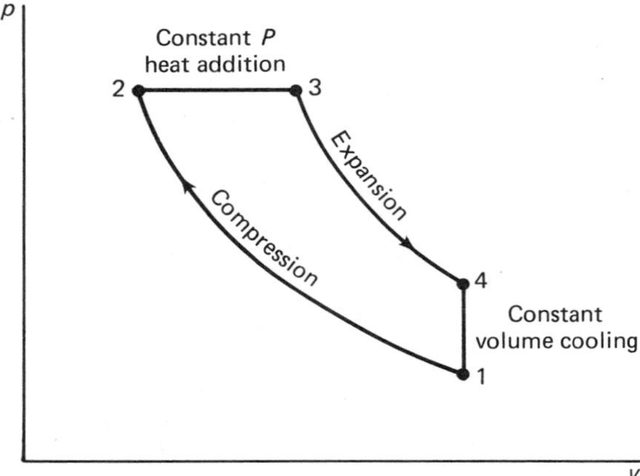

Figure 13.12 A typical pressure-cylinder volume diagram and air-standard p–v diagram for the basic diesel cycle.

For the isentropic compression (1-2) and expansion (3-4), the process equation is $Tv^{\gamma-1} =$ constant. That is,

$$\frac{T_2}{T_1} = \left(\frac{v_1}{v_2}\right)^{\gamma-1} \quad \text{and} \quad \frac{T_3}{T_4} = \left(\frac{v_4}{v_3}\right)^{\gamma-1} \tag{13-34}$$

With these, the ideal gas equation, and some luck, it can be shown that

$$\eta = 1 - \frac{1}{r^{\gamma-1}}\left[\frac{r_c^{\gamma} - 1}{\gamma(r_c - 1)}\right] \tag{13-35}$$

where

$$r_c = \frac{v_3}{v_2} \qquad \text{(Cutoff ratio)} \tag{13-36}$$

and

$$r = \frac{v_1}{v_2} \qquad \text{(Compression ratio)} \tag{13-37}$$

EXAMPLE 13.7. An internally reversible air-standard diesel cycle operates with inlet conditions of 15 psia and 80°F. The compression ratio is 15 and the cutoff ratio is 1.8. Compute the cycle efficiency and the work output per pound of air.

SOLUTION:

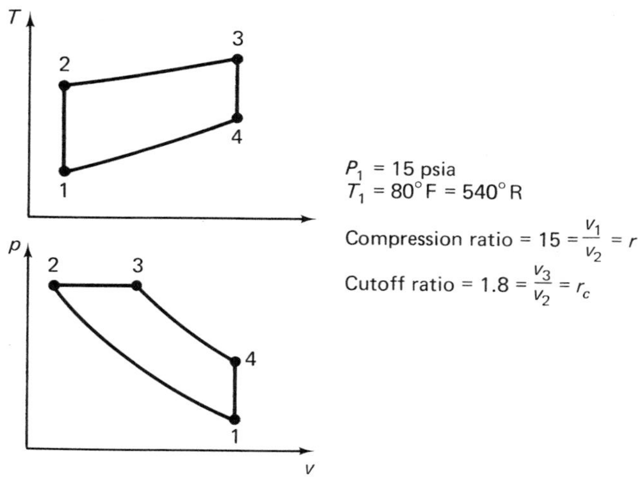

Assumptions implied in solution:

1. A closed cylinder of air executes the cycle.
2. Air is ideal gas.
   (a) Equation of state, $pV = mRT$.
   (b) c_p and c_v are constant and not functions of temperature.

Initial specific volume:

$$v_1 = \frac{RT_1}{p_1} = \frac{(53.3)(540)}{(15)(144)} = 13.3 \text{ ft}^3/\text{lbm}$$

Also, $v_1 = v_4$.

Properties of air at state 2:

$$v_2 = \frac{v_1}{r} = \frac{13.3}{15} = 0.89 \text{ ft}^3/\text{lbm}$$

$$T_2 = T_1 r^{\gamma - 1} = 540(15)^{0.4} = 1595°\text{R}$$

$$p_2 = \frac{RT_2}{v_2} = \frac{(53.3)(1595)}{(0.89)(144)} = 664 \text{ psia}$$

Also, $p_3 = p_2$.

Temperature at end of heat addition:

$$T_3 = \frac{p_3 v_3}{R} = \frac{p_3(1.8v_2)}{R} = \frac{(664)(1.8)(0.89)(144)}{(53.3)} = 2870°\text{R}$$

Temperature at end of isentropic expansion:

$$T_4 = T_3\left(\frac{v_3}{v_4}\right)^{\gamma-1} = 2870\left[\frac{(1.8)(0.89)}{(13.3)}\right]^{0.4} = 1230°\text{R}$$

Cycle efficiency:

$$\eta = \frac{W_{\text{net}}}{Q_{\text{in}}} = 1 - \frac{Q_{\text{rejected}}}{Q_{\text{in}}}$$

$$= 1 - \frac{T_4 - T_1}{T_3 - T_1}$$

$$= 1 - \frac{1230 - 540}{2870 - 1595} = 0.459$$

Heat added:

$$Q_{\text{in}} = U_3 - U_2 = c_v(T_3 - T_2) = 0.171(2870 - 1595)$$
$$= 218 \text{ Btu/lbm}$$

Net work:

$$W_{\text{net}} = \eta Q_{\text{in}} = (0.459)(218) = 100 \text{ Btu/lbm}$$

Alternatively, using Eq. (13-35) for efficiency,

$$\eta = 1 - \frac{1}{r^{\gamma-1}}\left[\frac{r_c^{\gamma} - 1}{\gamma(r_c - 1)}\right]$$

$$\eta = \frac{1}{(15)^{0.4}}\left[\frac{(1.8)^{1.4} - 1}{1.4(1.8 - 1)}\right]$$

$$\eta = 0.381$$

13-4(c) The Wankel engine

The Wankel engine, named for its inventor, Dr. Felix Wankel, who patented the basic mechanism in Germany in 1927, is an internal combustion engine that operates on the Otto cycle. But it is a rotary engine in contrast to the reciprocating piston engine commonly used in the past. To understand how the Wankel operates, refer to Fig. 13.13.

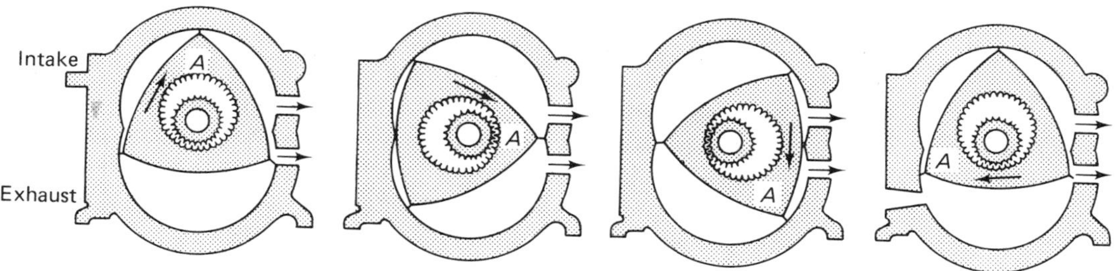

Figure 13.13 Schematic of Wankel operation. The shape of the chamber is called an epitrochoid.

Imagine following a point of contact (such as point A) around the cycle once, and consider what happens in the chamber just behind point A. As A passes the intake port (no valves needed), air is brought in. As A passes the fuel injection, fuel is added to chamber. As A moves well past the spark plug, the charge is compressed. Then the plug fires and burning and expansion take place until A uncovers the exhaust port.

The Wankel engine is said to have 40% fewer parts, is apparently simpler because of the absence of reciprocating motions, and is smaller and lighter than piston engines. Sealing is a more serious problem, however.

The Wankel engine is less efficient than conventional Otto engines. *And* it produces about *twice* the unburned hydrocarbon pollutants. Why then, is it getting so much attention? Because NO_x emissions are much less from the Wankel engine and much harder to control in piston engines. Also it is hoped that the Wankel engine will prove to be more adaptable to emissions control when there has been time enough for research and development.

The Wankel engine has all the inherent problems of high compression spark ignition internal combustion engines, plus a few of its own. Seals are one. Sealing at the apex of the rotor provides only a small contact area. Another problem is fuel economy. At present the Wankel engine is at least 10% less efficient than conventional piston engines, meaning higher operating costs and use of more increasingly precious oil resources. The Wankel engine also produces about twice as much unburned hydrocarbon pollutants. Thus this engine is no panacea for the social problems of automobile transportation.

The Wankel engine does have advantages, however, that are important enough to mean that its use will become widespread. It is smaller, lighter, and simpler to build than a piston engine of the same power. This means that there is both space and money available for the necessary pollution controls and it appears that the Wankel engine will be able to meet government antipollution specifications.

Japanese manufacturers were the first to market automobiles with Wankel engines. Of the United States producers, General Motors has purchased patent rights (for $50 million from Curtiss-Wright who bought them from Wankel and a German company for $2.1 million). Wankel engines are predicted to be in some smaller 1975 United States' cars.

It does not appear that the Wankel engine will completely replace conventional engines. The high cost of patent rights is one obstacle, the fuel economy question is another, and there *are* improvements (such as electronic fuel injection) that will help the piston engines compete. It appears that Wankel engines may dominate the smaller car market (though Volkswagen has no plans to use them) because for larger power ratings the Wankel engine itself becomes difficult and complex to build.

13-5 Other alternatives in the automobile-pollution battle

External combustion engines are better able to effect complete combustion than internal combustion engines. Thus gas turbines, steam engines, and a cycle called the *Stirling cycle* all are being given renewed attention by both government and industry.

Open cycle steam piston engines were used in automobiles in 1930. They were abandoned because they were large, heavy, difficult to start, and poor on gas mileage. Development in the 1960's was directed to such changes as closed cycles, use of Freon instead of water, and experimentation with turbines instead of pistons. Several small companies, with support from the government, are developing steam engines for use in typical United States' cars.

Another alternative receiving some attention is the Stirling cycle, shown in Fig. 13.14. Processes 1-2 and 3-4 are constant volume heat addition and rejection, respectively. Processes 2-3 and 3-4 are isothermal heat addition and rejection, respectively. The power producing process is 2-3. The attraction of the Stirling cycle is that it theoretically can be made to approach reversible efficiency. Thus if a practical engine can be designed, the fuel savings could be significant; however, the constant volume processes and the isothermal heating and cooling are complicated to produce.

The gas turbine is another external combustion engine used in some racing cars. The trouble with them is that they are efficient only when used in steady,

Figure 13.14 The Stirling cycle.

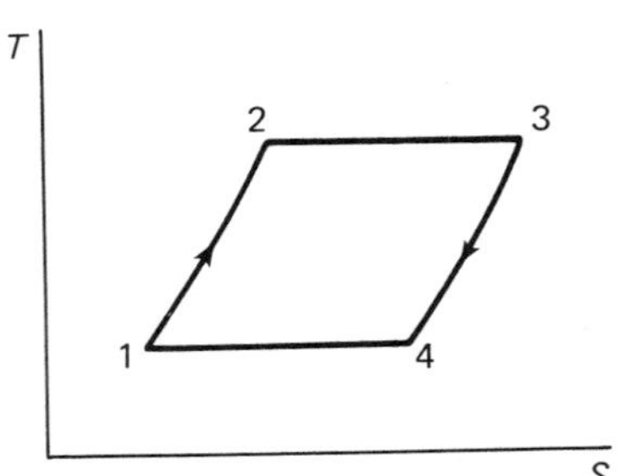

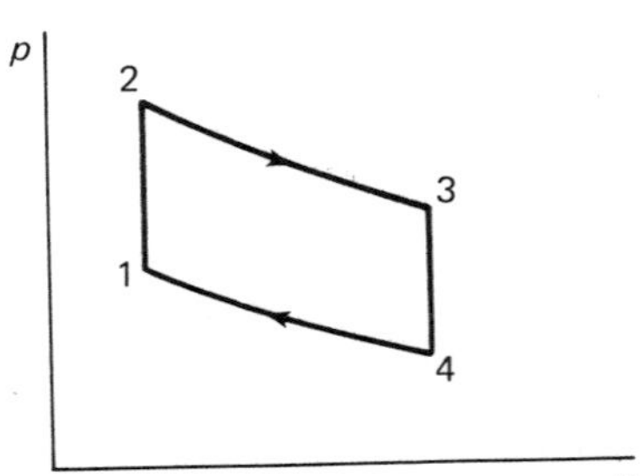

high-speed operation. In stop-and-go traffic, they become extremely inefficient, thus a great deal more development is needed for this external combustion engine also.

Electric cars are also under study and development. The electric car concept, of course, moves the pollution from the city to the power generating plant. The electricity can be generated by nuclear or solar energy, thus reducing dependence on fossil fuels. The trouble with lead–acid batteries, however, has always been their poor power-to-weight ratio. Research into metal–alkali batteries may provide a significant improvement. A number of companies in the United States are working on new batteries. It seems inevitable that electric powered transportation will be necessary sooner or later.

13-6 Summary

In this chapter, gas internal combustion engines are modeled and analyzed using the air-standard approximation. This allows approximate comparison of cycles with a minimum of labor. The air-standard model is a closed cycle with heat additions and rejections instead of combustion and exhaust as in the actual, open process.

The gas turbine is modeled as an air-standard Brayton cycle. For a reversible cycle, the efficiency (in terms of pressure ratio) is

$$\eta = 1 - \left(\frac{P_1}{P_2}\right)^{1-\gamma/\gamma}$$

Reheat and regeneration used in gas turbine cycles are analogous in their function to their use in steam cycles.

Inefficiencies in actual gas turbines include frictional effects, heat transfers, incomplete combustion, etc. In the air-standard cycle, one of these (friction in the adiabatic turbine and compressor) can be handled using an efficiency defined (for a turbine) as the work actually obtained divided by the work that would have been obtained in a reversible adiabatic process from the same inlet temperature and pressure to the same outlet pressure. That is,

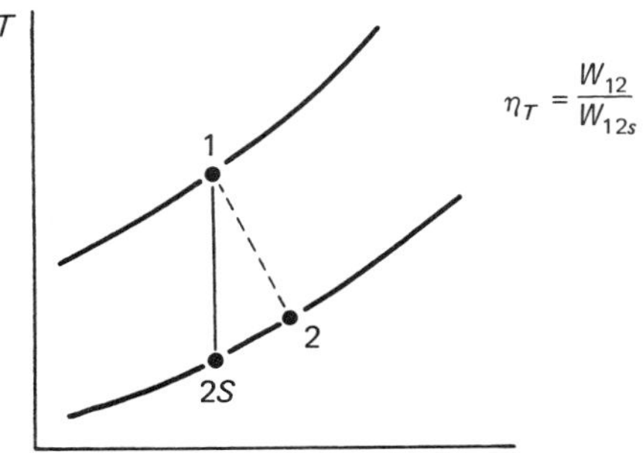

$$\eta_T = \frac{W_{12}}{W_{12s}}$$

Note that for a reversible, adiabatic process of an ideal gas

$$Tp^{(1-\gamma)/\gamma} = \text{constant}$$

The Otto cycle is the air-standard model of the reciprocating spark ignition internal combustion engine in common use in automobiles. Its air-standard efficiency is given in terms of compression ratio r.

$$\eta = 1 - (r)^{1-\gamma}$$

The Diesel cycle is also a reciprocating internal combustion engine but involves compression ignition instead of spark ignition. Its air-standard efficiency is given by

$$\eta = 1 - \frac{1}{(r)^{\gamma-1}}\left[\frac{r_c^{\gamma} - 1}{\gamma(r_c - 1)}\right]$$

where

$$r = \text{compression ratio}$$

and

$$r_c = \text{cutoff ratio}$$

The Wankel engine is a rotating spark ignition internal combustion engine.

Because of its high pollution effect, alternatives to the internal combustion engine are being sought. External combustion engines are less polluting but so far are not practical for large-scale use for other reasons. Ultimately, some alternatives to fossil fuel must be found.

Problems

13-1 Rework Example 13.1 assuming compressor and turbine efficiencies to be 0.90.

13-2 Rework Example 13.2 assuming compressor and turbine efficiencies to be 0.85.

13-3 Rework Example 13.3 assuming compressor and turbine efficiencies to be 0.80.

13-4 Rework Example 13.5 assuming compressor and turbine efficiencies to be 0.90.

13-5 The nozzle of a jet engine is provided with combustion products ($c_p = 0.22$ Btu/lbm-°F) at 80 psia and 1000°F. State assumptions and estimate the velocity at the nozzle exit.

13-6 The compressor of a jet engine has a pressure ratio of 5 and is 0.80 efficient. Inlet air temperature is 5°C. If the turbine is 0.90 efficient, how much work must be done by the combustion products in the turbine?

13-7 Without referring to the text, sketch the following air-standard cycles on p-v and T-S diagrams:

(a) Brayton.

(b) Otto.

(c) Diesel.

(d) Stirling.
(e) Wankel.

13-8 An ideal reciprocating compressor draws in 200 ft^3/min of air at 15 psia and 49°F and compresses it polytropically with $n = 1.35$ to 75 psia. Cooling water that removes heat from the air flows at a rate of 10.2 lb/min and undergoes a temperature rise of 10°F. Calculate the compressor power requirement in horsepower.

13-9 In a regenerative gas turbine cycle, air enters the compressor at 1 bar absolute and 27°C and leaves it at 5 bars absolute and 260°C. The temperature of air entering the combustion chamber is 427°C, entering the turbine, 870°C and leaving, 560°C. The lowest temperature in the surroundings is 27°C. Assuming no pressure drop in the regenerator or combustion chamber, calculate
(a) The cycle thermal efficiency.
(b) The compressor efficiency.
(c) The turbine efficiency.

13-10 For the cycle of Prob. 13-9, calculate per gram of air the degradation in the
(a) Compressor.
(b) Turbine.
(c) Regenerator.
(d) How much of the heat added to the cycle is available energy?
(e) How much of the heat rejected is available energy?

13-11 An ideal turbojet engine draws in air at 14.0 psia and 80°F with a velocity of 500 ft/sec at a rate of 32 lbm/sec. After passing through the compressor and the combustion chamber, the air enters the turbine at 60 psia and 1600°F. The velocity is 500 ft/sec at the compressor outlet, turbine inlet, and turbine exhaust. Calculate the velocity of the exhaust jet, assuming isentropic flow through the exhaust nozzle to a pressure of 14.0 psia.

13-12 A centrifugal compressor compresses air according to the process equation $pv^{1.30} =$ constant. The flow rate is 4 lbm/sec. Inlet conditions are 15 psia and 50°F. The pressure ratio is 3.0. The inlet velocity is 10 ft/sec and the outlet velocity is 600 ft/sec. Heat is removed at a rate of 15 Btu/sec.
(a) Calculate the compressor power requirement.
(b) Compare it with the requirement for a reversible, adiabatic compressor with the same inlet conditions and pressure ratio neglecting kinetic energy changes.

13-13 In a gas turbine plant working on the Brayton air cycle the air pressure and temperature before compression are, respectively, 15 psia and 80°F. The ratio of maximum pressure to minimum pressure is 6.00 and the temperature before expansion in the turbine is 1400°F. The turbine and compressor efficiencies are each 85%. Find
(a) The compressor shaft work per pound of air.
(b) The turbine shaft work per pound of air.
(c) The heat supplied per pound of air.
(d) The cycle efficiency.
(e) The turbine exhaust temperature.

13-14 Solve Prob. 13-13 if a regenerator of 80% effectiveness is added to the plant.

13-15 Solve Prob. 13-13 if the compression is divided into two steps, each of pressure ratio 2.45 and efficiency 85%, with intercooling to 80°F.

13-16 Solve Prob. 13-15 if a regenerator of 80% effectiveness is added to the plant.

13-17 For an air-standard Diesel cycle with compression ratio of 12, plot the efficiency as a function of cutoff ratio for cutoff ratios from 1 to 4.

13-18 In an air-standard Diesel cycle the compression ratio is 12 and compression begins at 15 psia and 120°F; the maximum temperature of the cycle is 2440°F. Find

(a) The heat supplied per pound of air.
(b) The work done per pound of air.
(c) The cycle efficiency.
(d) The temperature at the end of the isentropic expansion.
(e) The cutoff ratio.
(f) The maximum pressure of the cycle.

13-19 Plot the efficiency of the air-standard Otto cycle as a function of compression ratio for compression ratios from 4 to 18.

13-20 In an air-standard Otto cycle the compression ratio is 8 and compression begins at 80°F and 15 psia; the maximum temperature of the cycle is 2100°F. Find

(a) The heat supplied per pound of air.
(b) The work done per pound of air.
(c) The cycle efficiency.
(d) The temperature at the end of the isentropic expansion.
(e) The maximum pressure of the cycle.

13-21 What would be the advantages and disadvantages of helium as a working fluid for an Otto cycle? Of Freon-12 as the working fluid?

14

Mixtures of gases and air-water vapor mixtures

14-1 Basic assumptions

It is often necessary for engineers to work with mixtures of gases. Air is an obvious example, being about 78% nitrogen, 21% oxygen, and 1% other gases (mostly argon) by volume. Air also usually has some water vapor associated with it. Flue gases and gaseous mixtures for special processes also often require thermodynamic computations. In this chapter we shall deal only with mixtures of *ideal* gases. Mixtures of nonideal gases are extremely difficult to handle theoretically and are usually dealt with on an empirical basis. But since many real gases are nearly ideal under many practical circumstances, the ability to cope with mixtures of ideal gases is very important and useful.

In dealing with mixtures of gases we shall use unsubscripted letters to denote the properties of the total mixture and the subscripts $a, b, c, \ldots$ to denote properties of the components.

Consider a container of volume V at temperature T with a gaseous mixture of components $a, b, c, \ldots$. The basic assumption made in dealing with ideal gas mixtures is that each of the component gases occupies the total volume of the mixture *as if the others were not present*. Thus,

$$V_a = V_b = V_c = \cdots \tag{14-1}$$

Furthermore, all the components will be at the same temperature:

$$T = T_a = T_b = T_c = \cdots \tag{14-2}$$

The total mass of mixture (m) and the total moles of mixture (n) will be the sum of the masses and moles, respectively, of the components:

$$m = m_a + m_b + m_c + \cdots \tag{14-3}$$

and

$$n = n_a + n_b + n_c + \cdots \tag{14-4}$$

To distinguish between properties on a molar basis and those on a unit mass basis, we use a bar over the letter symbol to denote molar. Thus the gas constant per mole is designated as $\bar{R}$ and per pound it is R. Similarly $\bar{c}_p$ and $\bar{c}_v$ will designate specific heat per mole and c_p and c_v is the specific heat per pound. The same system will also be used for internal energy, enthalpy, entropy, etc.

The conversion from moles to pounds is done using molecular weight (MW), which has the units of pounds per pound-mole (lbm/lbmole). (*Note*: A pound-mole of a substance is an amount of pounds equal to its molecular weight. A pound-mole of oxygen is 32 lbm of oxygen.) For example, the molecular weight of oxygen is 32 lbm/lbmole and its molar specific heat is 7 Btu/lbmole-°R. Therefore,

$$\tfrac{7}{32} = 0.218 \text{ Btu/lbm-°R} = 0.913 \text{ j/gm-°K}$$

The total internal energy of the mixture will be the sum of the internal energies of the various components:

$$U = U_a + U_b + U_c + \cdots \tag{14-5}$$

or

$$mu = m_a u_a + m_b u_b + m_c u_c + \cdots \tag{14-6}$$

Finally, the total pressure of the mixture is the sum of the partial pressures of the components. This follows directly from our assumption that the gases behave independently of each other. Thus,

$$p = p_a + p_b + p_c + \cdots \tag{14-7}$$

The partial pressures of the components in a gaseous mixture are found from the equations of state. That is,

$$p_a = \frac{n_a \bar{R} T}{V} \qquad p_b = \frac{n_b \bar{R} T}{V} \qquad \cdots \tag{14-8}$$

Since we also have shown that

$$p = \frac{n\bar{R}T}{V} \tag{14-9}$$

we can divide these observations to obtain

$$\frac{p_a}{p} = \frac{n_a}{n} \qquad \frac{p_b}{p} = \frac{n_b}{n} \qquad \cdots \tag{14-10}$$

or

$$p_a = \frac{n_a}{n} p \qquad p_b = \frac{n_b}{n} p \tag{14-11}$$

Thus we see that partial pressures of components are proportional to the mole fraction (n_a/n) of the component in the mixture. This result is sometimes called Dalton's law.

With the basic premises above, we can show that a mixture of ideal gases is itself an ideal gas. The equation of state is written for each gas on a molar basis:

$$\begin{aligned} p_a V_a &= n_a \bar{R} T_a \\ p_b V_b &= n_b \bar{R} T_b \\ p_c V_c &= n_c \bar{R} T_c \\ \vdots \quad & \quad \vdots \end{aligned} \tag{14-12}$$

Adding these equations gives

$$(p_a + p_b + p_c + \cdots)V = (n_a + n_b + n_c + \cdots)\bar{R}T \tag{14-13}$$

or

$$pV = n\bar{R}T \tag{14-14}$$

where p, V, n, and T all apply to the mixture and $\bar{R}$ is the universal constant. Thus, the mixture also obeys the ideal gas equation of state.

Partial volume is defined as the fictitious volume a component would occupy by itself at the same temperature and total pressure of the mixture. We shall use a superscript prime (e.g., V') to denote partial volume. For a given component a, the partial volume is

$$V_a' = \frac{n_a \bar{R} T}{p} \tag{14-15}$$

Since the total volume is given by

$$V = \frac{n\bar{R}T}{p} \tag{14-16}$$

the partial volume is also

$$V_a = \frac{n_a}{n} V \tag{14-17}$$

EXAMPLE 14.1. Assume air consists of 79% nitrogen and 21% oxygen "by volume." Compute

1. The moles of nitrogen per mole of oxygen.
2. The partial pressure of oxygen and nitrogen if the total pressure is 1 atm.
3. The pounds of oxygen per pound of mixture.

(*Note: Percent by volume* is a commonly used term to denote the volume that would be occupied by a component at the same temperature and *total* pressure of the mixture.)

SOLUTION:
Translation to symbols and diagrams:

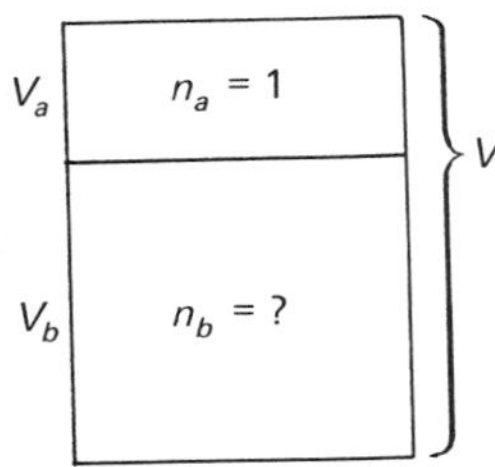

Let n_a = moles oxygen = 1.0
n_b = moles nitrogen
Let V contain $n = n_a + n_b$ moles of air at p and T.
Assume the oxygen and nitrogen are separated into their respective partial volumes.
Then $V_a' = 0.21\ V$
$V_b' = 0.79\ V$

Equations of state:

$$pV_a' = n_a \bar{R}T$$
$$pV_b' = n_b \bar{R}T$$

Dividing:

$$\frac{V_a'}{V_b'} = \frac{n_a}{n_b} \qquad n_b = \frac{n_a V_b'}{V_a'}\ (1)\left(\frac{0.79}{0.21}\right) = 3.76 \text{ moles}$$

Partial pressure:

$$p_a = \frac{n_a}{n} p = \left(\frac{1}{4.76}\right)(1) = 0.21 \text{ atm}$$

$$p_b = \frac{n_b}{n} p = \left(\frac{3.76}{4.76}\right)(1) = 0.79 \text{ atm}$$

Weight ratio:

$$\frac{m_b}{m_a + m_b} = \frac{n_b \, \text{MW}_b}{n_a \, \text{MW}_a + n_b \, \text{MW}_b} = \frac{(3.76)(28)}{(1)(32) + (3.76)(28)} = 0.77$$

$$= \frac{\text{lbm N}_2}{\text{lbm mixture}} \quad \text{or} \quad \frac{\text{kg N}_2}{\text{kg mixture}}$$

14-2 Properties of mixtures

Using the basic considerations in Sec. 14-1, we can easily show how the properties of an ideal gas mixture can be computed from the properties of the components. Using these properties, then, the mixture can be treated like any other ideal gas. We have already noted that the pressure p is given by

$$p = p_a + p_b + p_c + \cdots \tag{14-7}$$

and that T and V for the mixture and for all components are the same. Noting also that

$$U = U_a + U_b + U_c + \cdots \tag{14-5}$$

we can write

$$\begin{aligned} U_a &= m_a c_{v_a} T \\ U_b &= m_b c_{v_b} T \\ U_c &= m_c c_{v_c} T \\ &\vdots \end{aligned} \tag{14-18}$$

where the reference energy has been taken as zero at $T = 0$. Adding gives

$$U = (m_a c_{v_a} + m_b c_{v_b} + m_c c_{v_c})T \tag{14-19}$$

$$mc_v T = (m_a c_{v_a} + m_b c_{v_b} + m_c c_{v_c})T \tag{14-20}$$

or

$$c_v = \frac{m_a}{m} c_{v_a} + \frac{m_b}{m} c_{v_b} + \frac{m_c}{m} c_{v_c} + \cdots \tag{14-21}$$

and

$$c_p = \frac{m_a}{m} c_{p_a} \frac{m_b}{m} c_{p_b} \frac{m_c}{m} c_{p_c} \cdots \tag{14-22}$$

Thus we have the internal energy and specific heats of the mixture in terms of the component properties. To find the gas constant R for a mixture, we write

$$\begin{aligned} p_a V &= m_a R_a T \\ p_b V &= m_b R_b T \\ p_c V &= m_c R_c T \\ \vdots \quad & \quad \vdots \end{aligned} \tag{14-23}$$

Adding,

$$pV = (m_a R_a + m_b R_b + m_c R_c)T \tag{14-24}$$

But also

$$pV = mRT \tag{14-25}$$

Thus

$$mR = (m_a R_a + m_b R_b + m_c R_c) \tag{14-26}$$

$$R = \frac{m_a}{m} R_a + \frac{m_b}{m} R_b + \frac{m_c}{m} R_c + \cdots \tag{14-27}$$

The molecular weight of the mixture in terms of the molecular weights of the components is found by starting with

$$n = n_a + n_b + n_c \cdots \tag{14-4}$$

Noting that $n = m/\text{MW}$,

$$\frac{m}{\text{MW}} = \frac{m_a}{\text{MW}_a} + \frac{m_b}{\text{MW}_b} + \frac{m_c}{\text{MW}_c} + \cdots \tag{14-28}$$

or

$$\frac{1}{\text{MW}} = \frac{m_a}{m}\frac{1}{\text{MW}_a} + \frac{m_b}{m}\frac{1}{\text{MW}_b} + \frac{m_c}{m}\frac{1}{\text{MW}_c} + \cdots \tag{14-29}$$

Many problems are more easily handled on a molar basis. It is easy to show that

$$\bar{c}_v = \frac{n_a}{n}\bar{c}_{v_a} + \frac{n_b}{n}\bar{c}_{v_b} + \frac{n_c}{n}\bar{c}_{v_c} + \cdots \tag{14-30}$$

and

$$\bar{c}_p = \frac{n_a}{n}\bar{c}_{p_a} + \frac{n_b}{n}\bar{c}_{p_b} + \frac{n_c}{n}\bar{c}_{p_c} + \cdots \tag{14-31}$$

The general equation

$$\Delta S_{12} = mc_p \ln \frac{T_2}{T_1} - mR \ln \frac{P_2}{P_1} \tag{14-32}$$

is valid for mixtures as long as the composition of the mixture does not vary (as, for example, if one component should condense) during the process and, of course, as long as c_p, m, and R are the proper values for the mixture.

EXAMPLE 14.2. Compute the molecular weight of air (79% nitrogen, 21% oxygen, "by volume").

SOLUTION: From Example 14.1, we know that "air" contains 3.76 moles of N_2 per mole of O_2. Therefore, consider a system of 4.76 moles of air. The mass of O_2 will be

$$m_{O_2} = n_{O_2} MW_{O_2} = 1(32) = 32 \text{ lbm}$$
$$m_{N_2} = n_{N_2} MW_{N_2} = 3.76(28) = 105 \text{ lbm}$$

For the mixture,

$$m = 32 + 105 = 137 \text{ lbm} \qquad n = 4.76 \text{ lbmole}$$

$$MW = \frac{137}{4.76} = 28.8 \text{ lbm/lbmole} \quad (= 28.8 \text{ kg/kgmole})$$

Note: We shall use 29 for the molecular weight of "air" in the remainder of the book.

EXAMPLE 14.3. Compute the specific heats of "air" in Btu/lbm-°R and j/gm-°K for *both* N_2 and O_2:

$$\bar{c}_p = 7 \text{ Btu/lbmole-°R} = 29.1 \text{ j/gm-mole-°K}$$
$$\bar{c}_v = 5 \text{ Btu/lbmole-°R} = 20.8 \text{ j/gm-mole-°K}$$

SOLUTION: For the mixture,

$$\bar{c}_p = \frac{n_{O_2}}{n} \bar{c}_{pO_2} + \frac{n_{N_2}}{n} \bar{c}_{pN_2} = 7 \text{ Btu/lbmole-°R} = 29.1 \text{ j/gm-mole-°K}$$

$$\bar{c}_v = 5 \text{ Btu/lbmole-°R} = 20.8 \text{ j/gm-mole-°K}$$

On a per pound-mass basis,

$$c_p = \frac{\bar{c}_p}{MW} = \frac{7}{29} = 0.241 \text{ Btu/lbmole-°R}$$

$$c_v = \frac{c_v}{MW} = \frac{5}{29} = 0.172 \text{ Btu/lbmole-°R}$$

On a per gram basis,

$$c_p = \frac{29.1}{29} = 1.0 \text{ j/gm-°K}$$

$$c_v = \frac{20.8}{29} = 0.71 \text{ j/gm-°K}$$

EXAMPLE 14.4. The products of combustion of a certain fuel consist of 1 mole CO_2, 2 moles H_2O vapor, 1 mole O_2, and 11.3 moles N_2. Compute the ratio of specific heats ($\gamma = c_p/c_v$) for this mixture. For the components in Btu/lbm-°R,

$$c_{pO_2} = 0.218 \qquad c_{vO_2} = 0.158$$
$$c_{pN_2} = 0.250 \qquad c_{vN_2} = 0.179$$
$$c_{pH_2O} = 0.450 \qquad c_{vH_2O} = 0.340$$
$$c_{pCO_2} = 0.202 \qquad c_{vCO_2} = 0.156$$

SOLUTION: Assume, as a basis, 15.3 moles of the mixture. Then the masses of the various components will be

$$\begin{aligned} m_{O_2} &= 1(32) &&= 32 \text{ lbm} \\ m_{CO_2} &= 1(44) &&= 44 \text{ lbm} \\ m_{N_2} &= 11.3(28) &&= 316 \text{ lbm} \\ m_{H_2O} &= 2(18) &&= 36 \text{ lbm} \\ \text{Total mass} & &&= 428 \text{ lbm} \end{aligned}$$

For specific heats:

$$c_p = \frac{m_{O_2}}{m} c_{pO_2} + \frac{m_{N_2}}{m} c_{pN_2} + \frac{m_{H_2O}}{m} c_{pH_2O} + \frac{m_{CO_2}}{m} c_{pCO_2}$$

$$= \frac{32}{428}(0.218) + \frac{316}{428}(0.25) + \frac{36}{428}(0.450) + \frac{44}{428}(0.202)$$

$$c_p = 0.260 \text{ Btu/lbm-°R}$$

$$c_v = \frac{32}{428}(0.158) + \frac{316}{428}(0.179) + \frac{36}{428}(0.340) + \frac{44}{428}(0.156)$$

$$c_v = 0.189 \text{ Btu/lbm-°R}$$

$$\gamma = \frac{c_p}{c_v} = \frac{0.260}{0.189} = 1.37$$

Note: As we shall see in the next chapter, the assumption is often made in combustion processes that the products of combustion behave like air. In this example, though c_p and c_v are different from those of air ($c_{p_{air}} = 0.24$ and $c_{v_{air}} = 0.17$), the ratio is not affected much ($\gamma_{air} = 1.40$).

14-3 Mixing processes

Mixing occurs in many engineering processes and systems and, unless the things being mixed are exactly alike before mixing, is always a source of degradation. We shall consider first mixing in a closed system and then in open systems.

Refer to Fig. 14.1. We imagine two gases separated by a partition. In the general case, these need not be at the same initial temperatures or pressures.

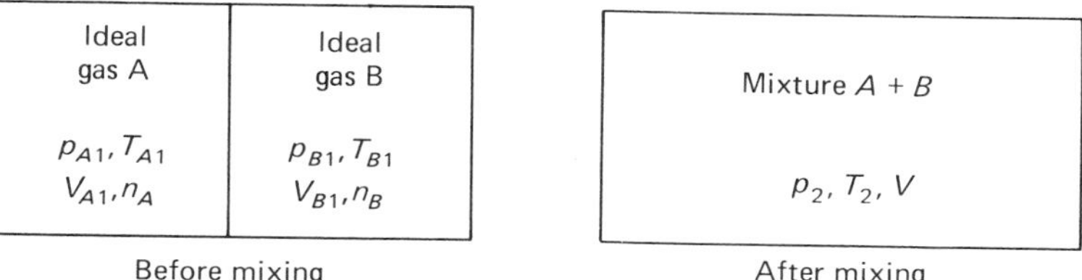

Figure 14.1 Mixing two ideal gases.

We do assume, however, that the system as a whole is isolated and rigid. The partition is removed and the gases mix. When equilibrium is established, we apply the laws:

First law:

$$U_{\text{initial}} = U_{\text{final}} \tag{14-33}$$

$$n_A \bar{U}_{A_1} + n_B \bar{U}_{B_1} = n_A \bar{U}_{A_2} + n_B \bar{U}_{B_2} \tag{14-34}$$

$$n_A \bar{c}_{v_A} T_{A_1} + n_B \bar{c}_{v_B} T_{B_1} = n_A \bar{c}_{v_A} T_2 + n_B \bar{c}_{v_B} T_2 \tag{14-35}$$

Given the initial conditions, Eq. (14-35) can be solved for the mixture temperature

$$T_2 = \frac{n_A \bar{c}_{v_A} T_{A_1} + n_B \bar{c}_{v_B} T_{B_1}}{n_A \bar{c}_{v_A} + n_B \bar{c}_{v_B}} \tag{14-36}$$

Equations of state:

For the mixture pressure,

$$p_2 V = (n_A + n_B)\bar{R}T_2 \tag{14-37}$$

For the partial pressures,

$$p_{A_2} V = n_A \bar{R} T_2 \tag{14-38}$$

$$p_{B_2} V = n_B \bar{R} T_2 \tag{14-39}$$

Second law:
For gas A,

$$\Delta S_A = n_A \bar{c}_{pA} \ln \frac{T_2}{T_{A_1}} - n_A \bar{R} \ln \frac{p_{A_2}}{p_{A_1}} \tag{14-40}$$

For gas B,

$$\Delta S_B = n_B \bar{c}_{pB} \ln \frac{T_2}{T_{B_1}} - n_B \bar{R} \ln \frac{p_{B_2}}{p_{B_1}} \tag{14-41}$$

From the entropy changes it is easy to compute the degradation for the entire adiabatic closed system:

$$D_{12} = T_0 \, \Delta S_{12} \tag{14-42}$$

EXAMPLE 14.5. One kilogram mole of O_2 and 3.76 kilogram mole of N_2 each at 27°C and atmospheric pressure are to be mixed adiabatically and at constant total volume. Compute the mixture temperature and pressure and the irreversibility of the mixing process.

SOLUTION:
Translation:

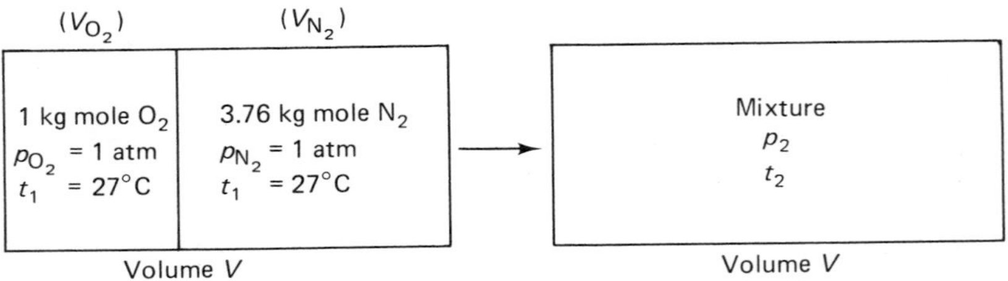

First law:

$$U_{\substack{\text{total}\\\text{initial}}} = U_{\substack{\text{total}\\\text{final}}}$$

$$n_{O_2} \bar{c}_{vO_2} T_1 + n_{N_2} \bar{c}_{vN_2} T_1 = n_{O_2} \bar{c}_{vO_2} T_2 + n_{N_2} \bar{c}_{vN_2} T_2$$

Thus

$$t_1 = t_2 = 27°\text{C}$$

Equation of state:
For p_2,

$$p_2 V = (n_{O_2} + n_{N_2})\bar{R}T_2$$

But

$$V = (V_{O_2})1 + (V_{N_2})1 = \frac{n_{O_2}\bar{R}T_1}{(p_{O_2})1} + \frac{n_{N_2}\bar{R}T_1}{(p_{N_2})1}$$

$$V = \frac{(n_{O_2} + n_{N_2})\bar{R}T_1}{p_1}$$

Thus

$$p_1 = \frac{(n_{O_2} + n_{N_2})\bar{R}T_2}{(n_{O_2} + n_{N_2})\bar{R}T_1} p_1 = p_1 = 1 \text{ atm}$$

For partial pressures,

$$(p_{O_2})_2 = \frac{1}{4.76}(1) = 0.21 \text{ atm}$$

$$(p_{N_2})_2 = \frac{3.76}{4.76}(1) = 0.79 \text{ atm}$$

Second law:

$$\Delta S_{O_2} = n_{O_2}\bar{c}_{pO_2} \ln\frac{T_2}{T_1} - n_{O_2}\bar{R}\ln\frac{(p_{O_2})_2}{(p_{O_2})_1}$$

$$= -n_{O_2}\bar{R}\frac{0.21}{1.0} = +(1)(8.312)\frac{0.21}{1.0} = 1.745 \text{ j/°K}$$

$$\Delta S_{N_2} = -n_{N_2}\bar{R}\frac{0.79}{1.0} = +(3.76)(8.312) = 24.7 \text{ j/°K}$$

$$D = T_0\,\Delta S_{\text{total}} = (300)(1.745 + 24.7) = 7934 \text{ j}$$

Mixing in steady flow open systems is handled in essentially the same way as closed systems except, of course, the energy equation is used for the first law. Otherwise there are no differences except that unless data are given on the flow conditions (pipe size, velocity, etc.), it is not possible to *calculate* the final mixture pressure as it is in closed systems.

EXAMPLE 14.6. A 1-lbm/min stream of O_2 at 2 atm pressure and 100°F is mixed in steady flow with a 2-lbm/min stream of air at 200°F and 1.5 atm. The mixture pressure is 1 atm. Compute the mixture temperature and the degradation.

SOLUTION:
Translation:

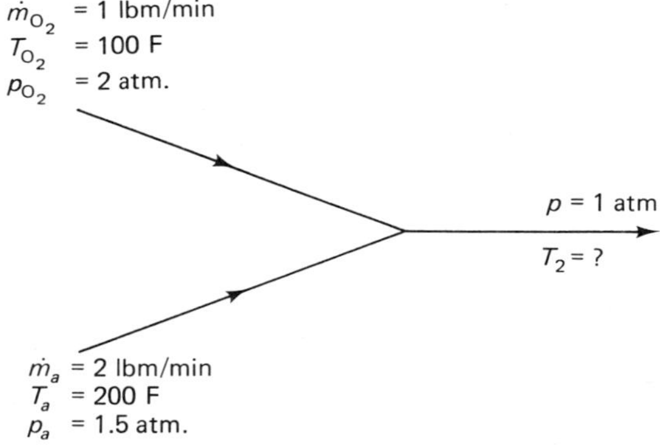

First law: Neglect kinetic and potential energies.

$$\dot{m}_{O_2}(h_{O_2})_1 + \dot{m}_a h_{a_1} = \dot{m}_{O_2}(h_{O_2}) + \dot{m}_a(h_{a_2})$$

$$\dot{m}_{O_2} c_{p_{O_2}}(T_{O_2})_1 + \dot{m}_a c_{p_a} T_{a_1} = (\dot{m}_{O_2} c_{p_{O_2}} + \dot{m}_a c_{p_a})T_2$$

$$T_2 = \frac{(1)(\frac{7}{32})(560) + (2)(0.241)(660)}{(1)(\frac{7}{32}) + (2)(0.241)}$$

$$T_2 = \frac{122 + 318}{0.218 + 0.482} = \frac{440}{0.700} = 630°\text{R} = 170°\text{F}$$

In the mixture:

$$\frac{n_{O_2}}{n} = \frac{(\frac{1}{32})}{(\frac{1}{32}) + (\frac{2}{29})} = \frac{1}{1 + (\frac{64}{29})} = 0.313$$

$$\frac{n_a}{n} = 0.687$$

Partial pressures:

$$p_{O_2} = 0.313 \text{ atm}$$
$$p_a = 0.687 \text{ atm}$$

Second law:

$$\Delta\dot{S}_{O_2} = \dot{m}_{O_2} c_{p_{O_2}} \ln\frac{630}{560} - \dot{m}_{O_2} R_{O_2} \ln\frac{0.313}{2}$$

$$= (1)(\tfrac{7}{32})(0.115) + (1)(\tfrac{2}{32})(1.86) = +0.141 \text{ Btu/°R}$$

$$\Delta\dot{S}_a = \dot{m}_a c_{p_a} \ln\frac{630}{660} - \dot{m}_a R_a \ln\frac{0.687}{1.5}$$

$$= -(2)(\tfrac{7}{29})(0.047) + (2)(\tfrac{2}{32})(0.78) = +0.0747 \text{ Btu/°R}$$

Degradation:

$$D = T_0\,\Delta\dot{S}_{\text{total}} = (540)(0.141 + 0.075) = 117 \text{ Btu/°R}$$

14-4 Separation processes

An interesting problem we can work using what we learned in Chap. 11 about work potential and what we now know about mixtures is that of finding the minimum work required to separate a mixture of gases into its components. As an example, consider the objective to be to separate "air" into its components at the same temperature and total pressure. That is, for each 4.76 moles of air at 1 atm and, say, 80°F, we wish to obtain 3.76 moles of nitrogen and 1.0 mole of oxygen, each at 1 atm and 80°F. What is minimum work required?

It is not necessary for us to ask how this process might be carried out in order to compute the minimum work. From Chap. 11, we remember that the minimum work required to produce a change of state is given by the change in work potential of the system. That is,

$$W_{\substack{\text{minimum}\\ \text{required}\\ 12}} = A_1 - A_2 \tag{14-43}$$

where for a closed system

$$A = U + P_0 V - T_0 S \tag{14-44}$$

In the process we are considering, $\Delta U = 0$ since the temperature is constant and the gases are assumed ideal. Also, $P_0 V = 0$ since the volume of the mixture will equal the sum of the volumes of the separated components at the same temperature and pressure. Thus we find simply

$$W_{\text{min}} = A_1 - A_2 = T_0(S_2 - S_1) \tag{14-45}$$

We have already discussed how to compute ΔS for the mixing process. Separation gives the same results except the sign. In our example,

$$\Delta S_{O_2} = -n_{O_2}\bar{R}\ln\frac{1}{0.21} = -(1)(2)(1.56) = -3.12 \text{ Btu/°R}$$

$$\Delta S_{N_2} = -n_{N_2}\bar{R}\ln\frac{1}{0.79} = -(3.76)(2)(0.237) = -1.78 \text{ Btu/°R}$$

EXAMPLE 14.7. A mixture of carbon dioxide and nitrogen is to be separated in steady flow. The mixture stream is at 1 atm pressure and 27°C and the mole fraction of CO_2 is 0.60. The separated streams are each to be at 1 atm and 27°C. Compute the minimum work required per mole of mixture.

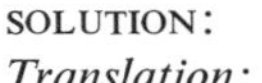
SOLUTION:

Translation:

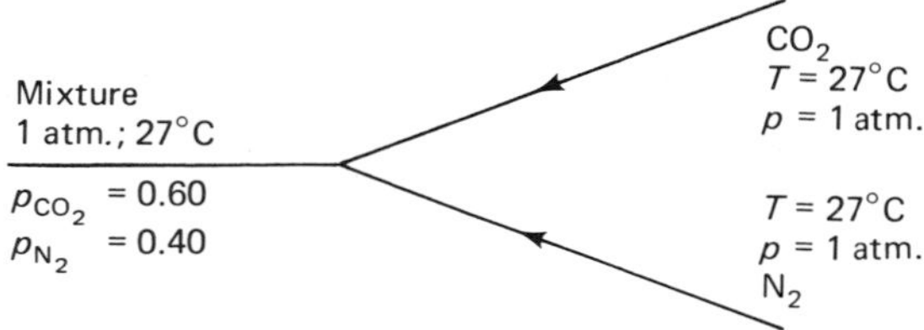

First law:

$\Delta H = 0$ since $T =$ constant and all gases are assumed ideal.

Second law:

$$W_{\substack{\text{minimum}\\\text{required}}} = A_2 - A_1$$

$$= (H - T_0 S)_2 - (H - T_0 S)_1$$

$$W_{\substack{\text{minimum}\\\text{required}}} = T_0(S_1 - S_2) = -T_0(\Delta S_{CO_2} + \Delta S_{N_2})$$

$$\Delta S_{CO_2} = -n_{CO_2} \bar{R} \ln \frac{1}{0.60} = -(0.60)(8.312)(0.511) = -2.55 \text{ j/°K}$$

$$\Delta S_{N_2} = -n_{N_2} \bar{R} \ln \frac{1}{0.40} = -(0.40)(8.312)(0.916) = -3.05 \text{ j/°K}$$

$$W_{\substack{\text{minimum}\\\text{required}}} = +(300)(2.55 + 3.05) = 1680 \text{ j}$$

14-5 Air-water mixtures

14-5(a) Unique properties

The atmospheric substance called *air* is actually an air–water vapor mixture. The amount of water present is quite variable and this variation is extremely important to human comfort ("It isn't the heat, it's the humidity"!). Therefore, of course, it is also important in air-conditioning processes. For engineering purposes, we treat this mixture as a mixture of *two* ideal gases, air and water vapor. In doing so we recognize that the *air* portion is not constant from place to place or time to time. That's partly what the air pollution problem is all about. But the variations from normal, though perhaps very important chemically and biologically, are not very significant in quantitative energy terms. Assuming that air is, by volume, 79% nitrogen and 21% oxygen, neglects about 1% by volume of other gases (mostly argon) and is the standard assumption made. The result is an ideal gas with the following properties:

$$\begin{aligned} MW_a &= 29 \text{ lbm/lbmole} \\ R_a &= 53.3 \text{ ft-lbf/lbm-°R} \\ c_{p_a} &= 0.241 \text{ Btu/lbm-°F} \\ c_{v_a} &= 0.172 \text{ Btu/lbm-°F} \end{aligned}$$

The water vapor associated with the air is at very low partial pressure even on very high humidity days. It is so low, in fact, that even as saturated vapor it may be considered to be an ideal gas with the following properties:

$$\begin{aligned} MW_{wv} &= 18 \\ R_{wv} &= 85.8 \text{ ft-lbf/lbm-°R} \\ c_{p_{wv}} &= 0.45 \text{ Btu/lbm-°R} \\ c_{v_{wv}} &= 0.34 \text{ Btu/lbm-°R} \end{aligned}$$

The fact that the amount of water vapor varies in engineering applications of air–water vapor mixtures leads to some rather special new properties that enable us to handle these mixtures fairly conveniently. These are relative humidity (ϕ), dew point (t_{dp}), specific humidity (ω), and wet bulb temperature (t_{wb}).

Relative humidity is defined as the actual partial pressure of the water vapor in the mixture divided by the pressure of *saturated* vapor at the same temperature. Refer to the skeleton property charts for H_2O shown in Fig. 14.2. The relative humidity is

$$\phi = \frac{P_{wv}}{P_g} \tag{14-46}$$

To find P_g, one merely looks up in the steam tables the value of the saturation pressure corresponding to the temperature of the air–water vapor mixture. Unfortunately relative humidity is not easily measured with accuracy and, though it is a term in common use, it is not very useful for engineering purposes.

If we assume $pV = mRT$ applies to the water vapor (as it does with considerable accuracy because of the very low pressure), then for a given volume of mixture

$$\phi = \frac{p_{wv}}{p_g} = \frac{m_{wv} RT}{v} \cdot \frac{v}{(m_{wv})_{\max} RT}$$

$$\phi = \frac{m_{wv}}{(m_{wv})_{\max}}$$

That is, relative humidity may be thought of as the ratio of mass of water vapor per unit volume of a mixture to the maximum amount the same volume of mixture could contain as vapor at the same temperature.

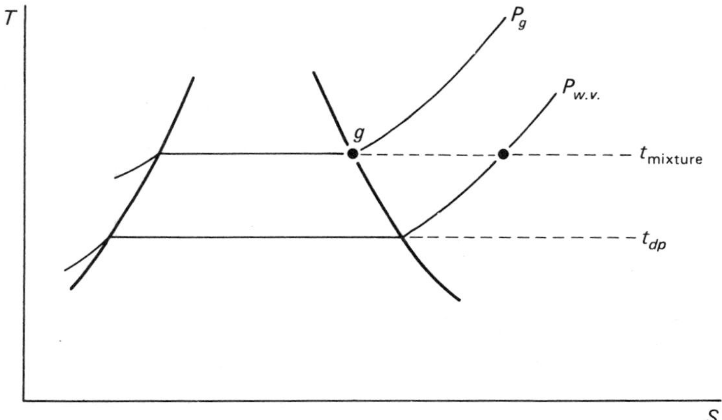

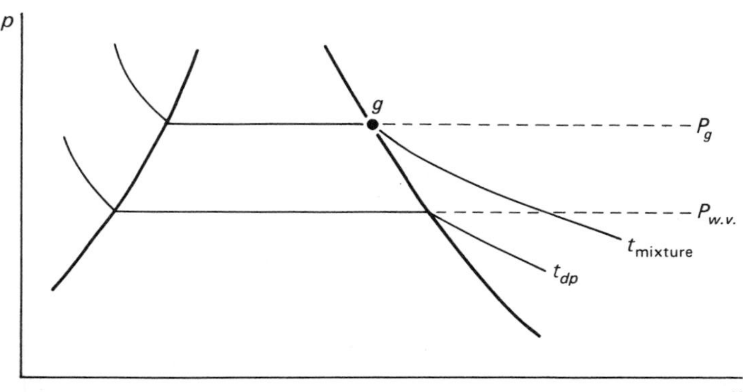

Figure 14.2 The dew point temperature shown on T–S and p–v plots.

The *dew point* is defined as the temperature at which condensation would begin if the mixture were cooled at constant pressure. On the T-S and p-v charts in Fig. 14.2 it is labeled as t_{dp}. In the tables, this is found as the saturation temperature corresponding to the partial pressure of the water vapor in the mixture.

The *specific humidity* is defined as the mass of water vapor in the mixture divided by the mass of air (*not* of mixture). That is, the mass of mixture is iven by

$$m = m_a + m_{wv} \tag{14-47}$$

and specific humidity is

$$\omega = \frac{m_{wv}}{m_a} \tag{14-48}$$

If we use the equation of state, we find

$$\omega = \frac{p_{wv} v / R_{wv} T}{p_a v / R_a T} = \frac{R_a}{R_{wv}} \frac{p_{wv}}{p_a} = 0.622 \frac{p_{wv}}{p_a} \tag{14-49}$$

$$\omega = 0.622 \frac{p_{wv}}{p - p_{wv}} \tag{14-50}$$

where p is the total pressure of the mixture. We shall see later when we deal with air-conditioning processes why specific humidity is used.

Sometimes specific humidity is reported in *grains* per pound of dry air. One pound-mass contains 7000 grains.

The *wet bulb temperature* is a number determined experimentally by passing a high velocity stream of the mixture over a thermometer kept wet by being wrapped in a cloth wick saturated with water. Evaporation of liquid from the wick reduces the temperature recorded. The wet bulb temperature is always intermediate between the dew point and the mixture temperature. It is important because it has been shown to be both theoretically and empirically related to the partial pressure of water vapor in the mixture. The empirical equation for partial pressure is

$$p_{wv} = p_{g_{wb}} - \frac{(p - p_{g_{wb}})(t - t_{wb})}{2800 - t_{wb}} \tag{14-51}$$

where p_{wv} = partial pressure of water vapor, psia
$p_{g_{wb}}$ = saturation pressure of H_2O at the wet bulb temperature, psia
p = total pressure of mixture, psia
t = (dry bulb) temperature, °F
t_{wb} = wet bulb temperature, °F

Wet bulb temperature is an experimental approximation to a theoretical parameter called the *temperature of adiabatic saturation*. The temperature of adiabatic saturation is the temperature an air–water vapor mixture will reach if it is saturated with water vapor in a completely adiabatic process. For practical purposes, the wet bulb temperature is a satisfactory approximation. As we shall see, it is related not only to the partial pressure of the water vapor but also to the enthalpy h^* of the mixture.

EXAMPLE 14.8. An air–water vapor mixture at 18 psia has a temperature of 100°F and a relative humidity of 60%. Compute the pound-mass of water vapor per pound-mass of air *and* per pound-mass of mixture. Also find the dew point.

SOLUTION: By definition,

$$\omega = \frac{m_{wv}}{m_a} = 0.622 \frac{p_{wv}}{p_a}$$

$$\phi = \frac{p_{wv}}{p_g}$$

From the tables, at 100°F, we find $p_g = 0.949$ psia. Thus

$$p_{wv} = 0.60(0.949) = 0.5694 \text{ psia}$$

and

$$p_a = 18.0 - 0.569 = 17.431 \text{ psia}$$

So

$$\omega = 0.622 \frac{0.5694}{17.431} = 0.0203 \frac{\text{lbm water vapor}}{\text{lbm air}}$$

and

$$\frac{\text{lbm water vapor}}{\text{lbm mixture}} = \frac{m_{wv}}{m_{wv} + m_a} = \frac{1}{1 + (m_a/m_{wv})} = \frac{1}{1 + (1/0.0203)}$$

$$= 0.0199 \frac{\text{lbm water vapor}}{\text{lbm mixture}}$$

The dew point is found from the tables. It is the saturation pressure corresponding to 0.5694 psia and lies between 83° and 84°F.

EXAMPLE 14.9. An air–water vapor mixture at 15 psia and 80°F has a wet bulb temperature of 60°F. Compute the relative humidity and specific humidity.

SOLUTION: The empirical relation for partial pressure of water vapor when wet bulb temperature is known is

$$p_{wv} = p_{g_{wb}} - \frac{(p - p_{g_{ab}})(t - t_{wb})}{2800 - t_{wb}}$$

From the tables, we find $p_{g_{ab}} = 0.256$ psia.

$$p_{wv} = 0.256 - \frac{(15.0 - 0.256)(80 - 60)}{(2800 - 60)}$$

$$= 0.256 - \frac{(14.744)(20)}{(2740)} = 0.256 - 0.108$$

$$p_{wv} = 0.148 \text{ psia}$$

Thus

$$\phi = \frac{p_{wv}}{(p_g)} = \frac{0.148}{0.507} = 0.292 = 29.2\%$$

and

$$\omega = 0.622 \frac{p_{wv}}{p_a} = (0.622)\left(\frac{0.148}{14.852}\right) = 0.00620$$

14-5(b) Enthalpy and specific volume

Because the amount of water vapor in systems handling air–water vapor mixtures is usually changing constantly, the weight of the mixture is also changing. This makes it inconvenient to put calculations on a per pound of mixture basis. What does not change is the (dry) air being handled or circulated. Hence the mass of (dry) air is usually taken as the basis. Instead of dealing with the enthalpy of the mixture per pound of mixture, we define an enthalpy of the mixture per pound of (dry) air, h^*. To illustrate this, suppose that a mixture contains m_a lbm of (dry) air and m_{wv} lbm of water vapor. Then the mass of mixture is

$$m = m_a + m_{wv} \tag{14-47}$$

The enthalpy of m is

$$H = m_a h_a + m_{wv} h_{wv} \tag{14-52}$$

We now define h^* such that

$$H = m_a h^* \tag{14-53}$$

Hence

$$m_a h^* = m_a h_a + m_{wv} h_{wv} \tag{14-54}$$

$$h^* = h_a + \frac{m_{wv}}{m_a} h_{wv} \tag{14-55}$$

$$h^* = h_a + \omega h_{wv} \tag{14-56}$$

An empirical relationship for h_{wv} that is sufficiently accurate for most air-conditioning engineering purposes is

$$h_{wv} = 1061.0 + 0.445t \tag{14-57}$$

where t is temperature (dry bulb) in Fahrenheit. Thus,

$$h^* = h_a + \omega(1061 + 0.445t) \tag{14-58}$$

Or, taking the air as an ideal gas with $h = 0$ at $t = 0°\text{F}$,

$$h^* = 0.241t + \omega(1061.0 + 0.445t) \tag{14-59}$$

This should make the usefulness of specific humidity apparent. As shown in Example 14.9, ω can be quickly computed from measurements of (dry bulb) temperature and wet bulb temperature. Thus, we easily can compute the enthalpy of the mixture *per pound of (dry) air* from a measurement of wet bulb and dry bulb temperatures. There could be little special usefulness in defining ω if we were interested in the enthalpy per unit of mixture but, as we have noted, air-conditioning analyses are much more practically done with the air only as the basis and this makes the specific humidity ω especially helpful.

EXAMPLE 14.10. Compute the enthalpy h^* of the air–water vapor mixture in Example 14.9.

SOLUTION:

$$h^* = 0.241t + \omega(1061 + 0.445t)$$
$$h^* = (0.241)(80) + 0.00625[1061 + 0.445(80)]$$
$$h^* = 19.28 + 6.85 = 26.1 \text{ Btu/lbm dry air}$$

EXAMPLE 14.11. An air–water mixture enters an air-conditioning unit at 80°F dry bulb and 60°F wet bulb temperatures. (See Examples 14.9 and 14.10.) It leaves at a dry bulb temperature of 60°F and a specific humidity of 0.0040. Compute the heat transferred and water removed *per pound of dry air.*

SOLUTION: From the steady flow energy equation, we find

$$\dot{Q}_{\text{out}} = \dot{m}_a(h^*_{\text{out}} - h^*_{\text{in}})$$
$$h^*_{\text{out}} = (0.241)(60) + 0.004[1061 + 0.445(60)]$$
$$h^*_{\text{out}} = 14.46 + 4.35 = 18.81 \text{ Btu/lbm dry air}$$

Thus

$$\frac{\dot{Q}}{\dot{m}_a} = 26.1 - 18.8 = 7.3 \text{ Btu/lbm dry air}$$

The water removed per pound of air is simply

$$\omega_{\text{in}} - \omega_{\text{out}} = 0.00625 - 0.0040 = 0.00225 \,\frac{\text{lb water vapor}}{\text{lb dry air}}$$

Specific humidity is also helpful in determining the volume of mixture to be handled, something that is needed to select fans and design ductwork. Consider a volume V of the mixture. Then

$$V = mv = m_a v^* = m_a v_a + m_{wv} v_{wv} \tag{14-60}$$

If the partial volumes are used,

$$m_a v^* = m_a v'_a + m_{wv} v'_{wv} \tag{14-61}$$

$$v^* = v'_a + \omega v'_{wv} \tag{14-62}$$

EXAMPLE 14.2. An air–water vapor mixture exists at 1 atm and 80°F dry bulb and 60°F wet bulb temperatures. Compute the specific volume per pound of dry air. (See Example 14.9).

SOLUTION:

$$v^* = v'_a + \omega v'_{wv}$$

Take $m_a = 1$ lbm as a basis. Then

$$v'_a = \frac{(53.3)(540)}{(15.00)(144)} = 13.3 \text{ ft}^3/\text{lbm}$$

$$v'_{wv} = \frac{(85.8)(540)}{(15.0)(144)} = 21.4 \text{ ft}^3/\text{lbm}$$

$$v^* = 13.3 + 0.00620(21.4)$$

$$v^* = 13.4 \text{ ft}^3/\text{lbm dry air}$$

14-5(c) Psychrometric charts

Charts called *psychrometric* charts have been developed to help reduce the computations in problems such as those in the preceding examples. They are published in a variety of styles but all have the basic features shown in the skeleton chart in Fig. 14.3. Charts are based on a specified total pressure, usually standard atmospheric (14.696 psia). Having taken the standard measurements of dry bulb temperature and wet bulb temperature, the charts may be used directly to find the relative humidity, the specific humidity,

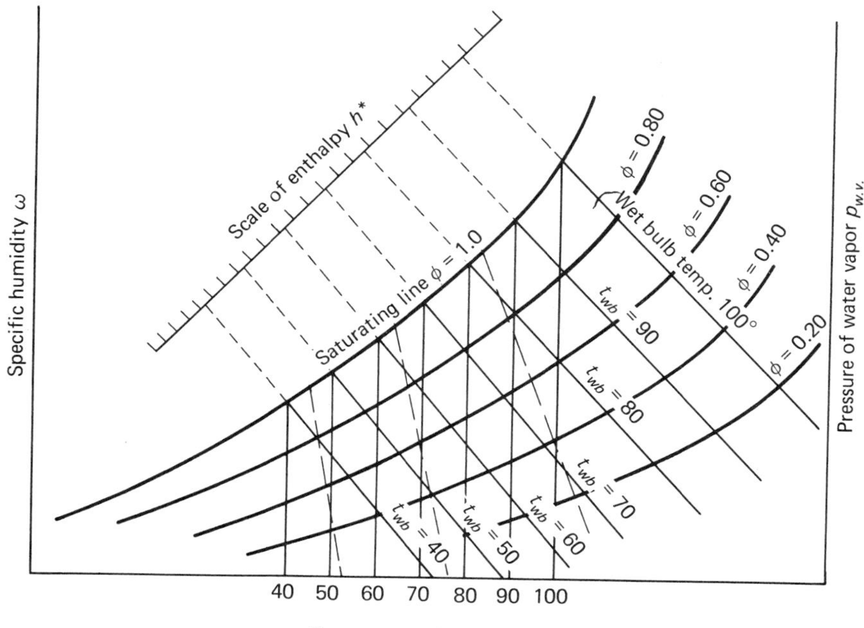

Figure 14.3 Skeleton psychrometric chart.

and the enthalpy h^*. Thus the charts are a great time and calculation saver but remember that they are approximate only.

A working chart is available in Appendix C.

EXAMPLE 14.13. Using the psychrometric chart in Appendix C, for a mixture with $t = 80°\text{F}$ and $t_{wb} = 60°\text{F}$, find ϕ, t_{dp}, specific volume per pound-mass of dry air, p_{wv}, and h^*.

SOLUTION: From the chart,

$$\phi = 0.30 = 30\%$$
$$t_{dp} \simeq 46\text{–}47°\text{F}$$
$$v = 13.74 \text{ ft}^3/\text{lbm dry air}$$
$$p_{wv} \simeq 0.14 \text{ psia}$$
$$h^* = 26.4 \text{ Btu/lbm dry air}$$

Note: Our previous calculation gave

$$\phi = 29.2\%$$
$$p_{wv} = 0.148 \text{ psia}$$
$$h^* = 26.1 \text{ Btu/lbm dry air}$$

14-6 Introduction to air conditioning

This section provides only the briefest introduction to air conditioning but this much does enable us to apply the material presented in this chapter as well as the first law in an interesting and useful way.

Reference is made to Fig. 14.4. There a space to be conditioned is shown schematically together with a unit to provide the necessary amount of properly conditioned air–water mixture. The system shown there is the typical summer cooling and dehumidifying system. In winter, it is usually necessary to heat and humidify in order to condition a space for maximum human comfort fully. The space is assumed to have a *space load* consisting of heat transfer ($\dot{Q}_s$) from the outside, from people, from lights and processes, etc., and of the energy associated with the moisture input ($m_{wv,s}$) from people or processes. Computation of space loads is a complex and important aspect of air-conditioning work but beyond our scope here.

It is desired to maintain the space at a temperature of t_r and a relative humidity of ϕ_r. These conditions could be specified by using $t_{r,wb}$ or ω_r but it is normal practice to use temperature and relative humidity. Space conditions are selected to meet the comfort needs of people and/or the operational needs of equipment or processes in the space. Another space condition that may also be required has to do with air velocity and distribution but we shall not consider that here. If there are people in the space, it is also often necessary

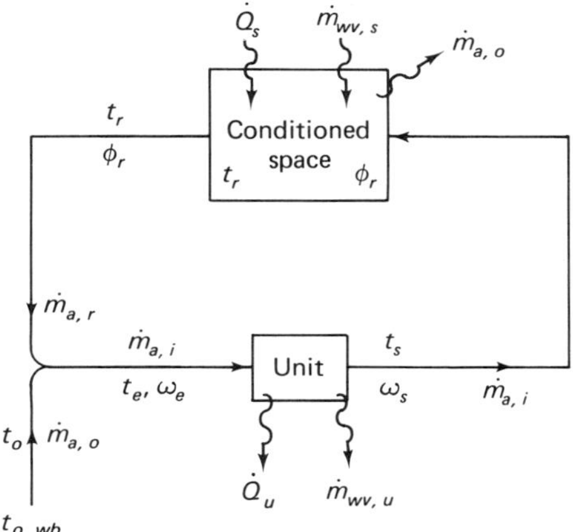

Figure 14.4 Schematic of typical air-conditioning system.

to introduce a certain amount of fresh outside air continuously into the system. In Fig. 14.4 this is done by bringing outside air $\dot{m}_{a,o}$ into the unit by mixing it with the return air $\dot{m}_{a,r}$. A mass balance requires that an equal amount ($\dot{m}_{a,o}$) of air be *ex*filtrated from the space. This is done by keeping a slightly positive air pressure in the space so that air leakage is out and not in. Such a procedure is the best way to control the space conditions but, for reasons of cost, it is not always used in smaller systems.

The return air ($\dot{m}_{a,r}$) is at t_r and ϕ_r and is mixed with the fresh air as shown. The mixture entering the unit is symbolized as $\dot{m}_{a,i}$ at t_e and ω_e. The unit removes $\dot{Q}_u$ of heat and $\dot{m}_{wv,u}$ of moisture from the mixture passing through and supplies it to the space at t_s and ω_s.

In most air-conditioning situations, the following quantities are either specified or determined by approximate analysis:

Space heat load:	$\dot{Q}_s$
Space moisture load:	$\dot{m}_{wv,s}$
Space conditions:	t_r, ϕ_r
Outside air needed:	$\dot{m}_{a,o}$
Outside air conditions:	t_o, ϕ_o
Temperature of supply air:	t_s

The temperature of supply air (t_s) is usually fixed by the minimum temperature of air that can be introduced into the space. For example, if people are present, extreme air that is too cold cannot be put in without creating discomfort. Outside air conditions are usually taken to be those hot and humid conditions that occur often enough to matter. They are really accepted by common practice and are usually in the range of 85 to 95°F temperature and 70 to 75°F wet bulb temperature.

With the quantities above given or computed, the air-conditioning problem is to find the state and rate of air to be supplied by the unit to the space (i.e., to find ω_s and $\dot{m}_{a,i}$) and to determine the unit loads ($\dot{Q}_u$ and $\dot{m}_{wv,u}$). With this information the engineer can then select a unit that will do the job and design the remainder of the system.

The problem thus has four unknowns: $\dot{m}_{a,i}$, ω_s, $\dot{Q}_u$, $\dot{m}_{wv,u}$. The first two unknowns are found by applying the steady flow continuity and energy equations simultaneously to the space. The second two unknowns are found by applying the same principles to the unit. The method is illustrated by the following example.

EXAMPLE 14.4. A room to be air conditioned has a space load of 100,000 Btu/hr and 20 lbm water vapor per hour. The room conditions are to be 80°F and 50% relative humidity. Outside air at design conditions of 95°F temperature and 75°F wet bulb temperature is to be brought into the unit as shown in Fig. 14.4 at a rate of 1000 ft³/min. The conditioned air supplied to the room is to be at a dry bulb temperature of 65°F. Compute the required mass rate and volume of conditioned air and its specific humidity. Also compute the unit load in tons. You may use the psychrometric chart.

SOLUTION:

Translation:

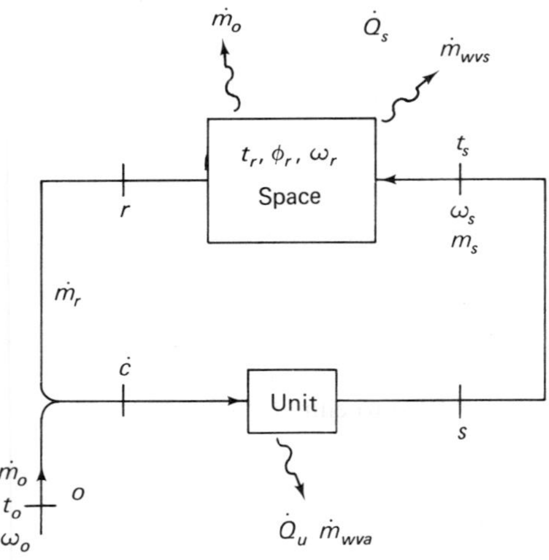

First law:

Space:

$$\dot{m}_s h_s^* + \dot{Q}_s = (\dot{m}_o + \dot{m}_r)h_r^* = \dot{m}_s h_r^*$$

From the psychrometric chart,

$$h_r^* = 31.1 \text{ Btu/lbm dry air}$$

Thus,

$$\dot{m}_s h_s^* + 100{,}000 = \dot{m}_s(31.1) \tag{14-63}$$

Water vapor mass balance:
Space:

$$\dot{m}_s \omega_s + \dot{m}_{wvs} = \dot{m}_s \omega_r$$

From the chart,

$$\omega_r = 76 \text{ grains/lbm dry air} = 0.0110 \frac{\text{lbm water vapor}}{\text{lbm dry air}}$$

Thus,

$$\dot{m}_s \omega_s + 20 = \dot{m}_s(0.0110) \tag{14-64}$$

Equations (14-63) and (14-64) may be solved simultaneously by trial and error using the psychrometric chart or by using

$$h_s^* = c_{pa} t_s + \omega_s(1061 + 0.445 t_s)$$

Using the latter, we find

$$h_s^* = (0.241)(65) + \omega_s[1061 + 0.445(65)]$$
$$h_s^* = 15.65 + \omega_s(1090)$$

Equation (14-63) becomes

$$\dot{m}_s(15.65 + 1090\omega_s) + 100{,}000 = \dot{m}_s(31.1) \tag{14-65}$$

Now Eq. (14-64) and (14-65) can be solved for $\dot{m}_s$ and ω_s.
From (14-64), we find

$$\omega_s = 0.0110 - \frac{20}{\dot{m}_s}$$

Inserting this into Eq. (14-65) to eliminate ω_s gives

$$\dot{m}_s\left[15.65 + 1090\left(0.0110 - \frac{20}{\dot{m}_s}\right)\right] + 100{,}000 = 31.1\dot{m}_s$$
$$15.65\dot{m}_s + 20.3\dot{m}_s - 11{,}990 + 100{,}000 = 31.1\dot{m}_s$$
$$4.85\dot{m}_s = 88{,}000$$
$$\dot{m}_s = 18{,}140 \text{ lbm/hr}$$

Going back to Eq. (14-68) to find ω_s,

$$\omega_s = 0.0110 - \frac{20}{16{,}100} = 0.00975 \frac{\text{lbm water vapor}}{\text{lbm dry air}} = 68.3 \frac{\text{grains}}{\text{lbm dry air}}$$

Before we can compute the unit load, we must find the conditions entering the unit. This involves a mixture of $\dot{m}_r$ and $\dot{m}_o$ where

$$\dot{m}_r + \dot{m}_o = \dot{m}_s$$

From the chart, we find

$$h_r^* = 31.1 \text{ Btu/lbm dry air}$$
$$h_o^* = 38.6 \text{ Btu/lbm dry air}$$

To find $\dot{m}_o$, we note that the volume flow rate is given as 1000 ft^3/min or 60,000 ft^3/hr From the chart, the specific volume per pound of dry air is 14.3 ft^3/lbm dry air. Thus,

$$\dot{m}_o = \frac{60{,}000}{14.3} = 4200 \text{ lbm/hr}$$

Now the energy equation applied to the mixer section gives

$$\dot{m}_o h_o^* + \dot{m}_r h_r = \dot{m}_i h_i^* = \dot{m}_s h_i^*$$

$$h_i^* = \frac{\dot{m}_o}{\dot{m}_s} h_o^* + \frac{\dot{m}_r}{\dot{m}_s} h_r^*$$

$$= \left(\frac{4200}{16{,}100}\right)(38.6) + \left(\frac{11{,}900}{16{,}100}\right)(31.1)$$

$$h_i^* = 10.0 + 23.1 = 33.1 \text{ Btu/lbm dry air}$$

An energy balance in the unit gives

$$\dot{Q}_u = \dot{m}_s(h_i - h_s) = 16{,}100(33.1 - h_s^*)$$

We know that $t_s = 65°\text{F}$ and $\omega_s = 68.3$ grains/lbm dry air so from the chart we can find h_s^*. It is

$$h_s^* = 26.5 \text{ Btu/lbm dry air}$$

Thus

$$\dot{Q}_u = 16{,}100(33.1 - 26.5)$$
$$\dot{Q}_u = 106{,}260 \text{ Btu/hr} = 21 \text{ tons}$$

14-7 Summary

A mixture of ideal gases is itself an ideal gas in which each component can be assumed to occupy the total volume without being influenced by the other components.

$$V = V_a = V_b = V_c = \cdots \tag{14-1}$$

The total pressure of the mixture is the sum of the partial pressures of the components.

$$p = p_a + p_b + p_c + \cdots \tag{14-7}$$

The partial pressure of each component is determined by its mole fraction.

$$p_a = \frac{n_a}{n} p \qquad p_b = \frac{n_b}{n} p \tag{14-11}$$

The ideal gas equation of state may be applied to the mixture or to each component.

$$p_a V = n_a \bar{R} T \tag{14-12}$$

$$pV = n\bar{R}T \tag{14-14}$$

Properties of mixtures may be computed from the known properties of the components.

$$c_v = \frac{m_a}{m} c_{va} + \frac{m_b}{m} c_{vb} + \frac{m_c}{m} c_{vc} + \cdots \tag{14-21}$$

$$\frac{1}{\text{MW}} = \frac{m_a}{m} \frac{1}{\text{MW}_a} + \frac{m_b}{m} \frac{1}{\text{MW}_b} + \frac{m_c}{m} \frac{1}{\text{MW}_c} + \cdots \tag{14-29}$$

Analysis of mixing processes is done by straightforward application of the first and second laws to the process. However, the general equation

$$\Delta s_{12} = mc_p \ln \frac{T_2}{T_1} - mR \ln \frac{p_2}{p_1} \tag{14-32}$$

is valid only for a mixture that does not change composition. Entropy changes can be computed from Eq. (14-32) in mixing situations only by applying it to the components separately, using their partial pressures in the mixture. The total entropy change is then obtained by summing.

Separation processes are often best analyzed by dealing with them in reverse, that is, as mixing processes.

Air–water vapor mixtures are treated as a mixture of two ideal gases: air and water vapor. Relative humidity is the ratio of the actual partial pressure of water vapor in the mixture to the saturation pressure (p_g) at the same temperature.

$$\phi = \frac{p_{wv}}{p_g} \tag{14-46}$$

The dew point is the temperature at which condensation would occur if the mixture were cooled at constant total pressure. The specific humidity is the ratio of the mass of water vapor in the mixture to the mass of air in the mixture.

$$\omega = \frac{m_{wv}}{m_a} \tag{14-48}$$

$$= 0.622 \frac{p_{wv}}{p - p_{wv}} \tag{14-50}$$

The enthalpy of an air–water vapor mixture is needed for convenience in air-conditioning work on a basis of unit mass of air and not of the mixture. On this basis,

$$h^* = h_a + \omega h_{wv} \tag{14-56}$$

$$= 0.241t + \omega(1061 + 0.445t) \tag{14-59}$$

Psychrometric charts are charts of properties of air–water vapor mixtures that are often used in lieu of calculation of properties. It must be noted that published charts are based on a total pressure of 14.696 psia.

Problems

14-1 Show that, for a homogeneous mixture of real gases,

$$\rho = \rho_1 + \rho_2 + \rho_3 + \cdots = \sum_i \rho_i$$

and thus

$$\frac{1}{v} = \frac{1}{v_1} + \frac{1}{v_2} + \frac{1}{v_3} + \cdots = \sum_i \frac{1}{v_i}$$

Is it necessary for the gases to be ideal or to obey Gibbs–Dalton law for these relations to be true?

14-2 A mixture of two ideal gases executes a reversible adiabatic process. Show that if each gas in this process is to execute an isentropic process also, it is necessary that they have equal values of the isentropic exponent, $\gamma(\gamma = c_p/c_v)$.

14-3 During the power stroke of an internal combustion engine, the products of combustion are composed of 78% N_2, 16% CO_2, 4% O_2, and 2% H_2O, *by mass*. At a certain point of the stroke, the temperature is 500°F, the pressure is 150 psia, and the volume is 0.10 ft^3. A short time later the temperature is 350°F and the pressure is 80 psia. Analysis of the indicator card for this period shows that the gases do 1100 ft-lb of work on the piston. Find the heat transfer during this time.

14-4 A city waterline carrying water at 45°F passes through a factory. The factory air temperature is 79°F. What is the maximum relative humidity that can be maintained and not have any condensation on the pipe? The barometer reads 29.92 in. Hg.

14-5 Typical conditions on a summer day in Pittsburgh, Pa., are 95°F dry bulb temperature and 50% relative humidity. On such a day, what is the minimum temperature to which water could be cooled in a cooling tower? The barometer reads 29.92 in. Hg.

14-6 The barometer reads 28.50 in. Hg (14.0 psia). The temperature and relative humidity are 80°F and 40%, respectively. Compute the specific humidity, ω.

14-7 Air at 32°F and 50% relative humidity is to be changed in an air conditioner to 70°F and 50% relative humidity. Water at 50°F is supplied in the spray chamber at a rate of 0.01 lb/lbm of dry air entering and drains from it at the same temperature. The barometer reads 29.92 in. Hg.

(a) Find the amount of excess water drained off per pound of dry air entering.

(b) Estimate the amount of heat that is supplied from an external source per pound of dry air.

14-8 A certain gas mixture is composed of 30% CO_2, 50% O_2, and 20% N_2, *by mass*. The mixture is initially at 70°F and 17 psia. After passing through the heat exchanger in steady flow, the mixture is at 110°F and 14.7 psia. Find the change in entropy per pound of mixture and the heat transferred per pound of mixture.

14-9 A steady stream of 1 lb/min of oxygen ($M = 32$) at 240°F and 1 atm is mixed with a steady stream of ethane (G_2H_6) ($M = 30$) of 2 lbm/min at 40°F and 1 atm pressure. The pressure of the mixture is 1 atm. Find the degradation in the mixing process. For oxygen, $c_{p*} = 7.0$ Btu/lbmole-°R. For ethane, $c_{p*} = 12.4$ Btu/lbmole-°R.

14-10 A room for processing very high temperature molybdenum is to be filled with inert monatomic argon (atomic weight = 40) at 16 psia. The heat from external and internal sources amounts to 20,000 Btu/hr and moisture from internal sources is added at a rate of 10 lb/hr (assumed to be saturated vapor at 60°F). The room is to be kept a 60°F with a relative humidity of 10%. A dryer is provided as shown, which, in conjunction with the air-conditioning unit, may be assumed to dry the circulating argon completely. Leaks in the system may be neglected. Compute the required rate of circulation (pounds per minute) through the air-conditioning unit and dryer. Also compute the required temperature of the argon entering the room.

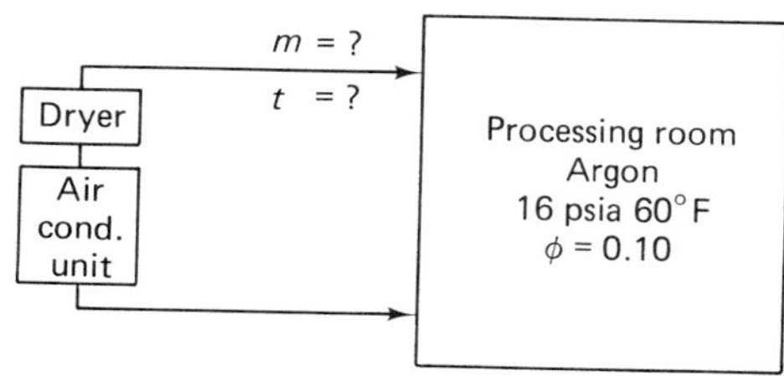

14-11 A room is to be air conditioned as shown in the sketch.

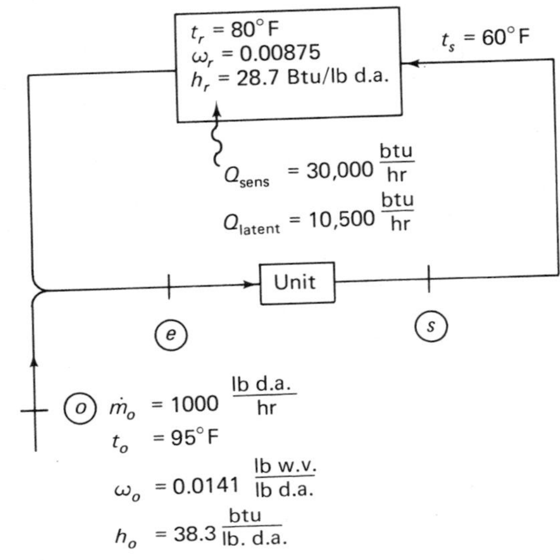

(a) Find the state and rate of supply air.
(b) Find the state of the mixture entering the unit.

14-12 A hospital operating room is to be air conditioned by using 100% outside air (i.e., none of the operating room air is to be recirculated through the air conditioner). The conditions to be maintained in the room are 80°F dry bulb temperature and 30% relative humidity. The load in the room (people and equipment) amounts to a heat addition of 25,000 Btu/hr and a moisture addition (assumed to be saturated vapor at 80°F) of 3 lb/hr. For comfort, the conditioned air entering the room must be at 65°F dry bulb temperature. Outside conditions are 95°F dry bulb temperature and 75°F wet bulb temperature. If the total air pressure is 14.7 psia, determine
(a) The specific humidity of the air entering the operating room.
(b) The air flow through the operating room in pounds per hour.
(c) The cooling load on the air conditioner in tons of refrigeration if the water leaving the air conditioner is assumed to be saturated at 65°F.

15

Combustion

15-1 Combustion reaction analysis

15-1(a)
When reactants are human

Combustion is the process of oxidizing or burning a fuel. The oxygen needed is usually supplied in air but it may be in other forms as it is in solid propellants. The fuel usually contains carbon and/or hydrogen that combine with the oxygen to form carbon and hydrogen oxides (i.e., CO_2, CO, H_2O, . . .). It is common to refer to the fuel and oxidizer as *reactants* and to the oxides and any other resulting materials as *products*. Thus the combustion process schematically is

$$\text{Reactants} \longrightarrow \text{products}$$
$$\text{Fuel} + \text{oxidizer} + \text{inerts} \longrightarrow \text{oxides} + \text{inerts}$$

Some examples of simple reactions that take place are the combustion of carbon

$$C + O_2 \longrightarrow CO_2$$

the combustion of hydrogen

$$H_2 + \tfrac{1}{2}O_2 \longrightarrow H_2O$$

and the combustion of hydrocarbons such as propane

$$C_3H_8 + 5O_2 \longrightarrow 3CO_2 + 4H_2O$$

The amount of oxygen in the relations above is called *stoichiometric oxygen.* The meaning of stoichiometric is that there is just enough O_2 to burn all the fuel. The coefficients in these equations are called the *stoichiometric coefficients.*

The combustion reaction relations above assume that just the right amount of pure oxygen is provided and that all the fuel is oxidized. This deviates from realism in at least three ways. First, as noted, the oxygen is usually provided as air and so there is also inert nitrogen present (in fact, 3.76 moles of N_2 for every mole of O_2). Second, with only the exact amount of oxygen present that is needed for complete combustion, it is very unlikely that there would be sufficient time or mixing in the combustion chamber for all the fuel and oxygen to meet and burn. Thus it is common to provide more oxygen than would be absolutely necessary for just complete combustion. Third, there may be some dissociation of compounds such as CO_2 into CO and O_2. This latter problem is beyond the scope of this treatment and is normally only a serious problem at very high temperatures. We can, however, handle the other two realisms.

First, let's consider the problem of using air instead of pure oxygen. For complete combustion of carbon, we obtain

$$C + O_2 + 3.76N_2 \longrightarrow CO_2 + 3.76N_2$$

For hydrogen,

$$H_2 + \tfrac{1}{2}O_2 + \tfrac{1}{2}(3.76)N_2 \longrightarrow H_2O + \tfrac{1}{2}(3.76)N_2$$

For a hydrocarbon, this time using octane as an example,

$$C_8H_{18} + 12.5O_2 + 12.5(3.76)N_2 \longrightarrow 8CO_2 + 9H_2O + 12.5(3.76)N_2$$

The amount of air required in the equations above is called the *stoichiometric air* or the *theoretical air.*

In order to accomplish complete combustion in a realistic system, it is necessary to provide an excess of air over the stoichiometric air. Usually from 20 to perhaps 100% *excess air* (or 120 to 200% of *stoichiometric* air) is needed. The following combustion relations illustrate differing amounts of excess air:

20% excess air:

$$C + (1.20)O_2 + (1.20)(3.76)N_2 \longrightarrow CO_2 + (0.20)O_2 + (1.20)(3.76)N_2$$

50% excess air:

$$CH_4 + 2(1.50)O_2 + 2(1.50)(3.76)N_2 \longrightarrow CO_2 + 2H_2O + O_2 + 2(1.50)(3.76)N_2$$

An insufficient amount of excess air will lead to incomplete combustion, perhaps resulting in CO or in unburned fuel.

Students should be able to set up these combustion relations and determine the coefficients from information given about the fuel and excess air. As an example, suppose we derive the combustion reaction for octane (C_8H_{18}) with 100% excess air (i.e., 200% stoichiometric air). We begin by writing the reactants and products without coefficients:

$$C_8H_{18} + O_2 + N_2 \longrightarrow CO_2 + H_2O + N_2$$

Taking 1 mole of the fuel as a basis, we now note that balance requires that there be 8 carbons and 18 hydrogens in the products. Thus

$$C_8H_{18} + O_2 + N_2 \longrightarrow 8CO_2 + 9H_2O + N_2$$

Next we see that balance of oxygen requires 16 from the $8CO_2$ and 9 from the $9H_2O$, or 25. That is 12.5 O_2's. But for each O_2 we know there will be 3.76 N_2's so

$$C_8H_{18} + 12.5O_2 + 12.5(3.76)N_2 \longrightarrow 8CO_2 + 9H_2O + 12.5(3.76)N_2$$

This is the stoichiometric air equation, however, and we have 100% *excess* air. That is, twice as much air as absolutely needed:

$$C_8H_{18} + 2(12.5)O_2 + (2)(12.5)(3.76)N_2 \longrightarrow 8CO_2 + 9H_2O + (2)(12.5)(3.76)N_2 + 12.5O_2$$

Note the unused 12.5 O_2's among the products. A check will now show every element balanced.

It is customary to define the *air–fuel* (A/F) ratio in combustion processes. It is the ratio of either the moles or mass of fuel to the corresponding amount of air. Thus we speak of either the molar or the mass air–fuel ratio. The stoichiometric or theoretical air–fuel ratio is the ratio when the stoichiometric or theoretical amount of air is used. For example, in the combustion of methane, the stoichiometric reaction relation is

$$C_3H_8 + 5O_2 + 5(3.76)N_2 \longrightarrow 3CO_2 + 4H_2O + 5(3.76)N_2$$

and the stoichiometric *molar* air–fuel ratio is

$$(A/F)_{molar} = \frac{5 + 5(3.76)}{1}$$

$$= 23.8 \frac{\text{moles of air}}{\text{mole of fuel}}$$

The stoichiometric *mass* air–fuel ratio is

$$(A/F)_{mass} = \frac{(23.8)(29)}{(1)(44)}$$

$$= 15.7 \frac{\text{lbm of air}}{\text{lbm of fuel}} \quad \text{or} \quad \frac{\text{kg of air}}{\text{kg of fuel}}$$

If 50% excess air (or 150% theoretical air) is used, the combustion relation for propane is

$$C_3H_8 + (1.50)(5)O_2 + (1.50)(5)(3.76)N_2 \longrightarrow 3CO_2 + 4H_2O + (0.50)(5)O_2 + (1.50)(5)(3.76)N_2$$

and the air–fuel ratios are

$$(A/F)_{\text{molar}} = \frac{(1.5)(5) + (1.5)(5)(3.76)}{1} = 35.7 \frac{\text{moles of air}}{\text{moles of fuel}}$$

$$(A/F)_{\text{mass}} = \frac{(35.7)(29)}{(1)(44)} = 23.5 \frac{\text{lbm of air}}{\text{lbm of fuel}} \quad \text{or} \quad \frac{\text{kg of air}}{\text{kg of fuel}}$$

EXAMPLE 15.1. Compute the molar and mass air–fuel ratio for the combustion of methane (CH_4) with 110% stoichiometric air.

SOLUTION:
Note: 110% stoichiometric air means 10% excess air.
The reaction relation is

$$CH_4 + (1.1)(2)O_2 + (1.1)(2)(3.76)N_2 \longrightarrow CO_2 + 2H_2O + (0.1)(2)O_2 + (1.1)(2)(3.76)N_2$$

The air–fuel ratios are

$$(A/F)_{\text{molar}} = \frac{(1.1)(2) + (1.1)(2)(3.76)}{(1)} = 10.47 \frac{\text{lbmoles of air}}{\text{lbmoles of fuel}}$$

$$(A/F)_{\text{mass}} = \frac{(10.45)(29)}{(1)(16)} = 19.0 \frac{\text{lbm of air}}{\text{lbm of fuel}} \quad \text{or} \quad \frac{\text{kg of air}}{\text{kg of fuel}}$$

15-1 (b) When products are known from Orsat data

It is often the case that an existing combustion process must be analyzed. The purpose may be to determine its efficiency, its pollution effects, etc. An experimental tool known as an *Orsat apparatus* is commonly used in such cases to obtain data about the composition of the products of combustion. The data are then used in calculations to determine the complete nature of the combustion process.

In an Orsat apparatus, a measured volume of the products of combustion is bubbled through a series of prepared chemical solutions, each of which absorbs one of the components in the mixture. The volume of gases absorbed in each solution can be measured so that the composition *by volume* of the combustion products is determined. For example, the Orsat apparatus yields data such as the following from the combustion of methane (CH_4) with air:

$$
\begin{aligned}
CO_2 &— \ \ 8.68\% \\
O_2 &— \ \ 4.82\% \\
CO &— \ \ 0.46\% \\
N_2 &— \ \underline{86.04\%} \\
& \quad 100.00\%
\end{aligned}
$$

Any water vapor in the products of combustion is condensed long before the sample of gases is analyzed at room temperature in the Orsat device. Thus the data exclude water vapor and give the volumetric analysis of the dry portion of the mixture only. This fact must be kept in mind when the data provided by an Orsat apparatus is used.

To use Orsat data to determine the details of the combustion process, the first step is to write the combustion equation using undetermined coefficients but assuming as a basis that 100 moles of dry products is involved. That is, for the example under consideration,

$$
\begin{aligned}
aCH_4 + bO_2 + cN_2 &\longrightarrow \\
&8.68CO_2 + 0.46CO + 4.82O_2 + 86.04N_2 + dH_2O
\end{aligned}
$$

Next, to determine the unknown coefficients (a, b, c, and d) a mass balance on each of the elements (C, H, O, and N) is performed. The resulting equations can be solved for the unknowns.

In the example above, a nitrogen balance gives $c = 86.04$. Since all the nitrogen comes from the air in this case, we know that $c = 3.76b$ and therefore $b = 86.04/3.76 = 22.8$. A balance of the carbon yields

$$a = 8.68 + 0.46 = 9.14$$

A balance of the hydrogen yields

$$4a = 2d \quad \text{or} \quad d = 2a = 18.28$$

The oxygen balance in this case can be used as a check on the other calculations:

$$
\begin{aligned}
2b &= 2(8.68) + 0.46 + 2(4.82) + 18.28 \\
b &= 22.87 \qquad \text{(checks)}
\end{aligned}
$$

15-2 Enthalpy of formation and enthalpy of combustion

As engineers, we are interested in the energy released in combustion processes. In order for us to make computations of that energy we must first understand the concept called *enthalpy* (or heat) *of formation*. A reference state of zero enthalpy for all *elements* (*not* compounds) has been selected at 77°F (25°C) and 1 atm pressure. The enthalpy of formation of a *compound* is then defined as its enthalpy at the same standard conditions (1 atm and 77°F).

For example, consider carbon dioxide:

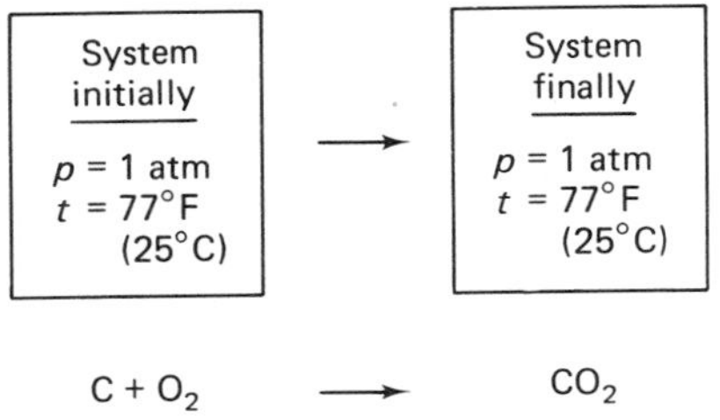

Experimentally it is found that the heat transfer for this process is $-169{,}293$ Btu/mole CO_2. The first law for a constant pressure process with no work other than ($p\,dv$)-type reduces to

$$d'Q - d'W = dU \tag{15-1}$$

$$d'Q - p\,dV = dU \tag{15-2}$$

$$d'Q = dV + p\,dU = dH \qquad (p = \text{constant}) \tag{15-3}$$

$$Q_{12} = \Delta H = H_2 - H_1 \qquad (p = \text{constant}) \tag{15-4}$$

We find the same result if we consider steady flow and neglect kinetic and potential energy terms. If we apply this result to the process above, we obtain

$$Q = H_p - H_r \tag{15-5}$$

where H_p is the enthalpy of the products and H_r is the enthalpy of the reactants. Since $H_r = 0$ at some reference temperature, we find

$$H_p = Q = -169{,}293 \text{ Btu/mole } CO_2$$

where $H_p = -169{,}293$ Btu is called the *enthalpy of formation* of CO_2 at the reference temperature. It is commonly given the symbol $\overline{h^0_{CO_2}}$, the overbar denoting the molar basis. Enthalpy of formation is really nothing more than the *enthalpy of compounds* at 77°F (25°C) and 1 atm when the enthalpy of *elements* is taken to be zero at these conditions. Table 15-1 lists the enthalpy of formation of a number of fuels.

Let us consider now a combustion process that takes place at constant pressure and in which the reactants and products enter and leave at 77°F and 1 atm pressure. The heat transfer will be designated by Q. If we write the first law for this process, we obtain, as before,

$$Q = H_p^0 - H_r^0 \tag{15-6}$$

Table 15.1 Enthalpies of formation

Substance	$\overline{h}^0$ (Btu/lbmole)	$\overline{h_0}$ (j/gm-mole)
Carbon, C, solid	0	0
Carbon monoxide, CO, gas	−47,550	−110,590
Carbon dioxide, CO_2, gas	−169,290	−393,741
Nitrogen, N_2, gas	0	0
Oxygen, O_2, gas	0	0
Water, H_2O, gas	−104,070	−242,050
Water, H_2O, liquid	−122,970	−286,000
Methane, CH_4, gas	−32,200	−74,890
Ethene, C_2H_4, gas	−22,490	−52,300
Ethane, C_2H_6, gas	−36,420	−84,710
Propane, C_3H_8, gas	−44,670	−103,900
Butane, C_4H_{10}, gas	−54,270	−126,220
Octane, C_8H_{18}, liquid	−107,530	−250,100

where the superscript zeros denote the standard conditions of 77°F (25°C) and 1 atm for products and reactants. Rewriting this on a per mole of fuel basis gives

$$\overline{Q} = \overline{H}_p - \overline{H}_r = \overline{H_{pr}^0} \tag{15-7}$$

where the overbar indicates the heat transfer per mole of fuel. The symbol $\overline{H_{pr}^0}$ is called the *enthalpy of combustion* and is simply the heat transfer in the combustion process per mole of fuel when the process is at constant pressure and both reactants and products are at 1 atm and 77°F (25°C). The corresponding symbol per unit mass of fuel is H_{pr}^0.

Since heat will be transferred *from* the system in a combustion process, H_{pr}^0 is negative. Care must be used, however, because different texts and references define different terms, some of which are the negative of the H_{pr}^0 as defined here. In particular, the term *heating value* is often used and taken to be positive but all books are not even consistent about this.

Different values are obtained for H_{pr}^0 depending on the phase state of the H_2O in the products. When the H_2O is taken to be liquid, the absolute numerical value of H_{pr}^0 is higher and this is called the *higher heating value*. When the H_2O is taken to be vapor, the absolute value of H_{pr}^0 is lower and this is called the *lower heating value*. Since in almost all practical cases the water vapor in the products is vapor, the lower value is the one that usually applies.

Values of the enthalpy of combustion in Btu/lbm for various fuels are listed in Table 15-2. To convert Btu/lbm to joules per gram, multiply by 2.326.

Table 15.2 Enthalpy of combustion of hydrocarbons at 25 °C. Reproduced from *Gas Tables* by Keenan and Kaye by permission of John Wiley & Sons, Inc., New York.

Hydrocarbon	*Composition*	*Liquid hydrocarbon* h_{RP} (Btu/lb)	*Gaseous hydrocarbon* h_{RP} (Btu/lb)
PARAFFIN SERIES			
Methane	CH_4	...	−21,502
Ethane	C_2H_6	...	−20,416
Propane	C_3H_8	−19,773[1]	−19,929
n-Butane	C_4H_{10}	−19,506[1]	−19,665
n-Pentane	C_5H_{12}	−19,340	−19,499
n-Hexane	C_6H_{14}	−19,233	−19,391
n-Heptane	C_7H_{16}	−19,157	−19,314
n-Octane	C_8H_{18}	−19,100	−19,256
n-Nonane	C_9H_{20}	−19,056	−19,211
n-Decane	$C_{10}H_{22}$	−19,020	−19,175

$$h_{RP} = (h_P - h_R)_{T,p}$$

where h_{RP} is the enthalpy of the gaseous products of combustion minus the enthalpy of the reactants, at 25°C and 1 atm, per pound of hydrocarbon.

For the liquid hydrocarbon, $$h_{RP} = \frac{-84{,}382 - 262{,}165n}{2.0160 + 14.0260n} \quad \text{for} \quad n > 5$$

For the gaseous hydrocarbon, $$h_{RP} = \frac{-85{,}244 - 264{,}288n}{2.0160 + 14.0260n} \quad \text{for} \quad n > 5$$

Hydrocarbon	*Composition*	*Liquid hydrocarbon* h_{RP} (Btu/lb)	*Gaseous hydrocarbon* h_{RP} (Btu/lb)
OLEFIN SERIES			
Ethene	C_2H_4	...	−20,276
Propene	C_3H_6	...	−19,683
1-Butene	C_4H_8	...	−19,483
1-Pentene	C_5H_{10}	...	−19,346
1-Hexene	C_6H_{12}	...	−19,262
1-Heptene	C_7H_{14}	...	−19,202
1-Octene	C_8H_{16}	...	−19,157
1-Nonene	C_9H_{18}	...	−19,122
1-Decene	$C_{10}H_{20}$	...	−19,094

For the gaseous hydrocarbon, $$h_{RP} = \frac{-35{,}242 - 264{,}288n}{14.0260n} \quad \text{for} \quad n > 5$$

Hydrocarbon	*Composition*	*Liquid hydrocarbon* h_{RP} (Btu/lb)	*Gaseous hydrocarbon* h_{RP} (Btu/lb)
ALKYLBENZENE SERIES			
Benzene	C_6H_6	−17,259	−17,446
n-Methylbenzene	C_7H_8	−17,424	−17,601
n-Ethylbenzene	C_8H_{10}	−17,596	−17,767
n-Propylbenzene	C_9H_{12}	−17,722	−17,887
n-Butylbenzene	$C_{10}H_{14}$	−17,823	−17,984
n-Amylbenzene	$C_{11}H_{16}$	...	−18,065

For the gaseous hydrocarbon, $$h_{RP} = \frac{-264{,}288n + 229{,}222}{14.0260n - 6.0480} \quad \text{for} \quad n > 5$$

In the several equations for h_{RP}, n denotes the number of carbon atoms in the formula of the hydrocarbon.

[1] At saturation pressure.

Note: To convert Btu/lbm to joules per gram, multiply by 2.326.

The enthalpy of formation data can often be used to compute enthalpy of combustion. For example, consider the combustion of methane at 1 atm and 77°F (25°C):

$$CH_4 + 2O_2 \longrightarrow CO_2 + 2H_2O$$

The first law:

$$\bar{Q} = \overline{H_p^0} - \overline{H_r^0} = \overline{H_{pr}^0}$$

Enthalpy of formation:

$$\begin{aligned}\overline{H_p^0} &= \overline{H_{CO_2}^0} + 2\overline{H_{H_2O}^0} \quad \text{(vapor)}\\ &= -169{,}290 - (2)(104{,}040)\end{aligned}$$

$$\overline{H_p^0} = -377{,}370 \text{ Btu/lbmole}$$

$$\begin{aligned}\overline{H_r^0} &= \overline{H_{CH_4}^0} + 2\overline{H_{O_2}^0}\\ &= -32{,}210 + 0 = -32{,}210 \text{ Btu/lbmole}\end{aligned}$$

Thus,

$$\overline{H_{pr}^0} = -377{,}370 + 32{,}210 = -345{,}160 \text{ Btu/lbmole}$$

To convert to metric units,

$$\overline{H_{pr}^0} = -345{,}160 \times 2.326 = -802{,}840 \text{ j/gm-mole}$$

15-3 Combustion processes

15-3(a) The general case

We are now in a position to deal with general constant pressure combustion processes in which the reactants and products do not necessarily enter and leave at the standard conditions of 1 atm pressure and 77°F. Though the calculations are not difficult conceptually, they are a bit complex because we have only data available for enthalpies at *standard* conditions of combustion for reactants *and* products at 1 atm and 77°F. Suppose, for example, we wish to analyze a combustion in which the reactants enter at 120°F and the products leave at 500°F as shown:

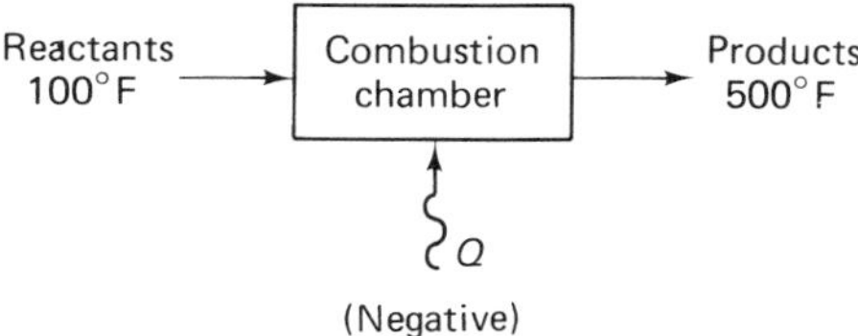

It is best to imagine this taking place in steps as follows:

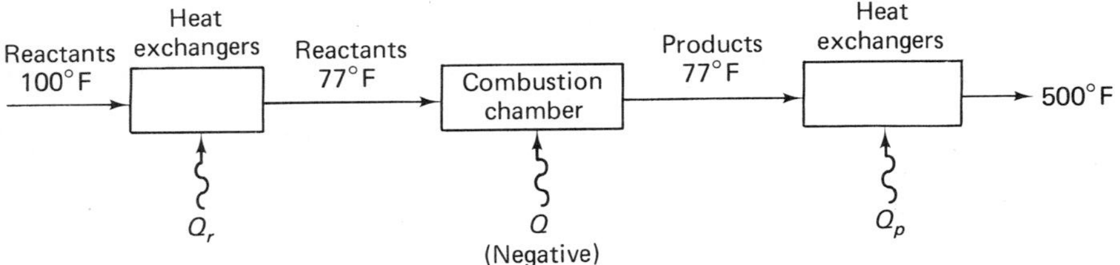

Writing the first law for the entire system gives

$$\sum \bar{Q} = \bar{H}_p - \bar{H}_r \tag{15-8}$$

$$\bar{Q}_r + \bar{Q} + \bar{Q}_p = \bar{H}_{p500} - \bar{H}_{r100} = \bar{H}_{pr}$$

We are interested in finding $\bar{H}_{pr}$ (not to be confused with $\overline{H_{pr}^0}$) for this system because that is the net heat transfer in the process.

Writing the first law for the first heat exchanger gives

$$\bar{Q}_r = \bar{H}_{r77} - \bar{H}_{r100}$$

For the combustor,

$$\bar{Q} = \bar{H}_{r77} - \bar{H}_{p77} = \overline{H_{rp}^0}$$

For the second heat exchanger,

$$\bar{Q}_p = \bar{H}_{p500} - \bar{H}_{p77}$$

To find numerical values for these quantities, we note the $Q = \overline{H_{rp}^0}$ is recorded data on enthalpy of combustion. For $\bar{Q}_r$, we know the specific heat of air and the specific heat of most fuels is assumed to be about 0.5 Btu/lbm-°R. For the products, we may use the methods described in Chap. 14 or we may use tables of properties. *Gas Tables* by Keenan and Kaye is the most complete. Excerpts are shown in Appendix E. The columns for 200 and 400% theoretical air are average properties for the products of combustion of a variety of hydrocarbon fuels with the specified amount of excess air.

EXAMPLE 15.2. Gaseous octane and 100% excess air enter a combustion chamber at 100°F and 1 atm. The products leave at 500°F. Compute the heat transfer per mole of octane.

SOLUTION: The reaction equation is

$$C_8H_{18} + 25O_2 + 94N_2 \longrightarrow 8CO_2 + 9H_2O + 94N_2 + 12.5O_2$$

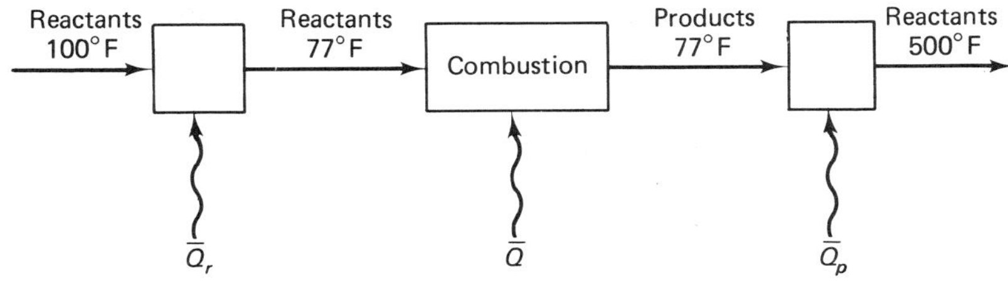

$$\bar{H}_{pr} = \bar{Q}_r + \bar{Q} + \bar{Q}_p = \bar{Q}_r + \overline{H^0_{pr}} + \bar{Q}_p$$

From the tables,

$$\overline{H^0_{pr}} = -2{,}371{,}400 \text{ Btu/lbmole}$$

For $\bar{Q}_r$ (noted that $\bar{c}_p = c_p(\text{MW})$,

$$\begin{aligned} \bar{Q}_r &= [nc_p(\text{MW})\,\Delta T]_{\text{fuel}} + (n\bar{c}_p\,\Delta T)_{\text{air}} \\ &= -(1)(0.5)(114)(23) - (119)(7)(23) \\ \bar{Q}_r &= -1310 - 19{,}200 = -20{,}510 \text{ Btu/lbmole of fuel} \end{aligned}$$

For $\bar{Q}_p$,

$$\begin{aligned} \bar{Q}_p &= (n\bar{c}_p\,\Delta T)_{CO_2} + (n\bar{c}_p\,\Delta T)_{H_2O} + (n\bar{c}_p\,\Delta T)_{N_2} + (n\bar{c}_p\,\Delta T)_{O_2} \\ &= (8)(0.202)(44)(423) + (9)(0.45)(18)(423) + (94)(0.25)(28)(423) \\ &\quad + (12.5)(0.22)(32)(423) \\ \bar{Q}_p &= 376{,}000 \text{ Btu/lbmole of fuel} \end{aligned}$$

Thus

$$\begin{aligned} \bar{H}_{pr} &= -20{,}500 - 2{,}371{,}400 + 376{,}000 \\ \bar{H}_{pr} &= -2{,}015{,}900 \text{ Btu/lbmole of fuel} \end{aligned}$$

Alternately, the gas tables could have been used to find $\bar{Q}_p$ or we might have computed the c_p for the product mixture.

15-3(b) Adiabatic flame temperature

The adiabatic flame temperature is the temperature of the products if *no* heat is transferred from the combustion system so that all the energy made available is used to increase the temperature of the products themselves. It is thus the maximum possible temperature of the products. In practice, even if the combustion could be made adiabatic, the adiabatic flame temperature would generally not be reached because of dissociation of the gaseous products. Still, adiabatic flame temperature is of interest as a limiting temperature for some design and theoretical purposes. It is computed as shown in the following example.

EXAMPLE 15.3. Compute the adiabatic combustion temperature for octane with 100% excess air if the fuel and air enter at 77°F and 1 atm.

SOLUTION:

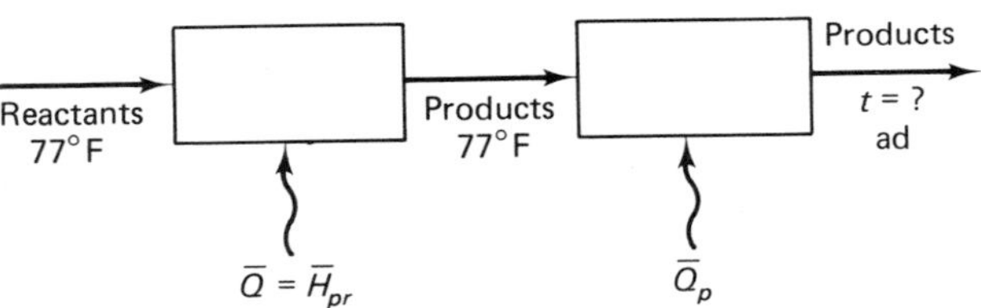

Since the reactants enter at 77°F and the combustion is adiabatic, the first law gives

$$\bar{Q}_p = -\overline{H^o_{pr}} = +2{,}371{,}400 \text{ Btu/lbmole of fuel}$$

The reaction is as in the preceding example. In this case, however, it is easiest for us to compute the specific heat of the products.

$$\bar{c}_p = \left(\frac{N_{CO_2}}{N}\right)\bar{c}_{pCO_2} + \left(\frac{N_{H_2O}}{N}\right)\bar{c}_{pH_2O} + \left(\frac{N_{N2}}{N}\right)\bar{c}_{pN_2} + \left(\frac{N_{O_2}}{N}\right)\bar{c}_{pO_2}$$

$$= \frac{8}{123.5}(0.202)(44) + \frac{9}{123.5}(0.45)(18) + \frac{94}{123.5}(7) + \frac{12.5}{123.5}(0.22)(32)$$

$$\bar{c}_p = 7.2 \text{ Btu/lbmole}$$

Therefore,

$$\bar{Q}_p = N\bar{c}_p(T_{\text{adiabatic}} - 77)$$

$$2{,}371{,}400 = (123.5)(7.2)(T_{\text{adiabatic}} - 77)$$

$$T_{\text{adiabatic}} = 2737°\text{F}$$

Note: The calculation above neglects dissociation of the products—not a very good assumption at these temperatures.

15-3(c) Dew point

The dew point of flue gases is important because any condensate that forms is likely to be highly corrosive. To compute the dew point, the partial pressure of the H_2O vapor in the products of combustion must first be obtained. This is found from

$$p_{H_2O} = \frac{n_{H_2O}}{n_{\text{total}}} p_{\text{total}}$$

Assuming that the air used for combustion is dry, we can find the mole fraction of H_2O from the reaction equation. Using the reaction given in Example 15.1, the mole fraction of H_2O is

$$\frac{n_{H_2O}}{n_{total}} = \frac{2}{1 + 2 + 0.2 + (2.2)(3.76)} = 0.175$$

Thus if the total pressure of the product gases is 1 atm, then the partial pressure of water vapor is

$$p_{H_2O} = 0.175(14.7) = 2.58 \text{ psia}$$

To determine the dew point we must now use steam tables. The dew point is the saturation temperature corresponding to the pressure of H_2O in the mixture. In the example above, the saturation temperature corresponding to 2.58 psia is found by interpolation in the table to be approximately 127°F. Thus, water vapor will not condense in this mixture unless it is cooled to 127°F or below.

In addition to the water vapor formed in the reaction process, there will also normally be some additional H_2O in the gases because the air used for combustion is not completely dry. Let us say, for example, that in the example above the specific humidity of the inlet air is $\omega = 0.01$ lbm water vapor/lbm dry air. To find the moles of water vapor per mole of dry air,

$$n_{H_2O} = 0.01\left(\frac{29}{18}\right) = 0.0161 \frac{\text{moles water vapor}}{\text{mole dry air}}$$

This amount of water vapor must now be added to the water vapor formed by combustion to find the total mole fraction of H_2O. That is,

$$n_{H_2O} = \frac{2 + 0.161}{1 + 2 + 0.2 + (2.2)(3.76) + 0.161} = 0.175$$

Now one proceeds as before to determine the partial pressure of water vapor and then the dew point.

15-4 Equilibrium

We have already shown that at constant temperature and pressure, equilibrium requires a minimum Gibbs function (see Chap. 11). That is, at equilibrium

$$(dG)_{T,p} = 0 \tag{15-9}$$

For mixtures of substances a, b, c, . . ., the Gibbs function will be

$$G = n_a\bar{g}_a + n_b\bar{g}_b + n_c\bar{g}_c + \cdots \tag{15-10}$$

In this chapter we have assumed that all reactions have gone to completion and that there has been no dissociation. Though this is often a fair first approximation, we should know how to check it and how to handle cases where it turns out to be a poor assumption. For this purpose, we must be able to find the equilibrium amounts of the various substances involved in a reaction. That is, suppose the reaction equation is

$$V_1 R_1 + V_2 R_2 \rightarrow V_3 P_1 + V_4 P_2 \tag{15-11}$$

where the V's are the stoichiometric coefficients. (There could be more or fewer terms on either side of the relation above.) We define a term ε called *degree of reaction* that indicates the extent to which the left-hand side has diminished and the right-hand side formed. Including ε in the equation, we write

$$V_1(1-\varepsilon)R_1 + V_2(1-\varepsilon)R_2 \rightarrow \varepsilon V_3 P_1 + \varepsilon V_4 P_2 \tag{15-12}$$

Note that if $\varepsilon = 0$, the reactants have done nothing; whereas if $\varepsilon = 1$, the reaction has gone to completion. Since ε will lie between 0 and 1 at equilibrium, all substances involved will be present and the Gibbs function for the mixture will be

$$G = V_1(1-\varepsilon)g_{R_1} + V_2(1-\varepsilon)g_{R_2} + \varepsilon V_3 g_{P_1} + \varepsilon V_4 g_{P_2} \tag{15-13}$$

Letting n be the number of moles of each substance,

$$G = n_1 g_{R_1} + n_2 g_{R_2} + n_3 g_{P_1} + n_4 g_{P_2} \tag{15-14}$$

$$\begin{aligned} dG = (g_{R_1}\,dn_1 + g_{R_2}\,dn_2 + g_{P_1}\,dn_3 + g_{P_2}\,dn_4) \\ + (n_1\,dg_{R_1} + n_2\,dg_{R_2} + n_3\,dg_{P_1} + n_4\,dg_{P_2}) = 0 \end{aligned} \tag{15-15}$$

The second parenthetical term on the right side of this expression is zero at equilibrium in a constant p and T process so

$$dG = g_{R_1}\,dn_1 + g_{R_2}\,dn_2 + g_{P_1}\,dn_3 + g_{P_2}\,dn_4 = 0 \tag{15-16}$$

We note that, by definition,

$$\begin{aligned} n_1 &= V_1(1-\varepsilon) & dn_1 &= -V_1\,d\varepsilon \\ n_2 &= V_2(1-\varepsilon) & dn_2 &= -V_2\,d\varepsilon \\ n_3 &= V_3\,\varepsilon & dn_3 &= V_3\,d\varepsilon \\ n_4 &= V_4\,\varepsilon & dn_4 &= V_4\,d\varepsilon \end{aligned} \tag{15-17}$$

so dG becomes

$$dG = (g_{R_1} V_1 + g_{R_2} V_2 + g_{P_1} V_3 + g_{P_2} V_4)\,d\varepsilon = 0 \tag{15-18}$$

or

$$g_{R_1} V_1 + g_{R_2} V_2 - g_{P_1} V_3 - g_{P_2} V_4 = 0 \tag{15-19}$$

For the Gibbs function, we start with the definition

$$dg = s\,dT - v\,dp \tag{15-20}$$

and integrate from a reference state of T_0 and $p = 1$ atm and find the Gibbs function *at the reference temperature* as a function only of p:

$$g - g^0 = \int_{T_0}^{T_0} s\,dT - \int_1^p v\,dp = -\int_1^p v\,dp \tag{15-21}$$

For an ideal gas,

$$\bar{g} - \overline{g^0} = -\bar{R}T \int_1^p \frac{dp}{p} = -\bar{R}T \ln p \qquad (p \text{ in atmospheres}) \tag{15-22}$$

Substituting this into Eq. (15-19),

$$-\bar{R}T(V_3 \ln p_{P_1} + V_4 \ln p_{P_2} - V_1 \ln p_{R_1} - V_2 \ln p_{R_2}) = V_3\overline{g^0_{P_1}} + V_4\overline{g^0_{P_2}} - V_1\overline{g^0_{R_1}} - V_2\overline{g^0_{R_2}} \tag{15-23}$$

The right-hand side of Eq. (15-23) is usually denoted by ΔG^0 and is sometimes called the free energy change of the reaction. The g^0 for each substance is its Gibbs function per mole at the standard conditions. Using this notation and writing the left side more elegantly, we find

$$-\bar{R}T \ln \frac{(p_{P_1})^{V_3}(p_{P_2})^{V_4}}{(p_{R_1})^{V_1}(p_{R_2})^{V_2}} = \Delta G^0 \tag{15-24}$$

Now the argument of the logarithm is also given a symbol and name of its own so that Eq. (15-24) becomes

$$-RT \ln K_P = \Delta G^0 \tag{15-25}$$

where K_P is called the *equilibrium constant* and is defined as

$$K_P = \frac{(p_{P_1})^{V_3}(p_{P_2})^{V_4}}{(p_{R_1})^{V_1}(p_{R_2})^{V_2}} \tag{15-26}$$

These results are known as the *law of mass action*. Values for K_P are tabulated in Table 15.3 as a function of temperature. K_P increases with temperature rather strongly.

Equation (15-25) can often be used to compute the equilibrium constant K_P since data on the Gibbs function of many substances are available in the literature. For example, ΔG^0 for H_2O gas is $-98{,}344$ Btu/lbmole at 1 atm and 298°K. The Gibbs function of elements is taken to be zero at these conditions. Thus if we wish to compute K_P for the equilibrium

$$H_2O \rightleftarrows H_2 + \tfrac{1}{2}O$$

Table 15.3 Logarithms to the base 10 of the equilibrium constant K_P[1]

For the reaction $aA + bB \rightleftarrows cC + dD$ the equilibrium constant K_P is defined as

$$K_P = \frac{p_C{}^c p_D{}^d}{p_A{}^a p_B{}^b}$$

where p is the partial pressure in atmospheres.

Temp. (°K)	$H_2 \rightleftarrows 2H$	$O_2 \rightleftarrows 2O$	$H_2O\,(g) \rightleftarrows H_2 + \frac{1}{2}O_2$	$H_2O\,(g) \rightleftarrows OH + \frac{1}{2}H_2$	$CO_2 \rightleftarrows CO + \frac{1}{2}O_2$	$CO_2 + H_2 \rightleftarrows CO + H_2O\,(g)$	$N_2 \rightleftarrows 2N$	$\frac{1}{2}O_2 + \frac{1}{2}N_2 \rightleftarrows NO$
298	−71.210	−80.620	−40.047	−46.593	−45.043	−4.996	−119.434	−15.187
400	−51.742	−58.513	−29.241	−33.910	−32.41	−3.169	−87.473	−11.156
600	−32.667	−36.859	−18.633	−21.470	−20.07	−1.432	−56.206	−7.219
800	−23.074	−25.985	−13.288	−15.214	−13.90	−0.617	−40.521	−5.250
1000	−17.288	−19.440	−10.060	−11.444	−10.199	−0.139	−31.084	−4.068
1200	−13.410	−15.062	−7.896	−8.922	−7.742	+0.154	−24.619	−3.279
1400	−10.627	−11.932	−6.344	−7.116	−5.992	+0.352	−20.262	−2.717
1600	−8.530	−9.575	−5.175	−5.758	−4.684	+0.490	−16.869	−2.294
1800	−6.893	−7.740	−4.263	−4.700	−3.672	+0.591	−14.225	−1.966
2000	−5.579	−6.269	−3.531	−3.852	−2.863	+0.668	−12.106	−1.703
2200	−4.500	−5.064	−2.931	−3.158	−2.206	+0.725	−10.370	−1.488
2400	−3.598	−4.055	−2.429	−2.578	−1.662	+0.767	−8.922	−1.309
2600	−2.833	−3.206	−2.003	−2.087	−1.203	+0.800	−7.694	−1.157
2800	−2.176	−2.475	−1.638	−1.670	−0.807	+0.831	−6.640	−1.028
3000	−1.604	−1.840	−1.322	−1.302	−0.469	+0.853	−5.726	−0.915
3200	−1.104	−1.285	−1.046	−0.983	−0.175	+0.871	−4.925	−0.817
3500	−0.458	−0.571	−0.693	−0.577	+0.201	+0.894	−3.893	−0.692
4000	+0.406	+0.382	−0.221	−0.035	+0.699	+0.920	−2.514	−0.526
4500	+1.078	+1.125	+0.153	+0.392	+1.081	+0.928	−1.437	−0.345
5000	+1.619	+1.719	+0.450	+0.799	+1.387	+0.937	−0.570	−0.298

[1] Values taken from or computed from "Selected Values of Chemical Thermodynamic Properties," Series III. National Bureau of Standards, Washington, D.C.

at 298°K (536°R) and 1 atm, we use

$$\ln K_P = -\frac{\Delta G^0}{\bar{R}T}$$

$$\Delta G^0 = g^0_{H_2} + \tfrac{1}{2}g^0_{O_2} = 98{,}344 \text{ Btu}$$

$$\ln K_P = -\frac{98{,}344}{(2)(536)} = -91.5$$

Converting this to base 10 so that the result can be compared with the value given in Table 15.3 gives

$$\log_{10} K_P = \frac{-91.5}{2.3} = -39.8$$

The table value is −40.0.

EXAMPLE 15.4. Consider the reaction

$$CO_2 \longrightarrow CO + \tfrac{1}{2}O_2$$

and compute the moles of each substance present at equilibrium at 1 atm and 500°K.

SOLUTION: From the table,

$$\log_{10} K_P = -25.0 \text{ at } 500°\text{K}$$

Thus

$$K_P = 10^{-25}$$

In terms of degree of reaction ε, the reaction is written

$$(1-\varepsilon)CO_2 \longrightarrow \varepsilon CO_2 + \tfrac{1}{2}\varepsilon O_2$$

Thus the total moles present will be

$$(1-\varepsilon) + \varepsilon + \tfrac{1}{2}\varepsilon = 1 + \tfrac{1}{2}\varepsilon$$

We define, for convenience, the mole fraction x as

$$x_a = \frac{n_a}{n}$$

so that from Dalton's law we can write

$$p_a = x_a p$$

Then

$$K_P = \frac{(p_{CO})^1(p_{O_2})^{1/2}}{(p_{CO_2})^1} = \frac{(x_{CO})^1(x_{O_2})^{1/2}}{(x_{CO_2})^1}(p)^1(p)^{1/2}(p)^1$$

Since $p = 1$, this is simply

$$K_P = \frac{(x_{CO})^1(x_{O_2})^{1/2}}{(x_{CO_2})^1}$$

Now in terms of ε, the mole fractions are

$$x_{CO} = \frac{\varepsilon}{1+\frac{1}{2}\varepsilon} \qquad x_{O_2} = \frac{\frac{1}{2}\varepsilon}{1+\frac{1}{2}\varepsilon} \qquad x_{CO_2} = \frac{1-\varepsilon}{1+\frac{1}{2}\varepsilon}$$

So

$$K_P = \frac{[\varepsilon/(1+\frac{1}{2}\varepsilon)][\frac{1}{2}\varepsilon/(1+\frac{1}{2}\varepsilon)]}{[(1-\varepsilon)/(1+\frac{1}{2}\varepsilon)]} = 10^{-25}$$

Solving this approximately for ε, we find

$$\varepsilon \simeq 1.4 \times 10^{-12.5}$$

The result indicates that dissociation of CO_2 into CO and O_2 is not a serious problem at 500°K (440°F) and will be even less at lower temperatures, unless, of course, you have a *lot* of CO_2 from, say, a lot of automobiles plus some CO from poor combustion . . . .

EXAMPLE 15.5. One mole of oxygen is heated to 4000°K and 1 atm pressure. Assuming the constituents are H_2 and H, determine the amount of each present.

SOLUTION: From Table 15.3, for the equilibrium

$$O_2 \rightleftarrows 2O$$

we find at 4000°K

$$\log_{10} K_P = +0.382$$

Thus

$$K_P = 10^{0.382} = 2.41$$

but

$$K_P = \frac{p_O^2}{p_{O_2}}$$

and

$$p_O + p_{O_2} = 1$$

Solving the two equations above simultaneously gives

$$p_O = 0.76 \text{ atm} \qquad p_{O_2} = 0.24 \text{ atm}$$

15-5 Summary

Combustion is oxidation (burning) of reactants (fuel) to form products of combustion. The amount of oxygen required to burn the fuel completely is called the *stoichiometric oxygen*. The amount of air required to provide the stoichiometric oxygen is called the *stoichiometric air* or *theoretical air*. In real situations, more air must be provided to achieve complete combustion because of incomplete mixing, finite burning rates, partial burning (as to CO), etc. The extra air is called *excess air*.

The air–fuel ratio is the ratio of either moles of air to moles of fuel or mass of air to mass of fuel.

Often the products of combustion are known from experimental Orsat analysis. The Orsat apparatus measures the composition of gaseous products by volume, excluding water vapor. With the Orsat data, it is possible to determine the actual reaction, amount of excess air, etc.

In order to make analysis of the energy released in combustion process, data are available on the enthalpy of formation of compounds at 77°F (25°C) and 1 atm. In addition, data are available on the enthalpy of combustion of substances when they are burned at constant pressure and reactants and products are both at 77°F (25°C).

The adiabatic flame temperature is the temperature the products of a combustion will reach if no heat is transferred to them.

The equilibrium constant for a reaction gives information on the degree to which the reaction goes to completion. It is a strong function of temperature as shown in Table 15.3. The equilibrium constant K_P and the Gibbs function are related by

$$-RT \ln K_P = \Delta G^0 \tag{15-25}$$

Problems

15-1 Propane (C_3H_8) is burned with dry air and an Orsat analysis of the products of combustion is as follows:

CO_2	8.00%
CO	1.00%
O_2	7.60%
N_2	83.40%

Determine the percent theoretical air, the air–fuel ratio, and the dew point of the mixture.

15-2 The products of combustion of a hydrocarbon fuel of unknown composition have the following Orsat analysis:

CO_2	9.00%
CO	0.60%
O_2	7.30%
N_2	83.10%

Determine the percent theoretical air and the air–fuel ratio.

15-3 Methane is burned at constant pressure with just enough oxygen to permit complete combustion. The equation of the reaction is

$$CH_4 + 2O_2 \longrightarrow CO_2 + H_2O$$

If the temperature and pressure of the final mixture are 100°F and 14.7 psia, find

(a) The partial pressure of water vapor in the products of combustion.
(b) The number of pounds of liquid water in the products of combustion.
(c) The volume of the products of combustion per pound of fuel, neglecting the volume of liquid water.

15-4 A mixture of octane vapor (C_8H_{18}) and air at 60°F and 75 psia burns at constant pressure in an insulated combustion chamber. Eighty pounds of air is supplied per pound of fuel. The constant pressure heat of combustion is 19,450 Btu/lb of octane burned at 60°F. Assume the properties of air and octane mixtures and of the products of combustion are identical with those of air. Calculate the final temperature. *State your assumptions.*

15-5 Hydrogen is supplied to a burner at 100°F and is completely burned with 20% excess air that is also supplied at 100°F. The products leave at 900°F. Determine the amount of heat released per pound of hydrogen.

15-6 Methane at 77°F is burned with 20% excess air supplied at 77°F in a steady flow process. The products leave at 800°F. For methane (CH_4) at 77°F, HHV (higher heating value) = 23,860 Btu/lb and LHV = 21,500 Btu/lb. Calculate

(a) The amount of heat released per pound of methane.

(b) The dew point of the products.

15-7 Determine the amount of heat released by the complete combustion of methane with 30% excess air if the methane is supplied at 100°F, air is supplied at 200°F, and the products leave at 800°F.

15-8 Determine the maximum adiabatic combustion temperature for the steady flow burning of propane (C_3H_8) with 100% excess air with propane and air supplied at 1 atm and 77°F. For propane at 77°F, LHV = 19,929 Btu/lb and HHV = 21,646 Btu/lb.

15-9 Carbon monoxide is burned completely without excess air in a steady flow process. The carbon monoxide and the air are supplied at 77°F. The heating value of carbon monoxide at 77°F is 4344 Btu/lb. Calculate the maximum adiabatic combustion temperature for this reaction.

15-10 Determine the equilibrium constant of the reaction

$$H_2O \rightleftarrows H_2 + \tfrac{1}{2}O_2$$

at 400°K and 1 atm using

$$\ln K_P = -\frac{\Delta G^0}{RT}$$

Compare with Table 15.3.

15-11 One mole of hydrogen is heated to 5000°K and 1 atm pressure. If H_2 and H are the only constituents present, determine the amount of each.

16

Relations among properties

16-1 Introduction

16-1(a) Why more property relations?

In Chap. 3 we discussed equations of state—the relationship among p, v, and T for substances. There and in later chapters we also discussed the compressibility (K), the coefficient of expansion (β), and the specific heats (c_p and c_v). Why more property relations? Partly because they are there, of course, and we are curious human beings; that is, we now have some new properties: u, h, s, g, and z. We are interested in what, if any, useful or important relationships exist among these properties and among these and our old ones.

But our interest is more than just human intellectual curiosity. Determining such relations among properties is far from useless. Equations of state and relations among properties are almost always important in the solution of practical problems in thermodynamics. If relations among properties exist for all substances based on the laws of thermodynamics and on mathematics, then they will certainly serve to reduce the amount of data taking and numerical work that might otherwise be necessary.

A second reason for more study of the thermodynamic relations among properties is to enable us to compute changes in certain unmeasurable properties from changes in the measurable ones. Note that while we have thermometers, pressure gauges, rulers, scales, and the like, we do *not* have any meters that measure internal energy, enthalpy, entropy, etc. The reason, of course, is that these latter properties are abstractions built on paper and pencil operations, deductions from inductively arrived at laws. As developed in Chap. 3, the logic of the process has a firm foundation in the operational definitions of the otherwise nonverbal fundamental concepts. We *do* want to be able to deal with these new properties that thermodynamics provides us with, however, and so we must be able to express them in other terms that *can* be measured. Thus, an important reason for this chapter is to develop expressions for our new properties such as u, h, s, g, and z in terms of p, v, T, and other easily measurable quantities such as c_p, c_v, β, and K.

The relations developed in this chapter are used in developing tables of properties too. In making up tables (such as steam tables, freon tables, etc.) not *all* points recorded are actual data. Only a relatively few data are taken and the tables are developed using empirical equations and property relations. Interested students should consult Sec. 12-8 in *Introduction to Thermodynamics: Classical and Statistical* by Sonntag and Van Wylen, and Appendix in the complete *Steam Tables* by Keenan, Keyes, Hill, and Moore.

16-1(b) Specific objectives

Since in general we want to be able to compute changes in our new and abstract properties from changes in physically measurable ones, we can state our objectives in the chapter very specifically as follows: to find expressions for changes in u, h, s, g, and z in terms of p, v, T, c_p, c_v, β, and K. We shall work only with substances that have two independent properties and we shall select these to be p and v or p and T or T and v. In other words we are trying to obtain expressions for the coefficients in terms of p, v, T, c_p, c_v, β, and K in the following set of equations:

$$ds = (\alpha)_1\, dp + (\alpha)_2\, dT \tag{16-1a}$$

$$ds = (\alpha)_3\, dp + (\alpha)_4\, dv \tag{16-1b}$$

$$ds = (\alpha)_5\, dv + (\alpha)_6\, dT \tag{16-1c}$$

$$du = (\alpha)_7\, dp + (\alpha)_8\, dT \tag{16-1d}$$

$$du = (\alpha_9)\, dp + (\alpha)_{10}\, dv \tag{16-1e}$$

$$du = (\alpha)_{11}\, dv + (\alpha)_{12}\, dT \tag{16-1f}$$

The properties h, g, and z have been omitted from Eq. (16-1) because if we have those, the others will be obtainable from their definitions:

$$h = u + pv \qquad dh = du + p\, dv + v\, dp \tag{16-2a}$$

$$z = u - Ts \qquad dz = du - T\, ds - s\, dT \tag{16-2b}$$

$$g = h - Ts \qquad dg = dh - T\, ds - s\, dT \tag{16-2c}$$

Thus it is the twelve α's in Eq. (16-1) that we would like to find from our efforts.

16-1(c) Review

From Chap. 4 students should remember that if $f(x, y, z) = 0$, then a chain results as follows:

$$\left(\frac{\partial x}{\partial y}\right)_z \left(\frac{\partial z}{\partial x}\right)_y \left(\frac{\partial y}{\partial z}\right)_x = -1 \tag{16-3}$$

and that

$$\left(\frac{\partial x}{\partial y}\right)_z = \frac{1}{(\partial y/\partial x)_z} \tag{16-4}$$

for any set of variables. We used the chain rule together with the definitions of β and K to find

$$\beta = \frac{1}{v}\left(\frac{\partial v}{\partial T}\right)_p \qquad K = -\frac{1}{v}\left(\frac{\partial v}{\partial p}\right)_T \qquad \frac{\beta}{K} = \left(\frac{\partial p}{\partial T}\right)_v \tag{16-5}$$

In addition, from mathematics we shall need to remember that if a variable P (designating a property now, not pressure) is a function of x, y, and z (two of which are independent), the following relations hold:

$$\left(\frac{\partial P}{\partial x}\right)_y = \left(\frac{\partial P}{\partial z}\right)_y \left(\frac{\partial z}{\partial x}\right)_y \tag{16-6}$$

$$\left(\frac{\partial P}{\partial x}\right)_y = \left(\frac{\partial P}{\partial z}\right)_x \left(\frac{\partial z}{\partial x}\right)_y = \left(\frac{\partial P}{\partial x}\right)_z \tag{16-7}$$

Also remember that under proper conditions that are met if P is a continuous function, the order of partial differentiation does not change the result. Thus

$$\frac{\partial^2 P}{\partial x\,\partial y} = \frac{\partial}{\partial y}\left[\left(\frac{\partial P}{\partial x}\right)_y\right]_x = \frac{\partial}{\partial x}\left[\left(\frac{\partial P}{\partial y}\right)_x\right]_y = \frac{\partial^2 P}{\partial y\,\partial x} \tag{16-8}$$

Finally, we shall make much use of the total derivative of a property P expressed as

$$P = P(x, y)$$

$$dP = \left(\frac{\partial P}{\partial x}\right)_y dx + \left(\frac{\partial P}{\partial y}\right)_x dy \tag{16-9}$$

In fact if we use the Eq. (16-9) together with the expressions containing the coefficients we are trying to find, we can express the coefficients above more explicitly. For example, Eq. (16-1a) is

$$ds = (\alpha)_1\, dp + (\alpha)_2\, dT$$

Taking p and T as the independent variables, $s = s(p, T)$, and differentiating, we find

$$ds = \left(\frac{\partial s}{\partial p}\right)_T dp + \left(\frac{\partial s}{\partial T}\right)_p dT \tag{16-10}$$

By comparison, then, since p and T are independent,

$$(\alpha)_1 = \left(\frac{\partial s}{\partial p}\right)_T \quad \text{and} \quad (\alpha)_2 = \left(\frac{\partial s}{\partial T}\right)_p \tag{16-11}$$

By analogy the other expressions we are seeking can be found. They are

$$(\alpha)_3 = \left(\frac{\partial s}{\partial p}\right)_v \quad \text{and} \quad (\alpha)_4 = \left(\frac{\partial s}{\partial v}\right)_p \tag{16-12a}$$

$$(\alpha)_5 = \left(\frac{\partial s}{\partial v}\right)_T \quad \text{and} \quad (\alpha)_6 = \left(\frac{\partial s}{\partial T}\right)_v \tag{16-12b}$$

$$(\alpha)_7 = \left(\frac{\partial u}{\partial p}\right)_T \quad \text{and} \quad (\alpha)_8 = \left(\frac{\partial u}{\partial T}\right)_p \tag{16-12c}$$

$$(\alpha)_9 = \left(\frac{\partial u}{\partial p}\right)_v \quad \text{and} \quad (\alpha)_{10} = \left(\frac{\partial u}{\partial v}\right)_p \tag{16-12d}$$

$$(\alpha)_{11} = \left(\frac{\partial u}{\partial v}\right)_T \quad \text{and} \quad (\alpha)_{12} = \left(\frac{\partial u}{\partial T}\right)_v \tag{16-12e}$$

16-2 Maxwell's relations

Starting with the very important combined first and second law equation

$$T\,ds = du + p\,dv \tag{16-13}$$

and incorporating the definitions for h, a, and g, from Eqs. (16-2), the following expressions are readily developed:

$$du = T\,ds - p\,dv \tag{16-14a}$$

$$dh = T\,ds + v\,dp \tag{16-14b}$$

$$dz = -s\,dT - p\,dv \tag{16-14c}$$

$$dg = -s\,dT + v\,dp \tag{16-14d}$$

But we also know that *du*, *dh*, *df*, and *dg* may be written as

$$du = \left(\frac{\partial u}{\partial s}\right)_v ds + \left(\frac{\partial u}{\partial v}\right)_s dv \tag{16-15a}$$

$$dh = \left(\frac{\partial h}{\partial s}\right)_p ds + \left(\frac{\partial h}{\partial p}\right)_s dp \tag{16-15b}$$

$$dz = \left(\frac{\partial z}{\partial T}\right)_v dT + \left(\frac{\partial z}{\partial v}\right)_T dv \tag{16-15c}$$

$$dg = \left(\frac{\partial g}{\partial T}\right)_p dT + \left(\frac{\partial g}{\partial p}\right)_T dp \tag{16-15d}$$

By a comparison of coefficients in Eq. 14(a) and 15(a), Eq. 14(b) and 15(b), etc., we can conclude that

$$\left(\frac{\partial u}{\partial s}\right)_v = T \qquad \left(\frac{\partial u}{\partial v}\right)_s = -p \tag{16-16a}$$

$$\left(\frac{\partial h}{\partial s}\right)_p = T \qquad \left(\frac{\partial h}{\partial p}\right)_s = v \tag{16-16b}$$

$$\left(\frac{\partial z}{\partial T}\right)_v = -s \qquad \left(\frac{\partial z}{\partial v}\right)_T = -p \tag{16-16c}$$

$$\left(\frac{\partial g}{\partial T}\right)_p = -s \qquad \left(\frac{\partial g}{\partial p}\right)_T = v \tag{16-16d}$$

Now taking the cross derivatives and remembering that

$$\frac{\partial^2 u}{\partial v\, \partial s} = \frac{\partial^2 u}{\partial s\, \partial v} \tag{16-17}$$

the equations known as Maxwell's relations result directly:

$$\left(\frac{\partial T}{\partial v}\right)_s = -\left(\frac{\partial p}{\partial s}\right)_v \tag{16-18a}$$

$$\left(\frac{\partial T}{\partial p}\right)_s = \left(\frac{\partial v}{\partial s}\right)_p \tag{16-18b}$$

$$\left(\frac{\partial s}{\partial v}\right)_T = \left(\frac{\partial p}{\partial T}\right)_v \tag{16-18c}$$

$$-\left(\frac{\partial s}{\partial p}\right)_T = \left(\frac{\partial v}{\partial T}\right)_p \tag{16-18d}$$

The great significance of these equations is that they relate entropy (s) to p, v, and T. This is certainly important in terms of our objectives here and the reader should notice that four of the coefficients (α's) for which we are looking appear in Maxwell's relations. Two of these, in fact, can be fully expressed in terms we desire from what we already have derived.

$$\alpha_5 = \left(\frac{\partial s}{\partial v}\right)_T = \left(\frac{\partial p}{\partial T}\right)_v = \frac{\beta}{K} \tag{16-19}$$

$$\alpha_1 = \left(\frac{\partial s}{\partial p}\right)_T = -\left(\frac{\partial v}{\partial T}\right)_p = -\beta v \tag{16-20}$$

16-3 A general method for deriving relations

16-3(a) The method using T and v as independent

In general, we may assume that the independent variables are T and p, T and v, or p and v. Each of these three cases can be analyzed and will produce results of interest. We shall present the general method using T and v as an example. Only the results from the other two cases will be presented, leaving as problems the details of those derivations.

1. Express ds in terms of the independent variables using Eq. (16-9):

$$ds = \left(\frac{\partial s}{\partial T}\right)_v dT + \left(\frac{\partial s}{\partial v}\right)_T dv \tag{16-21}$$

2. Also express ds using the $T\,ds$ equations (use the form that involves the most independent variables—either $T\,ds = du + p\,dv$ or $T\,ds = dh - v\,dp$):

$$ds = \frac{1}{T}\,du + \frac{1}{T}p\,dv \tag{16-22}$$

3. Using Eq. (16-9) as needed, put Eq. (16-21) completely in terms of the given independent variables by eliminating such terms as du, dh, and dg:

$$du = \left(\frac{\partial u}{\partial T}\right)_v dT + \left(\frac{\partial u}{\partial v}\right)_T dv \tag{16-23}$$

$$ds = \frac{1}{T}\left(\frac{\partial u}{\partial T}\right)_v dT + \frac{1}{T}\left[\left(\frac{\partial u}{\partial v}\right)_T + p\right] dv \tag{16-24}$$

4. Now compare and equate the coefficients of the independent variables from steps 1 and 3:

$$\left(\frac{\partial s}{\partial T}\right)_v = \frac{1}{T}\left(\frac{\partial u}{\partial T}\right)_v \tag{16-25}$$

$$\left(\frac{\partial s}{\partial v}\right)_T = \frac{1}{T}\left[\left(\frac{\partial u}{\partial v}\right)_T + p\right] \tag{16-26}$$

5. Apply the equivalence of the cross derivatives to the results of step 4:

$$\frac{\partial^2 s}{\partial v\, \partial T} = \frac{\partial^2 s}{\partial T\, \partial v} \tag{16-27}$$

$$\frac{\partial}{\partial v}\left[\frac{1}{T}\left(\frac{\partial u}{\partial T}\right)_v\right]_T = \frac{\partial}{\partial T}\left\{\frac{1}{T}\left[\left(\frac{\partial u}{\partial v}\right)_T + p\right]\right\}_v \tag{16-28}$$

6. Simplify and solve for the desired results:

$$\frac{\partial}{\partial v}\left[\frac{1}{T}\left(\frac{\partial u}{\partial T}\right)_v\right]_T = \frac{\partial}{\partial T}\left\{\frac{1}{T}\left[\left(\frac{\partial u}{\partial v}\right)_T + p\right]\right\}_v \tag{16-29}$$

$$\frac{1}{T}\frac{\partial^2 u}{\partial v\, \partial T} = -\frac{1}{T^2}\left[\left(\frac{\partial u}{\partial v}\right)_T + p\right] + \frac{1}{T}\frac{\partial^2 u}{\partial v\, \partial T} + \frac{1}{T}\left(\frac{\partial p}{\partial T}\right)_v \tag{16-30}$$

$$\left(\frac{\partial u}{\partial v}\right)_T + p = T\left(\frac{\partial p}{\partial T}\right)_v = T\frac{\beta}{K} \tag{16-31}$$

7. Inspect the intermediate steps above for possible useful relations. Note that in step 4 there is:

$$\left(\frac{\partial s}{\partial T}\right)_v = \frac{1}{T}\left(\frac{\partial u}{\partial T}\right)_v \tag{16-25}$$

Since $c_v = (\partial u/\partial T)_v$,

$$\frac{c_v}{T} = \left(\frac{\partial s}{\partial T}\right)_v \tag{16-32}$$

and we have another of the coefficients that we desire. Also by combining our final result

$$\left(\frac{\partial u}{\partial v}\right)_T = \frac{T\beta}{K} - p \tag{16-33}$$

with the intermediate result

$$\left(\frac{\partial s}{\partial v}\right)_T = \frac{1}{T}\left[\left(\frac{\partial u}{\partial v}\right)_T + p\right] \tag{16-34}$$

we can derive

$$\left(\frac{\partial s}{\partial v}\right)_T = \frac{\beta}{K} \qquad (16\text{-}35)$$

This, however, is not new to us because we obtained it more directly from Maxwell's relations. See Eq. (16-19).

16-3(b) Results taking q and T and p and v independently

Applying the general procedure above to the cases where p and T are independent leads to the following new results:

$$\left(\frac{\partial s}{\partial T}\right)_p = \frac{c_p}{T} \qquad (16\text{-}36)$$

$$\left(\frac{\partial u}{\partial p}\right)_T = pKv - T\beta v \qquad (16\text{-}37)$$

Note that we have already shown in Eq. (16-20) that

$$\left(\frac{\partial s}{\partial p}\right)_T = -\beta v \qquad (16\text{-}38)$$

If p and v are taken as independent variables, the results are

$$\left(\frac{\partial s}{\partial p}\right)_v = \frac{Kc_v}{\beta T} \qquad (16\text{-}39a)$$

$$\left(\frac{\partial s}{\partial v}\right)_p = \frac{c_v}{v\beta T} + \frac{\beta}{K} \qquad (16\text{-}39b)$$

$$\left(\frac{\partial u}{\partial v}\right)_p = \frac{c_v}{\beta v} + \frac{T\beta}{K} - p \qquad (16\text{-}39c)$$

Finally, notice that

$$\left(\frac{\partial u}{\partial p}\right)_v = \left(\frac{\partial u}{\partial T}\right)_v \left(\frac{\partial T}{\partial p}\right)_v = c_v\left(\frac{K}{\beta}\right) \qquad (16\text{-}40)$$

and

$$\left(\frac{\partial u}{\partial T}\right)_p = \left(\frac{\partial u}{\partial v}\right)_p \left(\frac{\partial v}{\partial T}\right)_p = \left(\frac{c_v}{\beta v} + \frac{T\beta}{K} - p\right)(\beta v) \qquad (16\text{-}41)$$

$$\left(\frac{\partial u}{\partial T}\right)_p = c_v + \frac{T\beta^2 v}{K} - \beta p v \qquad (16\text{-}42)$$

16-4 Compilation of results

In summary, the question marks we set out to find are

$$(\alpha)_1 = \left(\frac{\partial s}{\partial p}\right)_T = -v\beta \qquad (\alpha)_2 = \left(\frac{\partial s}{\partial T}\right)_p = \frac{c_p}{T} \tag{16-43}$$

$$(\alpha)_3 = \left(\frac{\partial s}{\partial p}\right)_v = \frac{Kc_v}{\beta T} \qquad (\alpha)_4 = \left(\frac{\partial s}{\partial v}\right)_p = \frac{c_v}{\beta v T} + \frac{\beta}{K} = \frac{c_p}{\beta v T} \tag{16-44}$$

$$(\alpha)_5 = \left(\frac{\partial s}{\partial v}\right)_T = \beta K \qquad (\alpha)_6 = \left(\frac{\partial s}{\partial T}\right)_v = \frac{c_v}{T} \tag{16-45}$$

$$(\alpha)_7 = \left(\frac{\partial u}{\partial p}\right)_T = pKv - T\beta v \qquad (\alpha)_8 = \left(\frac{\partial u}{\partial T}\right)_p = c_v + \frac{T\beta^2 v}{K} - pv \tag{16-46}$$

$$(\alpha)_9 = \left(\frac{\partial u}{\partial p}\right)_v = c_v \frac{K}{\beta} \qquad (\alpha)_{10} = \left(\frac{\partial u}{\partial v}\right)_p = \frac{c_v}{\beta v} + \frac{T\beta}{K} - p \tag{16-47}$$

$$(\alpha)_{11} = \left(\frac{\partial u}{\partial v}\right)_T = \frac{T\beta}{K} - p \qquad (\alpha)_{12} = \left(\frac{\partial u}{\partial T}\right)_v = c_v \tag{16-48}$$

Incorporating these into the complete equations gives

$$ds = c_v \frac{dT}{T} + \frac{\beta}{K} dv \tag{16-49}$$

$$ds = c_p \frac{dT}{T} - v\beta\, dp \tag{16-50}$$

$$ds = \frac{Kc_v}{\beta T} dp + \frac{c_p}{\beta v T} dv \tag{16-51}$$

$$du = c_v\, dT + \left(\frac{T\beta}{K} - p\right) dv \tag{16-52}$$

$$du = (c_p - pv\beta)\, dT + (pKv - T\beta v)\, dp \tag{16-53}$$

$$du = \frac{c_v K}{\beta} dp + \left(\frac{c_p}{\beta v} - p\right) dv \tag{16-54}$$

Equations (16-49) to (16-54) are the ones that we set out to find—expressions for ds and du in terms of β, K, p, v, T, and the specific heats.

One last very useful result can be obtained. Thus far we have found no relationship among our measurable properties themselves: β, K, c_p, c_v, p, v, and T. If a general relationship does exist, any experimental work can be reduced by use of the relation rather than another experiment. One *does* exist. The difference between the specific heats can be expressed in terms of the variables as follows:

$$c_p - c_v = \left(\frac{\partial h}{\partial T}\right)_p - \left(\frac{\partial u}{\partial T}\right)_p = \left(\frac{\partial u}{\partial T}\right)_p + p\left(\frac{\partial v}{\partial T}\right)_p - \left(\frac{\partial u}{\partial T}\right)_p \tag{16-55}$$

$$c_p - c_v = c_v + \frac{T\beta^2 v}{K} - \beta vp + p\beta v - c_v \tag{16-56}$$

$$c_p - c_v = \frac{T\beta^2 v}{K} \tag{16-57}$$

Notice that $(c_p - c_v)$ will be positive for all substances under all conditions.

In order to find complete expressions for enthalpy, Gibbs function, and Helmholtz function, we combine Eq. (16-52) to (16-54) with the definitions of h, g, and z. The results are shown below for enthalpy:

$$dh = c_p\, dT + (v - T\beta v)\, dp \tag{16-58}$$

$$dh = \left(v + \frac{c_v K}{\beta}\right) dp + \frac{c_p}{v\beta}\, dv \tag{16-59}$$

$$dh = \left(c_v + \frac{\beta v}{K}\right) dT + \left(\frac{T\beta}{K} - \frac{1}{K}\right) dv \tag{16-60}$$

16-5 Application to ideal gases

Some of the previous results can be used to prove that if $pv = RT$, then u and h are functions of temperature only. Starting with Eq. (16-32), we find

$$\left(\frac{\partial s}{\partial T}\right)_v = \frac{c_v}{T} \tag{16-32}$$

Now, as the order of differentiation may be reversed for these functions, we can write

$$\frac{\partial}{\partial v}\left[\left(\frac{\partial s}{\partial T}\right)_v\right]_T = \frac{\partial}{\partial T}\left[\left(\frac{\partial s}{\partial v}\right)_T\right]_v \tag{16-61}$$

$$\frac{1}{T}\left(\frac{\partial c_v}{\partial v}\right)_T = \frac{\partial}{\partial T}\left[\left(\frac{\partial s}{\partial v}\right)_T\right]_v \tag{16-62}$$

From Eq. (16-18c) we find

$$\left(\frac{\partial s}{\partial v}\right)_T = \left(\frac{\partial p}{\partial T}\right)_v$$

So

$$\frac{1}{T}\left(\frac{\partial c_v}{\partial v}\right)_T = \frac{\partial}{\partial T}\left[\left(\frac{\partial p}{\partial T}\right)_v\right]_v = \left(\frac{\partial^2 p}{\partial T^2}\right)_v \tag{16-63}$$

Now if $pv = RT$, then

$$\left(\frac{\partial^2 p}{\partial T^2}\right)_v = 0 \tag{16-64}$$

and so

$$\left(\frac{\partial c_v}{\partial v}\right)_T = 0 \tag{16-65}$$

That is, c_v is not a function of temperature. Since c_v is not, then c_p is also not. (Why?) Looking at Eq. (16-58), and noting that $\beta = 1/T$ for ideal gases, the result is

$$dh = c_p\, dT \tag{16-66}$$

And since $dh = du + d(pv)$,

$$du = c_p\, dT - R\, dT = c_v\, dT \tag{16-67}$$

With c_p and x_v being functions of temperature only (for ideal gases) then it is clear also that h and u are also functions of temperatures only.

16-6 Application to elastic systems

Though derived for systems in which pressure, volume, and temperature are the primary variables, the relations derived above have much wider usefulness. To illustrate how they are utilized in different situations, we shall here consider the thermodynamics of an elastic rod.

The situation we wish to explore is shown in the sketch. A bar of cross-sectional area A initially of length l_0 is stretched by a force F to a new length of $l_0 + dl$.

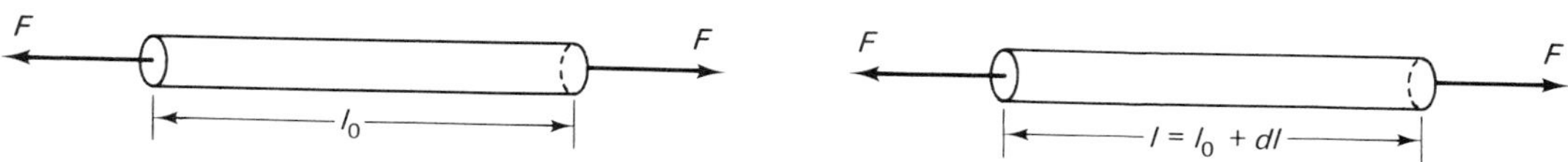

Let us first look at the mechanics of this situation. Students will remember that the stress σ is given by F/A and that the strain $d\varepsilon$ is expressed as dl/l_0. Young's modulus (Y) is defined as being the slope of the stress–strain curve under isothermal conditions. Thus, in our notation,

$$Y = \left(\frac{\partial \sigma}{\partial \varepsilon}\right)_T \tag{16-68}$$

Also, the coefficient of linear expansion (α) (not to be confused with the subscripted α's in Sec. 16-4) is defined as

$$\alpha = \left(\frac{\partial \varepsilon}{\partial T}\right)_\sigma \tag{16-69a}$$

Using the chain rule, we can easily show that

$$\left(\frac{\partial \sigma}{\partial T}\right)_\varepsilon = -\alpha Y \tag{16-70b}$$

The work done *on* the system is

$$d'W_{\text{on system}} = F\,dl = \sigma A\,dl = Al_0\,\sigma\,\frac{dl}{l_0} \tag{16-71}$$

$$= \sigma v\,d\varepsilon \quad \text{(per unit mass)} \tag{16-72}$$

Rewriting this as the work done *by* the system gives

$$d'W_v = -v\sigma\,d\varepsilon \tag{16-73}$$

Returning now to thermodynamics, the first law for the bar is

$$d'Q - d'W = du \quad \text{(per unit mass)}$$

$$d'Q + v\sigma\,d\varepsilon = du \tag{16-74}$$

Considering the process to be reversible ($dQ = T\,ds$) leads to a modified form of the $T\,ds$ equation for an elastic rod:

$$T\,ds = du - v\sigma\,d\varepsilon \tag{16-75}$$

Comparison of this result with the usual $T\,ds$ equation ($T\,ds = du + p\,dv$) shows that we might have simply written the result by analogy replacing p with σ and dv with $-v\,d\varepsilon$. With this hint we define an "enthalpy" for elastic systems as

$$h = u - v\sigma\varepsilon \tag{16-76}$$

The Gibbs and Helmholtz functions do not change:

$$z = u - Ts \tag{16-2b}$$

$$g = h - Ts \tag{16-2c}$$

Now Eqs. (16-76), (16-2b), and (16-2c) are analogous to the equations that lead to Eq. (16-13) and by manipulations similar to those in Sec. 16-2 they lead to a set of Maxwell's equations for an elastic rod:

$$\frac{1}{v}\left(\frac{\partial T}{\partial \varepsilon}\right)_s = \left(\frac{\partial \sigma}{\partial s}\right)_\varepsilon \tag{16-77a}$$

$$\left(\frac{\partial T}{\partial \sigma}\right)_s = -v\left(\frac{\partial \varepsilon}{\partial s}\right)_\sigma \tag{16-77b}$$

$$\left(\frac{\partial \sigma}{\partial T}\right)_\varepsilon = -\frac{1}{v}\left(\frac{\partial s}{\partial \varepsilon}\right)_T \tag{16-77c}$$

$$v\left(\frac{\partial \varepsilon}{\partial T}\right)_\sigma = \left(\frac{\partial s}{\partial \sigma}\right)_T \tag{16-77d}$$

Continuing by analogy with the previous development but using σ, ε, and T instead of p, v, and T, we can derive the following equation:

$$ds = \left(\frac{\partial s}{\partial T}\right)_\varepsilon dT + \left(\frac{\partial s}{\partial \varepsilon}\right)_T d\varepsilon \tag{16-78}$$

Multiplying by T,

$$T\,ds = T\left(\frac{\partial s}{\partial T}\right)_\varepsilon dT + T\left(\frac{\partial s}{\partial \varepsilon}\right)_T d\varepsilon \tag{16-79}$$

Remembering that in the more common system

$$c_v = T\left(\frac{\partial s}{\partial T}\right)_v \tag{16-80}$$

we define a specific heat at constant ε as

$$c_\varepsilon = T\left(\frac{\partial s}{\partial T}\right)_\varepsilon \tag{16-81}$$

Hence Eq. (16-78) becomes

$$ds = c_\varepsilon \frac{dT}{T} + \left(\frac{\partial s}{\partial \varepsilon}\right)_T d\varepsilon \tag{16-82}$$

Also from Eq. (16-77c) we find

$$\frac{1}{v}\left(\frac{\partial s}{\partial \varepsilon}\right)_T = -\left(\frac{\partial \sigma}{\partial T}\right)_\varepsilon \tag{16-83}$$

so that we have

$$ds = c_\varepsilon \frac{dT}{T} - v\left(\frac{\partial \sigma}{\partial T}\right)_\varepsilon d\varepsilon \tag{16-84}$$

In a similar way but starting with T and σ (analogous to T and p) as independent variables, we can derive

$$ds = \left(\frac{\partial s}{\partial T}\right)_\sigma dT + \left(\frac{\partial s}{\partial \sigma}\right)_T d\sigma \tag{16-85}$$

Defining a specific heat at constant σ,

$$c_\sigma = T\left(\frac{\partial s}{\partial T}\right)_\sigma \tag{16-86}$$

results in

$$ds = c_\sigma \frac{dT}{T} + \left(\frac{\partial s}{\partial \sigma}\right)_T d\sigma \tag{16-87}$$

But we also note that from Eq. (16-77d) and (16-69)

$$\left(\frac{\partial s}{\partial \sigma}\right)_T = v\left(\frac{\partial \varepsilon}{\partial T}\right)_\sigma = v\alpha \tag{16-88}$$

Thus we find

$$ds = c_\sigma \frac{dT}{T} + v\alpha\, d\sigma \tag{16-89}$$

To find an expression for the internal energy, we start with our $T\,ds$ equation rewriting it as

$$du = T\,ds + v\sigma\, d\varepsilon \tag{16-90}$$

Using Eq. (16-84) for ds,

$$du = c_\varepsilon\, dT + \left[v\sigma - vT\left(\frac{\partial \sigma}{\partial T}\right)_\varepsilon\right] d\varepsilon \tag{16-91}$$

Incorporating Eq. (16-70), we find

$$du = c_\varepsilon \, dT + (v\sigma + v\alpha YT) \, d\varepsilon \tag{16-92}$$

EXAMPLE 16.1. Determine how the temperature of an elastic system varies when it is stretched adiabatically.

SOLUTION: Assuming the stretching is done reversibly (i.e., slowly), then $ds = 0$. Since we are interested in the T, ε relationship, we select T and ε as the independent properties and hence start with Eq. (16-84)

$$ds = 0 = c_\varepsilon \frac{dT}{T} + v\alpha Y \, d\varepsilon$$

$$c_\varepsilon \frac{dT}{T} = -v\alpha Y \, d\varepsilon$$

$$c_\varepsilon \ln \frac{T}{T_0} = -v\alpha Y \, d\varepsilon$$

$$\frac{T}{T_0} = e^{-v\alpha Y/c_\varepsilon(\varepsilon - \varepsilon_0)}$$

Notice that Y, c_ε, and $(\varepsilon - \varepsilon_0)$ are all positive. Thus if α is also positive, the temperature will decrease when it is stretched adiabatically. Remember α is the coefficient of linear expansion:

$$\alpha = \left(\frac{\partial \varepsilon}{\partial T}\right)_\sigma$$

It is positive for metals and most substances but it is negative for materials like rubber. Unlike metals, rubber bands, in fact, increase in temperature when stretched.

16-7 Application to magnetic systems

Another interesting application of the results obtained earlier in this chapter is in the area of adiabatic demagnetization, a process used to obtain extremely low temperatures near absolute zero. We shall first derive some of the general relations for a magnetic system and then apply them to the adiabatic demagnetization process.

In Chap. 4 we found that magnetic work can be expressed as

$$dW = vH \, dM \tag{16-93}$$

where H is the magnetic field intensity (ampere turns per meter) and M is the magnetization (webers per square meter). Starting with the first law, we can derive our own $T\,ds$ equation for a magnetic system as follows:

$$dQ' - d'W = du$$

$$T\,ds - vH\,dM = du \tag{16-94}$$

$$T\,ds = du + vH\,dM \tag{16-95}$$

We have neglected volume changes and assumed a reversible process but the result is valid for all processes for the same reasons that the ordinary $T\,ds$ equations are valid: It consists only of properties.

Now we can derive expressions for ds and du in the same way we did earlier or we can work more simply by analogy. Then by analogy with Eq. (16-50) we find

$$ds = c_H\left(\frac{dT}{T}\right) + v\left(\frac{\partial M}{\partial T}\right)_H d\mathrm{H} \tag{16-96}$$

where c_H is a specific heat at constant field intensity analogous to c_p. Therefore by analogy to Eq. (16-55)

$$c_H = \left(\frac{\partial h}{\partial T}\right)_H = \left(\frac{\partial u}{\partial T}\right)_H + vH\left(\frac{\partial M}{\partial T}\right)_H \tag{16-97}$$

Cooling by adiabatic demagnetization makes use of substances called *paramagnetic salts* that follow Curie's law ($MT = \mathscr{C}H$) closely. For these substances, from the law itself, we find

$$\left(\frac{\partial M}{\partial H}\right)_T = \mathscr{C} \quad \text{and} \quad \left(\frac{\partial M}{\partial T}\right)_H = -\frac{\mathscr{C}H}{T^2} \tag{16-98}$$

(Note the $\mathscr{C}H$ here is *not* c_H.) Now if such a substance is magnetized to a large H, cooled at constant H to as low a temperature as available, and then demagnetized to $H = 0$ adiabatically (as far as possible), it will undergo a decrease in temperature. To see this, we look at Eq. (16-96) setting $ds = 0$ for an assumed reversible adiabatic process. Then

$$0 = c_H\frac{dT}{T} + v\left(\frac{\partial M}{\partial T}\right)_H dH \tag{16-99}$$

Incorporating Eq. (16-98) gives

$$c_H\frac{dT}{T} = \frac{v\mathscr{C}H}{T^2}\,dH \tag{16-100}$$

Integrating over an assumed drop from $H = H_1$ to $H = 0$, we find

$$\int_{T_1}^{T} T\,dT = \frac{v\mathscr{C}}{c_H}\int_{H_1}^{0} H\,dH \tag{16.101}$$

$$\frac{T^2 - T_1^2}{2} = -\frac{\mathscr{C}v}{c_H}\frac{H_1^2}{2} \tag{16-102}$$

$$T^2 = T_1^2 - \frac{\mathscr{C}v}{c_H}H_1^2 \tag{16-103}$$

$\mathscr{C}$, v, c_H, and H_1 are all positive so the new temperature T will be less than the initial T_1.

16-8 The Joule-Thomson coefficient

The Joule–Thomson coefficient is occasionally important. It is significant, for example, in some refrigeration systems and in the production of some liquefied gases. It is defined as the ratio of temperature change to pressure change in an isenthalic ($h = 0$) process. It is usually given the symbol μ.

$$\mu = \left(\frac{\partial T}{\partial p}\right)_h \tag{16-104}$$

Constant enthalpy processes (approximately) occur in valves, porous plugs, and capillary tubes. They are also called *throttling* processes. It matters a great deal, therefore, whether the Joule–Thomson coefficient is positive or negative for a given substance approaching a valve. If negative, its temperature will *rise* as it passes through the valve to the lower pressure. If μ is positive, the temperature will fall in a pressure drop. For refrigeration purposes, therefore, we would like to have positive Joule–Thomson coefficients.

The Joule–Thomson coefficient is shown visually by the slope of a constant enthalpy line on a T-p plot. Many substances have plots similar to the one shown schematically in Fig. 16.1. The dotted line joins points of zero slope of the enthalpy lines and is called the *inversion curve*. To the left of the inversion curve, the Joule–Thomson coefficient is positive. That is, throttling processes that start at some point like A and move to B (at lower pressure, of course) will have a temperature decrease. Processes that take place on the right of the curve (such as from C to D) will experience a temperature increase.

Liquid air can be made using the fact that the Joule–Thomson coefficient for air is positive and fairly large at pressures of 100 to 150 atm and temperatures around $-120°C$ (150°K). See Fig. 16.1, which shows constant enthalpy lines on a T-p plot. Throttling from this point takes the air into the two-phase region from which the liquid can be separated.

T
h = constant
Inversion line
A
B
C
D
h increasing
p

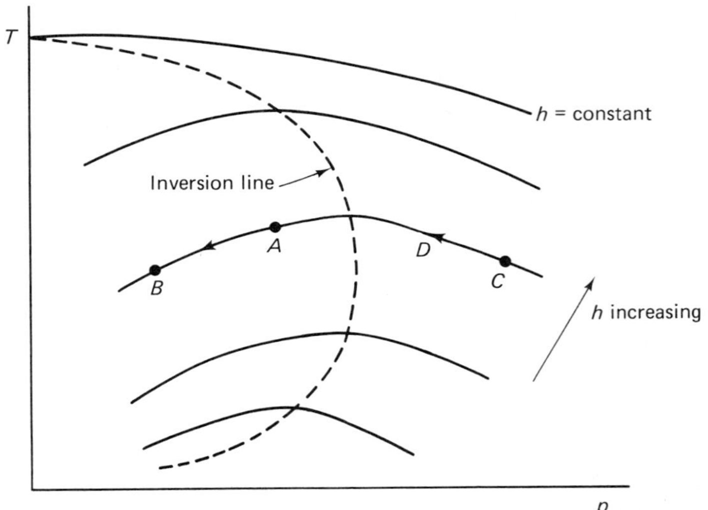

Figure 16.1 Inversion line for Joule–Thomson coefficient.

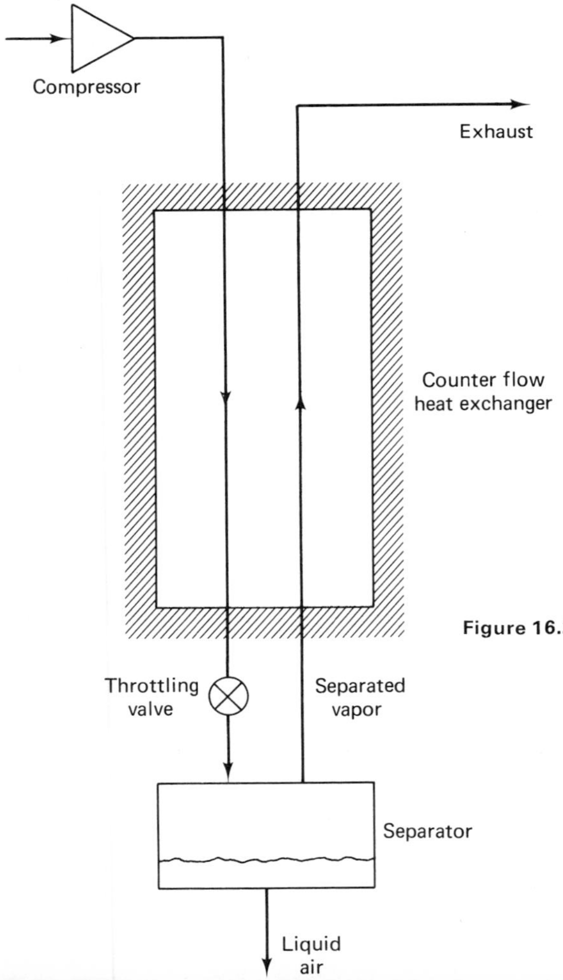

Compressor
Exhaust
Counter flow
heat exchanger
Throttling
valve
Separated
vapor
Separator
Liquid
air

Figure 16.2 Schematic diagram of liquid air process.

To *get* the air to that p and T, it is first compressed and then cooled back to atmospheric temperature. Then it is put through a counterflow heat exchanger as shown in Fig. 16.2. The very cold vapor leaving the separator is used to cool the inlet stream to the desired temperature.

16-9 The Clapeyron equation

Another useful result can be obtained using one of the Maxwell relations. We wish to derive an expression for the slope of the vapor pressure as shown on a p-T diagram like Fig. 16.3. The slope, of course, is given by dp/dT for a

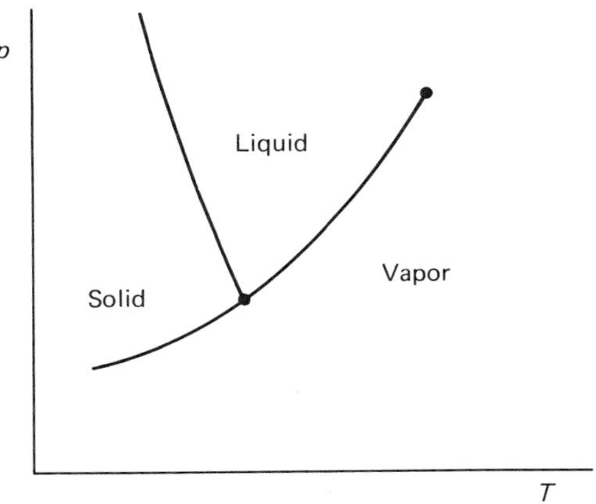

Figure 16.3 Typical p–T diagram.

two-phase mixture since the lines *are* the two-phase region. Taking $p = p(T, v)$, then

$$dp = \left(\frac{\partial p}{\partial T}\right)_v dT + \left(\frac{\partial p}{\partial v}\right)_T dv \tag{16-105}$$

But in the two-phase region $(\partial p/\partial v)_T = 0$ because p is a function *only* of temperature. Thus

$$\frac{dp}{dT} = \left(\frac{\partial p}{\partial T}\right)_v \qquad \text{Two-phase region} \tag{16-106}$$

We wish to make use of the Maxwell relation

$$\left(\frac{\partial p}{\partial T}\right)_v = \left(\frac{\partial s}{\partial v}\right)_T \tag{16-107}$$

We consider a state change in the two-phase region from saturated liquid to saturated vapor at constant temperature. Then

$$\left(\frac{\partial s}{\partial v}\right)_T = \frac{ds}{dv} = \frac{s_g - s_f}{v_g - v_f} = \frac{s_{fg}}{v_{fg}} \tag{16-108}$$

Therefore

$$\frac{dp}{dT} = \frac{s_{fg}}{v_{fg}} \tag{16-109}$$

but also

$$T\,ds = dh - v\,dp \tag{16-110}$$

Since p is also constant for the process under consideration, this becomes

$$T\,ds = dh \tag{16-111}$$

$$ds = \frac{dh}{T} \tag{16-112}$$

$$s_g - s_f = \frac{h_g - h_f}{T} \tag{16-113}$$

$$s_{fg} = \frac{h_{fg}}{T} \tag{16-114}$$

Combining with Eq. (16-109), we obtain the result called the *Clapeyron equation:*

$$\frac{dp}{dT} = \frac{h_{fg}}{Tv_{fg}} \tag{16-115}$$

Similar equations can be derived for the slope of the other vapor pressure lines.

At very low vapor pressures, if we assume that $v_g \simeq v_{fg}$ and the equation of state of the vapor is taken as $pv = RT$, then the Clapeyron equation becomes

$$\frac{dp}{dT} = \frac{h_{fg}}{v_{fg}T} = \frac{h_{fg}}{v_g T} = \frac{h_{fg}p}{RT^2} \tag{16-116}$$

This is sometimes called the Clausius–Clapeyron equation. Both it and the more basic Clapeyron equations are used to determine h_{fg} from vapor pressure–temperature data.

EXAMPLE 16.2. The following data are taken from the Freon-12 tables:

Saturation temperature (°F)	*Saturation pressure* (psia)	*Specific Volume* (ft³/lbm) v_f	v_g
−1	23.35	0.01102	1.641
0	23.85	0.01103	1.609
1	24.35	0.01104	1.577

Estimate the latent heat of evaporation h_{fg} at 0°F in Btu/lbm.

SOLUTION: From the Clapeyron equation,

$$h_{fg} = Tv_{fg}\frac{dp}{dT}$$

$$h_{fg} = (1.609 - 0.011)(460)\left(\frac{24.35 - 23.35}{2}\right)\left(\frac{144}{778}\right) = 68.1 \text{ Btu/lbm}$$

$$\textit{Units:}\quad \left(\frac{\text{ft}^3}{\text{lbm}}\right)(°\text{R})\left(\frac{1}{°\text{R}}\right)\left(\frac{\text{lbf}}{\text{in}^2}\right)\left(\frac{\text{in}^2}{\text{ft}^2}\right)\left(\frac{\text{Btu}}{\text{ft-lbf}}\right) \rightarrow \text{Btu/lbm}$$

16-10 Summary

Certain properties are directly measurable (p, v, T, c_p, c_v, β, K). Others (u, h, s, g, z) are computed by paper and pencil operations using data on the measurable ones and known relations among the properties. Many such relations have been developed in this chapter.

Among the most useful are Maxwell's relations, which relate entropy to p, v, and T but, in addition, how to compute changes in any abstract property from known changes in measurable properties is shown.

The Clapeyron equation expresses the slope of vapor pressure line of a p-T plot (dp/dT) in terms of data available.

$$\frac{dp}{dT} = \frac{h_{fg}}{Tv_{fg}} \tag{16-115}$$

Problems

16-1 A magnetic solid has an equation of state $CH = MT$. If this substance is magnetized isothermally from M_1 to $M_2 = 2M_1$, express the work required in terms of V, T, C, and M.

16-2 Show that if $CH = MT$, the heat per unit mass flowing into a paramagnetic substance when it is magnetized isothermally at T from $H = 0$ to H is given by

$$q_T = -\frac{CvH^2}{2T}$$

16-3 Compute the Joule–Thomson coefficient for an ideal gas.

16-4 Estimate the pressure under the blade of an ice skate. If $h_{if} \simeq 144$ Btu/lbm and $v_{if} \simeq 0.001$ ft^3/lbm, estimate the effect of this pressure on the melting point of the ice using the Clapeyron equation. Does this explain the film of liquid on which the skates ride?

16-5 Use the data for saturation pressure, saturation temperature and specific volume in the steam tables to estimate the latent heat of vaporization of water at 100 psia. Compare your result with the table value.

16-6 Determine the Joule–Thomson coefficient of water at 100 psia and 400°F.

PART III

An introduction to the kinetic theory of ideal gases

Part III—Chapter 17 can be studied as a unit or in parts in conjunction with Part I of this book. For those who wish to follow the latter course, the following table provides the key:

In Part 1		*In Part 2*
Section 4-5	is	Sections 17-1 and 17-2
Section 8-9	is	Sections 17-3 to 17-5

17

An introduction to the kinetic theory of ideal gases

17-1 Introduction

17-1 (a) What is kinetic theory?

Macroscopic thermodynamics deals with measurable properties, experimental laws, and relations that result therefrom. It makes no assumptions about the nature or structure of matter. It therefore has great generality.

Macroscopic thermodynamics also has great weaknesses, however. It uses the results of experiments but it does not help much in the decision about what experiments should be performed. It gives us relations among properties but says nothing about the absolute magnitudes of the properties themselves.

Kinetic theory is one of several microscopic approaches that can be taken to certain aspects of the thermodynamics of gases. It has usefulness beyond the realm of the ideal gas but we shall limit the discussion here to ideal gases. The kinetic theory of ideal gases makes the following assumptions about the nature of gases:

1. Gases are made up of an enormous number of very small particles called *molecules.*
2. The molecules are separated by distances very large in comparison to their own dimensions.
3. The molecules interact only through collisions.
4. The collisions between molecules or with the walls of a container are perfectly elastic.

5. The distribution of molecules and their directions of motion are completely uniform and random.
6. The usual macroscopic laws of mechanics apply to these molecules.

Some of these assumptions are easily justified from known data. The number of molecules in a kilogram-mole is over 6×10^{26} (Avogadro's number, N_0). Under conditions of standard atmospheric pressure and temperature, a kilogram-mole of gas occupies about 22.4 m^3. Converted to engineering units, this means that in a cubical volume 0.001 in. on a side, there would be about 5×10^{11} molecules, thus justifying assumption 1.

The size of a typical molecule can be appreciated as follows. If round, it would occupy a sphere with diameter of about 2×10^{-10} m. Working out the average molecular spacing gives a distance of about 10 times the diameter, fairly well justifying assumption 2.

The other assumptions cannot really be justified quantitatively. They serve to define the kinetic theory model of an ideal gas. The last assumption is incorrect as students of quantum mechanics will recognize.

17-1 (b) Why kinetic theory?

Kinetic theory has only limited usefulness but it does provide a great deal of microscopic insight into many thermodynamic properties. And it does so with relatively little cost in terms of effort. This makes it an excellent adjunct to an introductory study of thermodynamics. For a more sophisticated microscopic treatment, students should take a full course in statistical thermodynamics that deals with statistical mechanics and is based on quantum mechanical principles.

17-2 Equation of state of an ideal gas

17-2(a) Collisions with the walls

Consider a container of volume V having N molecules. The average density of molecules is then

$$n = \frac{N}{V} \tag{17-1}$$

The assumptions listed previously allow us to assume that n is a constant and applies as well to a small volume dV containing a small number of molecules dN. That is,

$$n = \frac{dN}{dV} \quad \text{or} \quad dN = n\,dV \tag{17-2}$$

We set out now to relate the pressure of the gas to the microscopic parameters. For this we must first determine the number of collisions that molecules make with the walls per unit time per unit area. Refer now to Fig. 17.1. It shows a portion of the wall of the container. On this portion, a section of area dA is taken as the base for a cylindrical volume tilted an angle θ from the vertical and rotated an angle ϕ from the x axis shown.

We shall now need some notation. The molecules may be moving in any direction with any speed. Let us designate the number of molecules with a speed between v and $v + dv$ as dN_v. Similarly, the number of molecules with direction between θ and $\theta + d\theta$ will be dN_θ and the number with direction between ϕ and $\phi + d\phi$ will be dN_ϕ.

In general, of course, a molecule will have a direction given by some θ and ϕ. To designate the number with velocity v and the direction given by θ and ϕ, we shall use $d^3N_{v\theta\phi}$.

Now consider the particular θ and ϕ of the cylinder of Fig. 17.1. Note also that we make the cylinder length equal to $v\,dt$. We wish to find the number of molecules with direction θ and ϕ and with velocity v that strike dA in time dt. Then by integration over all θ, all ϕ, and all v, we can obtain the total collisions with dA. Now if a molecule has direction $\theta\phi$ and velocity v and if it is in the cylinder, it will strike dA. If not in the cylinder, it will not strike dA. Other molecules will also strike dA in time dt but we shall handle these by integration. The point is that in our notation all $\theta\phi v$ molecules *in* the cylinder strike dA in time dt.

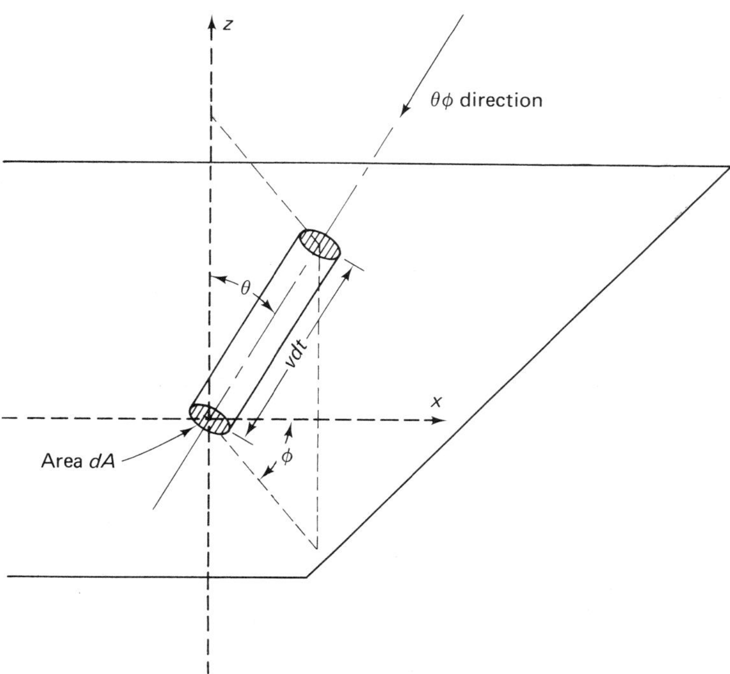

Figure 17.1

Next we must find the number of $\theta\phi v$ molecules in the cylinder. First let us find the number of $\theta\phi$ molecules in the cylinder. Remember that all directions of molecules are equally likely. Imagine therefore that the vectors representing the directions of all the molecules are brought together with their origins at a point. They will then all point outward and be uniformly distributed. There are a total of N of them.

To find the number with a particular direction $\theta\phi$, we imagine a sphere drawn with the point above as the center and all vectors extended to intersect the sphere. In terms of θ and ϕ, a unit of surface area on that sphere is given by

$$dS = r^2 \sin\theta \, d\theta \, d\phi$$

where r is the radius of the sphere. See Fig. 17.2.

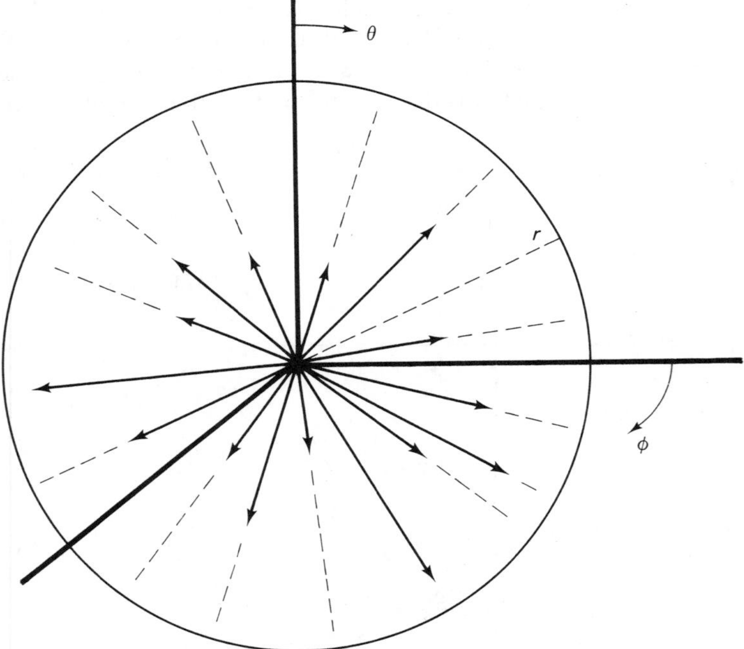

Figure 17.2

Since the directions are random and thus evenly distributed around the sphere, the number intersecting dS will be to the total number as dS is to the total surface area of the sphere. That is,

$$\frac{d^2 N_{\theta\phi}}{N} = \frac{dS}{4\pi r^2} \tag{17-3}$$

or

$$d^2N_{\theta\phi} = \frac{N}{4\pi} \sin\theta \, d\theta \, d\phi \tag{17-4}$$

It is simply a matter of notation to see that the number of molecules with velocity v is given by dN_v and that the density of such molecules is dn_v.

Since the molecules are randomly or evenly distributed throughout the container, we note that the fraction of all the $\theta\phi$ particles that will have a particular velocity v is dN_v/N. That is,

$$d^3N_{\theta\phi v} = d^2N_{\theta\phi} \frac{dN_v}{N} \tag{17-5}$$

Combining this with Eq. (17-1) gives

$$d^3N_{\theta\phi v} = \frac{dN_v}{4\pi} \sin\theta \, d\theta \, d\phi \tag{17-6}$$

Put on a per unit volume basis, this becomes

$$d^3n_{\theta\phi v} = \frac{dn_v}{4\pi} \sin\theta \, d\theta \, d\phi \tag{17-7}$$

Equation (17-7) expresses the number of molecules per unit volume in the container with a direction $\theta\phi$ and a velocity v. The volume of the cylinder shown in Fig. 17.1 is

$$dv_{\text{cyl}} = \cos\theta \cdot v \cdot dt \, dA \tag{17-8}$$

and so the number of $\theta\phi v$ molecules in the cylinder is

$$\text{No. of molecules in cylinder} = d^3n_{\theta\phi v} \, dv_{\text{cyl}} = \frac{v \, dn_v}{4\pi} \sin\theta \cos\theta \, d\theta \, d\phi \, dA \, dt \tag{17-9}$$

We have already argued that the number of $\theta\phi v$ molecules in the cylinder is the number striking the wall in time dt. So, dividing by dA and dt,

$$\text{No. of collisions per unit area per unit time} = \frac{v \, dn_v}{4\pi} \sin\theta \cos\theta \, d\theta \, d\phi \tag{17-10}$$

To find the total number of collisions with the wall by molecules with a velocity v, we integrate θ over 0 to $\pi/2$ and ϕ over 0 to 2π. The result is

$$\text{No. of collisions per unit time per unit area} = \tfrac{1}{4} v \, dn_v \tag{17-11}$$

The total number of collisions is found by integrating Eq. (17-11) over all v. As yet, however, we do not have a relation between dn_v and v so this cannot be carried out. We can only write

$$\text{Total collisions per unit time per unit area} = \tfrac{1}{4}\int_0^\infty v\,dn_v \qquad (17\text{-}12)$$

From the definition of an average, we find the average speed $\bar{v}$ is

$$\bar{v} = \frac{\int v\,dN_v}{\int dN_v} = \frac{\int v\,dN_v}{N} = \frac{\int v\,dn_v}{n} \qquad (17\text{-}13)$$

and this can also be written as

$$\text{Total collisions per unit time per unit area} = \tfrac{1}{4}n\bar{v} \qquad (17\text{-}14)$$

17–2(b) Pressure

We now wish to relate microscopic considerations to our macroscopic concept pressure. To do this we use the Newtonian principle that force equals rate of change of momentum.

Suppose a molecule with mass m having direction θ and velocity v strikes the wall and bounces off according to our assumptions. Then its change in momentum on the z direction (normal to the wall) will be

$$\text{Change in } z \text{ momentum of one } \theta v \text{ molecule} = 2mv\cos\theta \qquad (17\text{-}15)$$

Considering again the cylinder of Fig. 17.1, the number of collisions with dA in time dt of $\theta\phi v$ molecules is

$$\frac{1}{4\pi}\,v\,dn_v \sin\theta\cos\theta\,d\theta\,d\phi\,dA\,dt \qquad (17\text{-}16)$$

Thus the change in momentum of all $\theta\phi v$ molecules striking dA in time dt is the product of Eq. (17-15) and (17-16):

$$\text{Change in momentum of all } \theta\phi v \text{ molecules striking } dA \text{ in time } dt$$
$$= \frac{1}{2\pi}\,mv^2\,dn_v \sin\theta\cos^2\theta\,d\theta\,d\phi\,dA\,dt \qquad (17\text{-}17)$$

Integrating over θ and ϕ as before gives

$$\text{Change in momentum of all } v \text{ molecules striking } dA \text{ in time } dt$$
$$= \tfrac{1}{3}mv^2\,dn_v\,dA\,dt \qquad (17\text{-}18)$$

$$\text{Change in momentum of all molecules striking } dA \text{ in time } dt$$
$$= \tfrac{1}{3}m\,dA\,dt\int v^2\,dn \qquad (17\text{-}19)$$

Now involving Newton's second law and noting that pressure equals force/area:

$$\text{Force} = \frac{d}{dt} \quad \text{(momentum)} \tag{17-20}$$

$$p\,dA\,dt = \tfrac{1}{3}m\,dA\,dt \int v^2\,dn_v \tag{17-21}$$

$$p = \tfrac{1}{3}m \int v^2\,dn \tag{17-22}$$

The definition of the root mean square is

$$\overline{v^2} = \frac{\int v^2\,dN_v}{\int dN_v} = \frac{\int v^2\,dN}{N} = \frac{\int v^2\,dn}{n} \tag{17-23}$$

So

$$p = \tfrac{1}{3}mn\overline{v^2} \tag{17-24}$$

Note that the left side of Eq. (17-24) is a macroscopic property p whereas the right side contains only microscopic parameters. This is the first of several expressions to be developed that relate the macroscopic and microscopic worlds of thermodynamics and kinetic theory.

17-2(c) Temperature

Using the fact that $n = N/v$, the expressions for p become

$$pV = \tfrac{1}{3}mN\overline{v^2} \tag{17-25}$$

Our macroscopic equation of state for an ideal gas is

$$pV = MRT \tag{17-26}$$

where M is mass and R is the gas constant per unit mass. We can convert MR to microscopic terms using molecular mass (M_0) and Avogadro's number N_0 as follows:

$$MR = \frac{M}{M_0} \cdot M_0 R = \overline{M}\overline{R} = \frac{N}{N_0} \cdot N_0 k = Nk \tag{17-27}$$

where $\overline{M}$ = number of moles
$\overline{R}$ = gas constant per mole
k = gas constant per molecule = 1.38×10^{-23} j/°K
N = number of molecules

Thus we obtain

$$pV = NkT \tag{17-28}$$

Equating Eq. (17-25) and (17-28), we find

$$NkT = \tfrac{1}{3}N m\overline{v^2} \tag{17-29}$$

or

$$T = \frac{1}{3}\frac{m\overline{v^2}}{k} \tag{17-30}$$

Again we have derived an expression relating a macroscopic property to microscopic terms. Notice that temperature is related to the *speed* of the molecules in an important way.

It is interesting to rearrange Eq. (17-30) as follows:

$$\tfrac{1}{2}m\overline{v^2} = \tfrac{3}{2}kT \tag{17-31}$$

Now the left side is the average *energy* of a molecule. Evidently then kT is also energy. It is especially interesting that the energy of a molecule given by kT is independent of mass. This says that at the same temperature the average *energy* of molecules in various gases is the same.

17-3 Equipartition of energy

Based on our assumptions, the molecules with which we have been dealing have only translational, or kinetic, energy. We can therefore write the internal energy of the gas in a container of N particles as

$$U = N\bar{u}^* = N\tfrac{1}{2}m\overline{v^2} = \tfrac{3}{2}nkT \tag{17-32}$$

or

$$u^* = \tfrac{1}{2}m\overline{v^2} = \tfrac{3}{2}kT \tag{17-23}$$

where u^* is the average total internal energy of a molecule. Now if the vector velocity is resolved into its x, y, and z components, it is easy to show that

$$\overline{v^2} = \overline{v_x^2} + \overline{v_y^2} + \overline{v_z^2} \tag{17-34}$$

and so

$$\bar{u} = \tfrac{1}{2}m\overline{v_x^2} + \tfrac{1}{2}m\overline{v_y^2} + \tfrac{1}{2}m\overline{v_z^2} \tag{17-35}$$

Because of the assumption that all directions are equally likely, it is reasonable to assume that the total energy of the gas is *equally divided* among the three degrees of freedom (x, y, and z). Notice that this assumption results in

$$\tfrac{1}{2}m\overline{v_x^2} = \tfrac{1}{2}\overline{v_y^2} = \tfrac{1}{2}m\overline{v_z^2} = \tfrac{1}{2}kT \tag{17-36}$$

and that each translational degree of freedom has energy equal to $\frac{1}{2}kT$.

The special case above is often extended or generalized to systems of gases with more than just translational energy. The *equipartition* theory assumes that the energy associated with each active degree of freedom of a molecule is $\frac{1}{2}kT$. For our simple gas with translation only, we find

$$U = \tfrac{3}{2}NkT \tag{17-37}$$

or in general

$$U = \frac{f}{2}NkT \tag{17-38}$$

where f is the active degrees of freedom of the molecules. Remembering that $Nk = MR$, we find

$$U = \frac{f}{2}MRT \tag{17-39}$$

If we were to add heat at constant volume to the gas, the first law gives

$$U_2 - U_1 = M\bar{c}_v(T_2 - T_1) = \frac{f}{2}M\bar{R}(T_2 - T_1) \tag{17-40}$$

Thus

$$\bar{c}_v = \frac{f}{2}\bar{R} \tag{17-41}$$

Since we know that $c_p - c_v = R$ or $\bar{c}_p - \bar{c}_v = \bar{R}$, this gives

$$\bar{c}_p = \left(\frac{f}{2} + 1\right)\bar{R} = \left(\frac{f+2}{5}\right)\bar{R} \tag{17-42}$$

and, for the ratio of specific heats γ,

$$\gamma = \frac{\bar{c}_p}{\bar{c}_v} = \frac{f+2}{f} \tag{17-43}$$

Applying Eq. (17-43), we see that $f = 3$ for a monatomic gas; then for all monatomic gases, according to this theory,

$$\bar{c}_p = \tfrac{5}{2}\bar{R} \qquad \bar{c}_v = \tfrac{3}{2}\bar{R} \qquad \gamma = \tfrac{5}{3} = 1.67 \tag{17-44}$$

For diatomic gases, a dumbbell model is used as shown in the sketch. Now such a molecule can have three active translational degrees of freedom and two rotational ones (rotation about z is assumed negligible). It also has

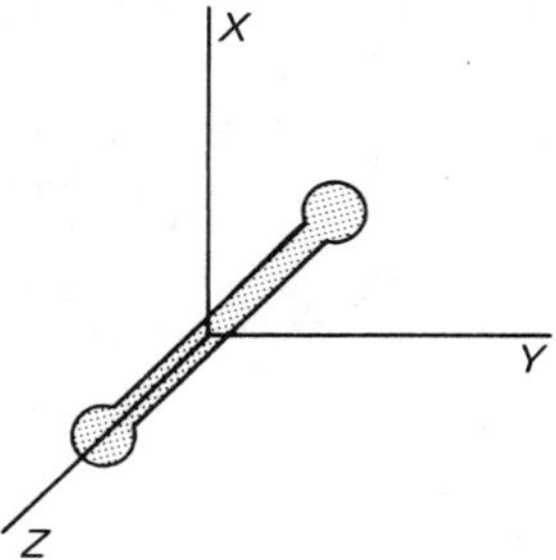

two vibrational degrees of freedom because its bonds are not rigid and it can vibrate along its axis. This vibration gives two degrees of freedom—one for potential energy and one for kinetic. Thus for such a molecule, $f = 7$, and we find

$$\bar{c}_p = \tfrac{7}{2}\bar{R} \qquad \bar{c}_v = \tfrac{5}{2}\bar{R} \qquad \gamma = \tfrac{7}{5} = 1.40 \tag{17-45}$$

Data for some monatomic and diatomic gases and others are found in Table 17-1. This shows the general agreement with the theory. Usually, if the assumption is that a gas is ideal, then the theory above will be valid for finding the specific heats. At low temperatures, the diatomic molecule's degrees of freedom become "inactive" and f falls to 5 and then to 3. Also at very high temperatures other vibrational degrees of freedom may become active and there may also be an electron contribution to specific heat. But the theory is very often a quick and sure way to find the specific heat of an ideal gas under typical engineering circumstances.

Table 17-1 Some molar specific heats of gases at room temperature

Gas	c_p	c_v	$\gamma = c_p/c_v$
H_2	6.84	4.88	1.40
O_2	7.04	5.01	1.40
N_2	7.00	4.90	1.43
CO	7.00	4.98	1.41
Cl_2	8.24	6.04	1.36
A	5.04	3.02	1.67
He	5.04	3.04	1.66
CO_2	8.80	6.76	1.30
NH_3	8.96	6.84	1.31

For a solid, the equipartition theory predicts a total energy of

$$U = 3NkT \qquad (17\text{-}46)$$

based on an energy of kT for each of three vibrational degrees of freedom. This gives a specific heat of

$$\bar{c}_v = 3\bar{R} \qquad (17\text{-}47)$$

and is in fair agreement with experimental results for solids at relatively high temperatures.

17-4 The distribution of molecular velocities

So far we have not developed any information about the distribution of molecular velocities. That is, we do not know dn_v as a function of v. We have written our various results in terms of $\bar{v}$ or v^2 but we cannot compute these averages. In this section we shall derive a velocity distribution equation.

To find the desired result we first construct a "velocity space" by setting up a rectangular coordinate system with v_x, v_y, and v_z as the coordinates.

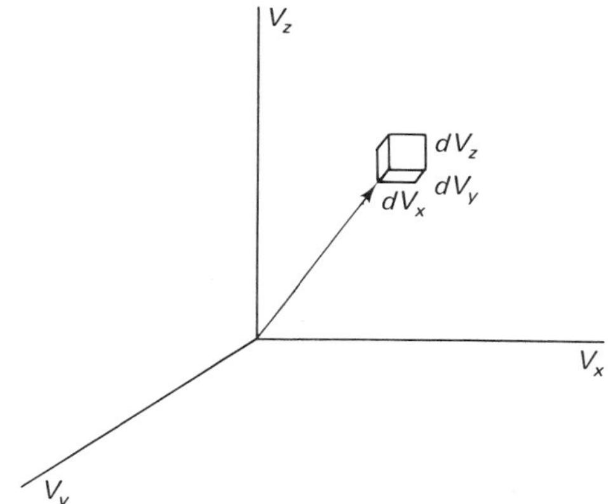

Figure 17.3

See Fig. 17.3. As in Fig. 17.2 we move all the velocity vectors for the molecules to the origin of this velocity space system except that now, however, we do not extend the vectors as needed to intersect any sphere of radius r. We leave them as they are and imagine their arrow tips defining a point in space. If we can express the density of these points in terms of v_x, v_y, and v_z, note that we shall have accomplished what we wished because that density will be the density of molecules with a particular velocity.

Now consider a small element $dv_x \cdot dv_y \cdot dv_z$ at a distance v from the origin. We note that geometry gives

$$v^2 = v_x^2 + v_y^2 + v_z^2 \tag{17-48}$$

Next consider the fraction of particles that have a velocity between v_x and $v_x + dv_x$. This fraction can be designated as

$$\frac{dN_{vx}}{N}$$

and it will depend directly on the size of d_{vx} chosen and it will be some function of v_x (not of v_y or v_z because the x component of a molecule's velocity is not to be dependent on its y or z component).

$$\frac{dN_{vx}}{N} = f(v_x)\, d_{vx} \tag{17-49}$$

In a similar manner we can write

$$\frac{dN_{vy}}{N} = f(v_y)\, d_{vy} \tag{17-50}$$

and

$$\frac{dN_{vz}}{N} = f(v_z)\, d_{vz} \tag{17-51}$$

The function f is the same in each case but we do not know yet what it is.

Now the fraction of particles with velocity v_x which also has velocity v_y is given by

$$\frac{d^2N_{v_xv_y}}{dN_{vx}}$$

and the fraction of the total number which has velocity v_y is given by

$$\frac{dN_{vy}}{N} = f(v_y)\, d_{vy} \tag{17-52}$$

Now since the molecules' velocities are evenly distributed, we can state that

$$\frac{\text{No. of particles with } v_y}{\text{Total no. of particles}} = \frac{\text{No. of particles with } v_x \text{ and with } v_y}{\text{Total no. of particles with } v_x}$$

Thus

$$\frac{dN_{vy}}{N} = \frac{d^2N_{v_xv_y}}{dN_{vx}} \tag{17-53}$$

or

$$d^2N_{v_xv_y} = dN_{v_x} f(v_y)\, d_{v_y} \tag{17-54}$$

$$d^2N_{v_xv_y} = Nf(v_x)f(v_y)\, dv_x\, dv_y \tag{17-55}$$

In the same way we can find

$$d^3N_{v_xv_yv_z} = Nf(v_x)f(v_y)f(v_z)\, dv_x\, dv_y\, dv_z \tag{17-56}$$

This is the number of points in the little element $dv_x\, dv_y\, dv_z$. The number of these points per unit volume or their "density" is

$$\frac{d^3n_{v_xv_yv_z}}{dv_x\, dv_y\, dv_z} = \rho \tag{17-57}$$

The total derivative of ρ is

$$d\rho = \frac{\partial \rho}{\partial v_x} dv_x + \frac{\partial \rho}{\partial v_y} dv_y + \frac{\partial \rho}{\partial v_z} dv_z \tag{17-58}$$

and

$$\frac{\partial \rho}{\partial v_x} = Nf'(v_x)f(v_y)f(v_z) \tag{17-59}$$

where

$$f'(v_x) = \frac{d}{dv_x} f(v_x) \tag{17-60}$$

Finding $\partial\rho/\partial v_y$ and $\partial\rho/\partial v_z$ in the same way and combining the results above gives

$$d\rho = \frac{f'(v_x)}{f(v_x)} dv_x + \frac{f'(v_y)}{f(v_y)} dv_y + \frac{f'(v_z)}{f(v_z)} dv_z \tag{17-61}$$

Remembering that $v^2 = v_x^2 + v_y^2 + v_z^2$, by differentiation, we find

$$v\, dv = v_x\, dv_x + v_y\, dv_y + v_z\, dv_z \tag{17-62}$$

Now let us restrict ourselves to changes that take place with v constant. v_x, v_y, and v_z may change but the distance from the origin, and hence ρ, will be kept constant. That is,

$$dv = 0 \quad \text{and} \quad d\rho = 0 \tag{17-63}$$

Employing the method of Lagrangian multipliers to find the value of ρ subject to the constraint that $v^2 = v_x^2 + v_y^2 + v_z^2$, we now multiply Eq. (17-62) by a constant λ and add to Eq. (17-61). The result is

$$\left[\frac{f'(v_x)}{f(v_x)} + \lambda v_x\right] dv_x + \left[\frac{f'(v_y)}{f(v_y)} + \lambda v_y\right] dv_y + \left[\frac{f'(v_z)}{f(v_z)} + \lambda v_z\right] dv_z = 0 \quad (17\text{-}64)$$

Since v_x, v_y, and v_z are independent, each coefficient above must be zero. That is,

$$\frac{f'(v_x)}{f(v_x)} + \lambda v_x = 0 \quad (17\text{-}65)$$

and similarly for the v_y and v_z terms. Integration gives

$$\frac{f'(v_x)\, dv_x}{f(v_x)} = -\lambda v_x\, dv_x \quad (17\text{-}66)$$

or

$$\ln f(v_x) = -\frac{\lambda v_x^2}{2} + C \quad (17\text{-}67)$$

where C is the integration constant. This may be rewritten as

$$f(v_x) = Ce^{-\lambda v_x{}^2/2} \quad (17\text{-}68)$$

It is common practice now to rewrite this using α and β as follows:

$$f(v_x) = \alpha e^{-\beta^2 v_x{}^2} \quad (17\text{-}69)$$

Similarly we can derive

$$f(v_y) = \alpha e^{-\beta^2 v_y{}^2} \quad (17\text{-}70)$$

$$f(v_z) = \alpha e^{-\beta^2 v_z{}^2} \quad (17\text{-}71)$$

We have written [Eq. (17-56)] that the number of points in an element $dv_x\, dv_y\, dv_z$ is

$$d^3N_{v_x v_y v_z} = Nf(v_x)f(v_y)f(v_z)\, dv_x\, dv_y\, dv_z \quad (17\text{-}56)$$

so this now becomes

$$d^3N_{v_x v_y v_z} = N\alpha^3 e^{-\beta^2(v_x{}^2 + v_y{}^2 + v_z{}^2)}\, dv_x\, dv_y\, dv_z \quad (17\text{-}72)$$

or

$$d^3N_{v_x v_y v_z} = N\alpha^3 e^{-\beta^2 v^2}\, dv_x\, dv_y\, dv_z \quad (17\text{-}73)$$

and the density of points is

$$\rho = N\alpha^3 e^{-\beta^2 v^2} \tag{17-74}$$

This is known as the Maxwell velocity distribution function. But before it is of much value to us, we must find the physical meaning of α and β. To do this, it helps first to find the *speed* distribution (as distinguished from the *velocity* distribution above).

Note that all points a given distance v from the origin in velocity space have the same *speed*. The volume of a thin spherical shell of thickness dv and radius v will be $dV = 4\pi v^2\,dv$. We have found the number per unit volume to be ρ so the number of points in such a shell (dN_v) is

$$dN_v = \rho\,dV \tag{17-75}$$

$$= 4\pi N\alpha^3 v^2 e^{-\beta^2 v^2}\,dv \tag{17-76}$$

To find α and β, we next note that the total number of particles must add up to N. That is,

$$N = \int_0^\infty dN_v \tag{17-77}$$

$$N = 4\pi N\alpha^3 \int_0^\infty v^2 e^{-\beta^2 v^2}\,dv \tag{17-78}$$

This integral can be evaluated as

$$\int_0^\infty v^2 e^{-\beta^2 v^2}\,dv = \frac{\sqrt{\pi}}{4\beta^3} \tag{17-79}$$

and so we find α in terms of β:

$$\alpha^3 = \beta^3 \pi^{-3/2} \tag{17-80}$$

and

$$dN_v = \frac{4N}{\sqrt{\pi}}\,\beta^2 v^2 e^{-\beta^2 v^2}\,dv \tag{17-81}$$

To find β, we shall compute v^2 and make use of the result already obtained that

$$\frac{1}{2}\,m\overline{v^2} = \frac{3}{2}\,kT \tag{17-31}$$

$$\overline{v^2} = \frac{\int_0^\infty v^2\,dN_v}{N} = \frac{4\beta^3}{\sqrt{\pi}} \int_0^\infty v^4 e^{-\beta^2 v^2}\,dv \tag{17-82}$$

$$\overline{v^2} = \frac{4\beta^3}{\sqrt{\pi}} \frac{3\sqrt{\pi}}{8\beta^5} = \frac{3}{2\beta^2} \tag{17-83}$$

$$\beta^2 = \frac{3}{2} \frac{1}{\overline{v^2}} = \frac{m}{2kT} \tag{17-84}$$

Notice that once again we have an equation that relates the macroscopic world with the microscopic term β. Thus the distribution of speeds is given by

$$\frac{dN_v}{dv} = \frac{4N}{\sqrt{\pi}} \left(\frac{m}{2kT}\right)^{3/2} v^2 e^{-mv^2/2kT} \tag{17-85}$$

Equation (17-87) is plotted in Fig. 17.4 qualitatively to show how it varies with temperature. Note that at low temperatures there are large numbers of molecules with low velocity whereas at high T's the distribution is flatter.

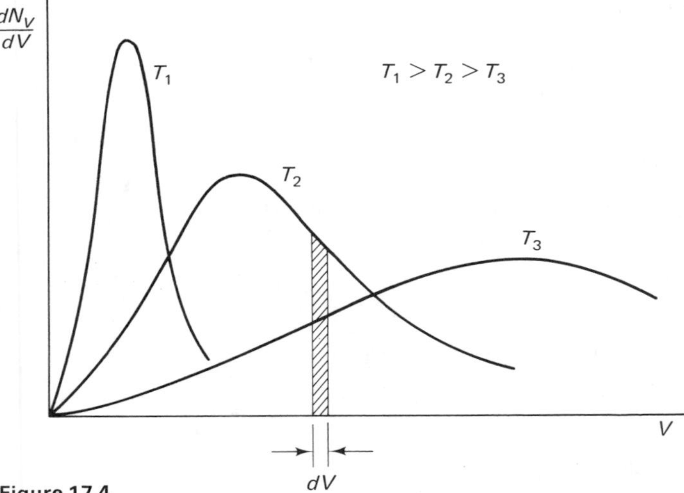

Figure 17.4

Students encountering this distribution for the first time are sometimes confused by the notation. Note that the slice dv wide shown in Fig. 17.4 has a height of dN_v/dv. Therefore its area is dN_v and gives the number of particles with a velocity between v and $v + dv$. The area under the curve is thus the number of particles.

To establish the *velocity* distribution, we go back to Eq. (17-72) and include our results for α and β. The result is

$$\frac{d^3 N_{v_x v_y v_z}}{dv_x\, dv_y\, dv_z} = \left(\frac{N}{\pi^{3/2}}\right)\left(\frac{m}{2kT}\right)^{3/2} e^{-mv^2/2kT} \tag{17-86}$$

For the distribution of velocities in a particular direction, say, x, we find

$$\frac{dN_{v_x}}{dv_x} = \left(\frac{N}{\sqrt{\pi}}\right)\left(\frac{m}{2kT}\right)^{1/2} e^{-mv_x^2/2kT} \tag{17-87}$$

The results above can now be used to compute the average and rms speeds. To find the average speed,

$$\bar{v} = \frac{\int_0^\infty v\, dN_v}{N} = \frac{4}{\sqrt{\pi}}\left(\frac{m}{2kT}\right)^{3/2} \int_0^\infty v^3 e^{-mv^2/2kT}\, dv \tag{17-88}$$

Using the information in Fig. 17.5, this becomes

$$\bar{v} = \sqrt{\frac{8kT}{\pi m}} = \sqrt{2.55\frac{kT}{m}} \tag{17-89}$$

We already know that the rms speed is

$$v_{\text{rms}} = \sqrt{\overline{v^2}} = \sqrt{\frac{3kT}{m}} \tag{17-90}$$

Students may be interested in the fact that the speed of sound in an ideal gas is found to be

$$v_{\text{sound}} = \sqrt{\frac{\gamma kT}{m}} \tag{17-91}$$

where $\gamma = 1.67$ for monatomic and 1.4 for diatomic gases.

17-5 Some applications

A number of interesting problems can be worked with the results obtained so far. For example, one can now derive equations for the viscosity and for the diffusion coefficient of an ideal gas. The mean free path—that is, the average distance a molecule travels between collisions—can also be computed. It is found to be

$$\lambda = \frac{1}{\sqrt{2}}\frac{1}{\sigma n} \tag{17-92}$$

where λ = mean free path
σ = cross-sectional area of the moving molecule
n = density of molecules per unit volume

For air under normal conditions the mean free path is of the order of 10^{-6} cm.

We know that when a gas in a container is allowed to escape through a *macroscopic* hole (e.g., a centimeter or so in diameter) that its escape is governed by the laws of compressible fluid flow. You will learn that if the pressure ratio exceeds about 2, for most gases a limiting condition will be reached when the velocity of the gas through the passage reaches its sonic velocity. And the hole is large enough that the gas escaping is a true sample of the gas inside.

What happens, however, when the hole is *micro*scopic? We shall define microscopic to mean of the order of magnitude of the mean free path. In such a case, macroscopic considerations are of no value but the equations just derived are. We know, for example, that the number of particles striking a wall per unit time, per unit area, is given by

$$\frac{1}{4}\int v\,dn_v$$

and we have an expression for the speed distribution dn_v. If we assume that all the particles that "strike" a hole of area A pass through it, then we can find the rate at which gas escapes such a container.

No. of molecules striking per unit area per unit time

$$= \frac{1}{4}\int_0^\infty v\,dn_v = \frac{1}{4V}\int_0^\infty v\,dN_v = \frac{N\bar{v}}{4V} \qquad (17\text{-}93)$$

Incorporating the expression for $\bar{v}$ derived in Eq. (17-89), we find

$$\text{No. of molecules striking per unit area per unit time} = \frac{N}{4V}\sqrt{\frac{8kT}{\pi m}} \qquad (17\text{-}94)$$

Remembering that $pV = NkT$, this can be reduced to

$$\text{No. of molecules striking per unit area per unit time} = \frac{p}{\sqrt{2\pi mkT}} \qquad (17\text{-}95)$$

These expressions enable us to compute the rate of escape of a gas through a "small" hole.

Students will find study of the following examples to be helpful preparation for working some of the problems at the end of the chapter.

EXAMPLE 17.1. Show that the average energy of the molecules that escape from a "small" hole is *not* $\frac{3}{2}kT$ but $2kT$.

SOLUTION: The average energy of the molecules that escape a hole will be given by

$$\text{Average energy} = \frac{\text{total energy}}{\text{number of molecules that escape}}$$

Per unit time per unit area, we know the number that escape is

$$\text{No. of molecules that escape} = \frac{N}{4V}\sqrt{\frac{8kT}{\pi m}}$$

The total energy of those that escape must be found by integration. The number escaping per unit time per unit area with a velocity v is

$$\frac{1}{4V} v \, dN_v$$

The energy of these is $(\frac{1}{2}mv^2)$ so the energy leaving is

$$E_v = \frac{1}{4V} v \, dN_v \cdot \frac{1}{2} mv^2 = \frac{m}{8V} v^3 \, dN_v$$

Integrating over all velocities to find the total energy,

$$E = \frac{m}{8V}\int_0^\infty v^3 \, dN_v$$

$$= \frac{m}{8V}\frac{4N}{\sqrt{\pi}}\left(\frac{m}{2kT}\right)^{3/2}\int_0^\infty v^5 e^{-mv^2/2kT} \, dv$$

$$f(n) = \int_0^\infty x^n e^{-ax^2} dx$$

n	$f(n)$	n	$f(n)$
0	$\frac{1}{2}\left(\frac{\pi}{a}\right)^{\frac{1}{2}}$	4	$\frac{3}{8}\left(\frac{\pi}{a^5}\right)^{\frac{1}{2}}$
1	$\frac{1}{2a}$	5	$\frac{1}{a^3}$
2	$\frac{1}{4}\left(\frac{\pi}{a^3}\right)^{\frac{1}{2}}$	6	$\frac{15}{16}\left(\frac{\pi}{a^7}\right)^{\frac{1}{2}}$
3	$\frac{1}{2a^2}$	7	$\frac{3}{a^4}$

Figure 17.5

Using Figure 17.5 to determine the integral, we find

$$E = \frac{m}{8V}\frac{4N}{\sqrt{\pi}}\left(\frac{m}{2kT}\right)^{3/2}\left(\frac{2kT}{m}\right)^3$$

Thus the average energy is given by

$$\text{Average energy} = \frac{(m/8V)(4N/\sqrt{\pi})(2kT/m)^{3/2}}{(N/4V)(8kT/\pi m)^{1/2}}$$

This reduces to simply

$$\text{Average energy} = 2kT$$

EXAMPLE 17.2. A tank of volume V develops a small hole of area A. The region surrounding the tank is kept at very low pressure so that leakage back into the tank is negligible. Derive an expression for the pressure in the tank as a function of time, A, V, and $\bar{v}$. Assume T is constant.

SOLUTION: To solve this problem, we note that the rate at which molecules *leave* the tank is given by

$$\frac{N\bar{v}A}{4V}$$

where N is the number of molecules in the tank at any time. But the rate at which molecules leave the tank can also be expressed as

$$-\frac{dN}{dt}$$

Equating these gives

$$-\frac{dN}{dt} = \frac{N\bar{v}A}{4V}$$

$$\int_{N_0}^{N} \frac{dN}{N} = -\frac{\bar{v}A}{4V}\int_0^t dt$$

$$\ln\frac{N}{N_0} = -\frac{\bar{v}tA}{4V}$$

$$N = N_0\, e^{-A\bar{v}t/4V}$$

17-6 Summary

Based on the kinetic theory model of an ideal gas, the number of collisions per unit time of molecules with its container wall per unit area is

$$\frac{1}{4}\int_0^{\infty} v\, dn_v = \frac{1}{4}n\bar{v} = \frac{N\bar{v}}{4V} \qquad (17\text{-}96)$$

The pressure exerted by the gas on the wall is

$$p = \tfrac{1}{3}mn\overline{v^2} \tag{17-97}$$

The equation of state is

$$pV = NkT \tag{17-28}$$

From combining Eqs. (17-97) and (17-28), we find

$$\tfrac{1}{2}m\overline{v^2} = \tfrac{3}{2}kT \tag{17-31}$$

The principle of equipartition of energy assumes that the energy of the gas is divided equally among its degrees of freedom so that each degree of freedom has $\frac{1}{2}kT$. This leads to

$$\gamma = \frac{f+2}{f} = \frac{\bar{c}_p}{\bar{c}_v} \tag{17-43}$$

where f is the number of degrees of freedom of the gas. For a monatomic gas, $f = 3$; for a diatomic, $f = 5$.

The distribution of molecular *speeds* is given by

$$dN_v = \frac{4N}{\sqrt{\pi}}\left(\frac{m}{2kT}\right)^{3/2} v^2 e^{-mv^2/2kT}\, dv \tag{17-85}$$

The distribution of molecular velocities is given by

$$dN_{v_x} = \frac{N}{\sqrt{\pi}}\left(\frac{m}{2kT}\right)^{1/2} e^{-mv_x^2/2kT}\, dv_x \tag{17-87}$$

From these, the following results are obtained:

$$\bar{v} = \left(\frac{8kT}{\pi m}\right)^{1/2} \tag{17-89}$$

$$v_{\text{rms}} = \sqrt{\overline{v^2}} = \left(\frac{3kT}{m}\right)^{1/2} \tag{17-90}$$

The number of molecules striking the wall per unit time per unit area can also be written as

$$\frac{p}{(2\pi mkT)^{1/2}} \tag{17-95}$$

The average energy of molecules that escape a microscopic hole is $2kT$.

Problems

17-1 Estimate c_p and c_v for the following gases at low pressures:
(a) Oxygen.
(b) Hydrogen.
(c) Nitrogen.
(d) Argon.
(e) Carbon dioxide.

17-2 For oxygen at 300°K and 1 atm pressure, compute $\bar{v}$, v_{rms}, $\bar{v}^2$, and the velocity of sound.

17-3 Repeat Prob. 17-2 for oxygen at 10°K.

17-4 Repeat Prob. 17-2 for oxygen at 600°K.

17-5 Repeat Prob. 17-2 for argon at 300°K and 1 atm.

17-6 Compute the number of nitrogen molecules in a cubical box whose sides are $\frac{1}{16}$ in. long if the temperature is 300°K and the pressure is 1 atm.

17-7 How many molecules strike one side of the wall of the container in Prob. 17-6 per unit time?

17-8 What is the average energy of oxygen molecules at 300°K? What is the average energy of oxygen molecules that escape a small hole at 300°K?

17-9 A container of gas is separated into equal parts by a partition that contains a very small hole.

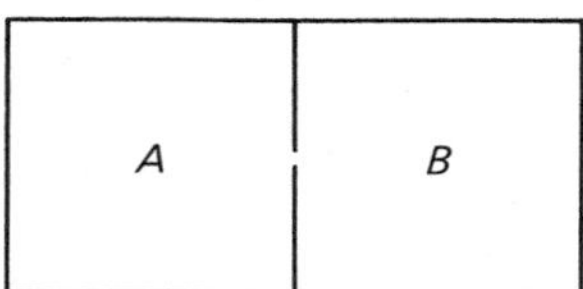

The two sides of the container are maintained at different temperatures.
(a) Derive an expression for the ratio of the pressures in the two sides in terms of the temperatures.
(b) Assuming $T_A > T_B$, derive an expression for the rate at which heat must be supplied to or from side A to maintain it at constant temperature.

17-10 A tank, maintained at constant temperature T, contains a porous plug at one end that consists of very fine holes through which a gas is to be withdrawn. The total hole area is A. If the tank initially contains N_A/N_0 moles of gas A and N_B/N_0 moles of gas B, derive an expression for
(a) The composition.
(b) The pressure in the tank as a function of time.
The initial tank pressure is designated by p_0. Express your result for composition as the ratio N_A/N_B. (N_0 is Avogadro's number.)

17-11 An evacuated tank in a constant pressure environment develops a microscopic leak. Derive an expression for the pressure in the tank as a function of time.

17-12 A container of volume $2V$ is divided into two equal parts. The left side contains a gas at 300°K. The right side is evacuated. Temperature is kept constant. A microscopic hole develops in the partition between the two sides. Derive expressions for the pressure in both sides as functions of time.

Appendices

Appendix A

UNITS USED IN THIS BOOK[1]

Both English and metric systems of units are used in this book. Since all nations, including the United States, are converting to a metric system, engineering students must be familiar with both systems during the transition period.

The particular metric system being adopted is the *Système International d'Unites* (International System of Units), also known as SI units. Table A-1 shows some of the basic SI and English units.

Table A-1 Some Basic English and SI Units

Quantity name	*Symbol*	*English unit*	*SI unit*
length	l	foot (ft)	meter (m)
mass	M	slug (slug)	kilogram (kg)
		pound-mass (lbm)	
time	θ	second (sec)	second (sec)
temperature	T	Rankine (°R)	Kelvin (°K)
	t	Fahrenheit (°F)	Celsius (°C)
electric current	I	ampere (amp)	ampere (amp)

Force is a derived unit using Newton's second law and the basic units in Table A-1. Newton's second law is written as

$$F = \frac{1}{g_0} Ma$$

[1] Electrical units are discussed in Chapter 6. See especially Table 6–1.

Table A-2 shows various values of g_0 depending on the units used for mass and force.

Table A-2 Units of Force and Mass in $F = (1/g_0)(Ma)$

Mass	*Acceleration*	*Force*	g_0
slug (slug)	feet per second per second (ft/sec^2)	pound-force (lbf)	1 slug-ft/lbf-sec^2
pound-mass (lbm)	feet per second per second (ft/sec^2)	pound-force (lbf)	32.2 lbm-ft/lbf-sec^2
kilogram (kgm)	meter per second per second (m/sec^2)	newton (N)	1 kg-m/N-sec^2
kilogram (kgm)	meter per second per second (m/sec^2)	kilogram-force (kgf)	9.8 kg-m/kgf-sec^2
gram (gm)	centimeter per second per second (cm/sec^2)	dyne (dyne)	1 gm-cm/dyne-sec^2

In this book, we normally (that is, unless otherwise specified) use mass in pound-mass (lbm), kilograms (kg), or grams (gm). We normally use force in either pound-force (lbf) or newtons (N). Meters (m) [and centimeters (cm)] and feet (ft) are used for length, and the Rankine (°R), Fahrenheit (°F), Kelvin (°K), and Celsius (°C) temperature scales are used.

In addition to the basic units listed in Table A-1 and A-2, the derived units used in this book are listed in Table A-3. The table omits routine quantities, such as area (m^2 or ft^2), and quantities common to both the English and SI systems, such as angular velocity (radians/sec).

Table A-3 Some Derived English and SI Units

Quantity name	*Symbol*	*English unit*	*SI unit*
force	F	pound-force (lbf)	newton (N) or kilogram-force (kgf)
pressure	p	pound-force per square foot (lbf/ft^2)	Pascal (Pa) or Newton per square meter (N/m^2 or bar (10^5 N/m^2)
energy-work-heat	E, Q, W	British thermal unit (Btu) or foot-pound-force (ft-lbf)	joule (j) (1 N-m)
power	p	British thermal unit per second (Btu/sec) or foot-pound-force per second (ft-lbf/sec)	watt (1 j/sec)

Students and practicing engineers will occasionally need to convert quantities from English to SI units or vice versa. For assistance in making such conversions, and also for conversions within each system, Table A-4 is presented.

Table A-4 Some Conversion Factors

Quantity	Given unit (abbreviated)	× Conversion factor	= New unit (abbreviated)
length	ft	× 0.3048	= m
	mi	× 1.609	= km
	in.	× 2.540	= cm
area	ft^2	× 0.0929	= m^2
	in^2.	× 6.4516	= cm^2
volume	U.S. gal	× 3.785	= liter
	U.S. quart	× 0.9463	= liter
	liter	× 1.0×10^{-3}	= m^3
mass	slug	× 14.594	= kg
	lbm	× 0.4536	= kg
	slug	× 32.2	= lbm
density	lbm/ft^3	× 16.02	= kg/m^3
	slug/ft^3	× 515.4	= kg/m^3
	gm/cm^3	× 1000	= kg/m^3
	lbm/ft^3	× 0.016	= gm/cm^3
specific volume	ft^3/lbm	× 0.06243	= m^3/kg
	ft^3/lbm	× 62.43	= cm^3/gm
	cm^3/gm	× 0.016	= ft^3/lbm
velocity	ft/sec	× 0.3048	= m/sec
	mil/hr (mph)	× 0.447	= m/sec
acceleration	ft/sec^2 (fps)	× 0.3048	= m/sec^2
volume flow rate	ft^3/sec	× 0.0283	= m^3/sec
	gal/hr	× 1.051×10^{-6}	= m^3/sec
mass flow rate	lbm/sec	× 0.4536	= kg/sec
	lbm/min	× 7.56×10^{-3}	= kg/sec
force	lbf	× 0.4536	= kgf
	lbf	× 4.448	= N
	kgf	× 9.807	= N
	dyne	× 1.0×10^{-5}	= N
pressure	lbf/ft^2	× 47.88	= N/m^2 (Pascal)
	N/m^2	× 1.0×10^{-5}	= bars
	lbf/in.^{2} (psi)	× 0.0703	= kgf/cm^2
	lbf/in.^{2} (psi)	× 0.06895	= bars
	in. of H_2O	× 0.002491	= bars
	in. of Hg	× 0.03377	= bars
	atm	× 14.696	= lbf/in.^{2} (psi)
	atm	× 1.013	= bars
	atm	× 1.033	= kgf/cm^2
pressure	mm of Hg	× 0.001333	= bars
	dyne/cm^2	× 0.10	= N/m^2 (Pascal)
	kgf/m^2	× 9.807	= N/m^2 (Pascal)
	gmf/cm^2	× 98.07	= N/m^2 (Pascal)
	kgf/cm^2	× 0.9807	= bars
temperature	°K	× 1.8	= °R
energy, work, heat	Btu	× 1055	= j
	ft-lbf	× 1.356	= j
	Btu	× 778	= ft-lbf
	watt-sec	× 1.0	= j
	cal	× 4.187	= j
	kw-hr	× 3.6×10^6	= j
	kw-hr	× 3413	= Btu
	hp-hr	× 2545	= Btu
	cal	× 3.968×10^{-3}	= Btu
	cal	× 3.088	= ft-lbf
power	Btu/hr	× 0.293	= watts
	Btu/sec	× 1054	= watts
	ft-lbf/min	× 0.0226	= watts
	hp	× 745.7	= watts
	j/sec	× 1.0	= watts
specific heat	Btu/lbm-°R	× 4187	= j/kg-°K
	Btu/lbmole-°R	× 4187	= j/kgmole-°K
	Btu/lbm-°R	× 1.0	= cal/gm-°K
specific energy	Btu/lbm	× 2.326	= j/gm
	Btu/lbm	× 0.555	= cal/gm

Appendix B

THERMODYNAMIC PROPERTIES OF WATER—ENGLISH UNITS[1]

Table B-1 Saturation: Temperatures

		Specific	Volume	Internal Energy			Enthalpy			Entropy		
Temperature (°F) t	Pressure (psia) p	Saturated Liquid ft³/lbm v_f	Saturated Vapor ft³/lbm v_g	Saturated Liquid Btu/lbm u_f	Evaporated Btu/lbm u_{fg}	Saturated Vapor Btu/lbm u_g	Saturated Liquid Btu/lbm h_f	Evaporated Btu/lbm h_{fg}	Saturated Vapor Btu/lbm h_g	Saturated Liquid Btu/lbm-°R s_f	Evaporated Btu/lbm-°R s_{fg}	Saturated Vapor Btu/lbm-°R s_g
32.018	0.08866	0.016022	3302	0.00	1021.2	1021.2	0.01	1075.4	1075.4	0.00000	2.1869	2.1869
40	0.12166	0.016020	2445	8.02	1015.8	1023.9	8.03	1070.9	1078.9	0.01617	2.1430	2.1592
50	0.17803	0.016024	1704.2	18.06	1009.1	1027.2	18.06	1065.2	1083.3	0.03607	2.0899	2.1259
60	0.2563	0.016035	1206.9	28.08	1002.4	1030.4	28.08	1059.6	1087.7	0.05555	2.0388	2.0943
70	0.3632	0.016051	867.7	38.09	995.6	1033.7	38.09	1054.0	1092.0	0.07463	1.9896	2.0642
80	0.5073	0.016073	632.8	48.08	988.9	1037.0	48.09	1048.3	1096.4	0.09332	1.9423	2.0356
90	0.6988	0.016099	467.7	58.07	982.2	1040.2	58.07	1042.7	1100.7	0.11165	1.8966	2.0083
100	0.9503	0.016130	350.0	68.04	975.4	1043.5	68.05	1037.0	1105.0	0.12693	1.8526	1.9822
110	1.2763	0.016166	265.1	78.02	968.7	1046.7	78.02	1031.3	1109.3	0.14730	1.8101	1.9574
120	1.6945	0.016205	203.0	87.99	961.9	1049.9	88.00	1025.5	1113.5	0.16465	1.7690	1.9336
130	2.225	0.016247	157.17	97.97	955.1	1053.0	97.98	1019.8	1117.8	0.18172	1.7292	1.9109
140	2.892	0.016293	122.88	107.95	948.2	1056.2	107.96	1014.0	1121.9	0.19851	1.6907	1.8892
150	3.722	0.016343	96.99	117.95	941.3	1059.3	117.96	1008.1	1126.1	0.21503	1.6533	1.8684
160	4.745	0.016395	77.23	127.94	934.4	1062.3	127.96	1002.2	1130.1	0.23130	1.6171	1.8484
170	5.996	0.016450	62.02	137.95	927.4	1065.4	137.97	996.2	1134.2	0.24732	1.5819	1.8293
180	7.515	0.016509	50.20	147.97	920.4	1068.3	147.99	990.2	1138.2	0.26311	1.5478	1.8109
190	9.343	0.016570	40.95	158.00	913.3	1071.3	158.03	984.1	1142.1	0.27866	1.5146	1.7932
200	11.529	0.016634	33.63	168.04	906.2	1074.2	168.07	977.9	1145.9	0.29400	1.4822	1.7762
210	14.125	0.016702	27.82	178.10	898.9	1077.0	178.14	971.6	1149.7	0.30913	1.4508	1.7599
220	17.188	0.016772	23.15	188.17	891.7	1079.8	188.22	965.3	1153.5	0.32406	1.4201	1.7441
240	24.97	0.016922	16.327	208.36	876.9	1085.3	208.44	952.3	1160.7	0.35335	0.3609	1.7143
260	35.42	0.017084	11.768	228.64	861.8	1090.5	228.76	938.8	1167.6	0.38193	1.3044	1.6864
280	49.18	0.017259	8.650	249.02	846.3	1095.4	249.18	924.9	1174.1	0.40986	1.2504	1.6602
300	66.98	0.017448	6.742	269.52	830.5	1100.0	269.73	910.4	1180.2	0.43720	1.1984	1.6356
400	247.1	0.018638	1.8661	374.27	742.4	1116.6	375.12	826.8	1202.0	0.56672	0.9617	1.5284
450	422.1	0.019433	1.1011	428.6	690.9	1119.5	430.2	775.4	1205.6	0.6282	0.8523	1.4806
500	680.0	0.02043	0.6761	485.1	632.3	1117.4	487.7	714.8	1202.5	0.6888	0.7448	1.4335
550	1044.0	0.02175	0.4249	544.9	563.7	1108.6	549.1	641.6	1190.6	0.7497	0.6354	1.3851
600	1541.0	0.02363	0.2677	609.9	480.1	1090.0	616.7	549.7	1166.4	0.8130	0.5187	1.3317
650	2205.0	0.02673	0.16206	685.0	368.7	1053.7	695.9	423.9	1119.8	0.8831	0.3820	1.2651
700	3090.0	0.03666	0.07438	801.7	145.9	947.7	822.7	167.5	990.2	0.9902	0.1444	1.1346
705.44	3204.0	0.05053	0.05053	872.6	0	872.6	902.5	0	902.5	1.0580	0	1.0580

[1] Condensed from *Steam Tables* by Keenan, Keyes, Hill, and Moore (1969) by permission of authors and John Wiley & Sons, Inc., New York.

Table B-2 Saturation: Pressures

Pressure (psia) p	Temperature (°F) t	Specific Volume: Saturated Liquid ft³/lbm v_f	Specific Volume: Saturated Vapor ft³/lbm v_g	Internal Energy: Saturated Liquid Btu/lbm u_f	Internal Energy: Evaporated Btu/lbm u_{fg}	Internal Energy: Saturated Vapor Btu/lbm u_g	Enthalpy: Saturated Liquid Btu/lbm h_f	Enthalpy: Evaporated Btu/lbm h_{fg}	Enthalpy: Saturated Vapor Btu/lbm h_g	Entropy: Saturated Liquid Btu/lbm-°R s_f	Entropy: Evaporated Btu/lbm-°R s_{fg}	Entropy: Saturated Vapor Btu/lbm-°R s_g
0.08866	32.02	0.016022	3302.0	0.00	1021.2	1021.2	0.01	1075.4	1075.4	0.00000	2.1869	2.1869
0.10	35.02	0.016021	2946.0	3.02	1019.2	1022.2	3.02	1073.7	1076.7	0.00612	2.1702	2.1764
0.20	53.15	0.016027	1526.3	21.22	1007.0	1028.2	21.22	1063.5	1084.7	0.04225	2.0736	2.1158
0.50	79.56	0.016071	641.5	47.64	989.2	1036.9	47.65	1048.6	1096.2	0.09250	1.9443	2.0368
1.0	101.70	0.016136	333.6	69.74	974.3	1044.0	69.74	1036.0	1105.8	0.13266	1.8453	1.9779
3.0	141.43	0.016300	118.72	109.38	947.2	1056.6	109.39	1013.1	1122.5	0.20089	1.6852	1.8861
5.0	162.21	0.016407	73.53	130.15	932.9	1063.0	130.17	1000.9	1131.0	0.23486	1.6093	1.8441
10	193.19	0.016590	38.42	161.20	911.0	1072.2	161.23	982.1	1143.3	0.28358	1.5041	1.7877
14.696	211.99	0.016715	26.80	180.10	897.5	1077.6	180.15	970.4	1150.5	0.31212	1.4446	1.7567
20	227.96	0.016830	20.09	196.19	885.8	1082.0	196.26	960.1	1156.4	0.33580	1.3962	1.7320
30	250.34	0.017004	13.748	218.84	869.2	1088.0	218.93	945.4	1164.3	0.36821	1.3314	1.6996
40	267.26	0.017146	10.501	236.03	856.2	1092.3	236.16	933.8	1170.0	0.39214	1.2845	1.6767
50	281.03	0.017269	8.518	250.08	845.5	1095.6	250.24	924.2	1174.4	0.41129	1.2476	1.6589
60	292.73	0.017378	7.177	262.06	836.3	1098.3	262.25	915.8	1178.0	0.42733	1.2170	1.6444
70	302.96	0.017478	6.209	272.56	828.1	1100.6	272.79	908.3	1181.0	0.44120	1.1909	1.6321
80	312.07	0.017570	5.474	281.95	820.6	1102.6	282.21	901.4	1183.6	0.45344	1.1679	1.6214
90	320.31	0.017655	4.898	290.46	813.8	1104.3	290.76	895.1	1185.9	0.46442	1.1475	1.6119
100	327.86	0.017736	4.434	298.28	807.5	1105.8	298.61	889.2	1187.8	0.47439	1.1290	1.6034
120	341.30	0.017886	3.730	312.27	796.0	1108.3	312.67	878.5	1191.1	0.49201	1.0966	1.5886
140	353.08	0.018024	3.221	324.58	785.7	1110.3	325.05	868.7	1193.8	0.50727	1.0688	1.5761
160	363.60	0.018152	2.836	335.63	776.4	1112.0	336.16	859.8	1196.0	0.52078	1.0443	1.5651
180	373.13	0.018273	2.533	345.68	767.7	1113.4	346.29	851.5	1197.8	0.53292	1.0223	1.5553
200	381.86	0.018387	2.289	354.9	759.6	1114.6	355.6	843.7	1199.3	0.5440	1.0025	1.5464
250	401.04	0.018653	1.8448	375.4	741.4	1116.7	376.2	825.8	1202.1	0.5680	0.9594	1.5274
300	417.43	0.018896	1.5442	393.0	725.1	1118.2	394.1	809.8	1203.9	0.5883	0.9232	1.5115
400	444.70	0.019340	1.1620	422.8	696.7	1119.5	424.2	781.2	1205.5	0.6218	0.8638	1.4856
500	467.13	0.019748	0.9283	447.7	671.7	1119.4	449.5	755.8	1205.3	0.6490	0.8154	1.4645
600	486.33	0.02013	0.7702	469.4	649.1	1118.6	471.7	732.4	1204.1	0.6723	0.7742	1.4464
800	518.36	0.02087	0.5691	506.6	608.4	1115.0	509.7	689.6	1199.3	0.7110	0.7050	1.4160
1000	544.75	0.02159	0.4459	538.4	571.5	1109.9	542.4	650.0	1192.4	0.7432	0.6471	1.3903
1500	596.39	0.02346	0.2769	605.0	486.9	1091.8	611.5	557.2	1168.7	0.8082	0.5276	1.3359
2000	636.00	0.02565	0.18813	662.4	404.2	1066.6	671.9	464.4	1136.3	0.8623	0.4238	1.2861
2500	668.31	0.02860	0.13059	717.7	313.4	1031.0	730.9	360.5	1091.4	0.9131	0.3196	1.2327
3000	695.52	0.03431	0.08404	783.4	185.4	968.8	802.5	213.0	1015.5	0.9732	0.1843	1.1575
3203.6	705.44	0.05053	0.05053	872.6	0	872.6	902.5	0	902.5	1.0580	0	1.0580

Table B-3 Vapor

p psia (t Saturated) Temperature (°F)	1.0(101.70)				10(193.19)				14.696(211.99)			
	v ft^3/ lbm	*u* Btu/ lbm	*h* Btu/ lbm	*s* Btu/ lbm-°R	*v*	*u*	*h*	*s*	*v*	*u*	*h*	*s*
Saturated	333.6	1044.0	1105.8	1.9779	38.42	1072.2	1143.3	1.7877	26.80	1077.6	1150.5	1.7567
200	392.5	1077.5	1150.1	2.0508	38.85	1074.7	1146.6	1.7927	26.29	1073.2	1144.7	1.7479
300	452.3	1112.0	1195.7	2.1150	44.99	1110.4	1193.7	1.8592	30.52	1109.6	1192.6	1.8157
400	511.9	1147.0	1241.8	2.1720	51.03	1146.1	1240.5	1.9171	34.67	1145.6	1239.9	1.8741
500	571.5	1182.8	1288.5	2.2235	57.04	1182.2	1287.7	1.9690	38.77	1181.8	1287.3	1.9263
600	631.1	1219.3	1336.1	2.2706	63.03	1218.9	1335.5	2.0164	42.86	1218.6	1335.2	1.9737
700	690.7	1256.7	1384.5	2.3142	69.01	1256.3	1384.0	2.0601	46.93	1256.1	1383.8	2.0175
800	750.3	1294.9	1433.7	2.3550	74.98	1294.6	1433.3	2.1009	51.00	1294.4	1433.1	2.0584
900	809.9	1333.9	1483.8	2.3932	80.95	1333.7	1483.5	2.1393	55.07	1333.6	1483.4	2.0967

p (t Saturated) Temperature (°F)	20(277.96)				60(292.73)				100(327.86)			
	v	*u*	*h*	*s*	*v*	*u*	*h*	*s*	*v*	*u*	*h*	*s*
Saturated	20.09	1082.0	1156.4	18.378	7.177	1098.3	1178.0	1.6444	4.434	1105.8	1106.5	1188.7
200	19.191	1071.4	1142.5	1.7113	6.047	1057.1	1124.2	1.5680				
300	22.36	1108.7	1191.5	1.7805	7.260	1101.3	1181.9	1.6496	4.228	1093.1	1171.4	1.5822
400	25.43	1145.1	1239.2	1.8395	8.353	1140.8	1233.5	1.7134	4.934	1136.2	1227.5	1.6517
500	28.46	1181.5	1286.8	1.8919	9.399	1178.6	1283.0	1.7678	5.587	1175.7	1279.1	1.7085
600	31.47	1218.4	1334.8	1.9395	10.425	1216.3	1332.1	1.8165	6.216	1214.2	1329.3	1.7582
700	34.47	1255.9	1383.5	1.9834	11.440	1254.4	1381.4	1.8609	6.834	1252.8	1379.2	1.8033
800	37.46	1294.3	1432.9	2.0243	12.448	1293.0	1431.2	1.9022	7.445	1291.8	1429.6	1.8449
900	40.45	1333.5	1483.2	2.0627	13.452	1332.5	1481.8	1.9408	8.053	1331.5	1480.5	1.8838

p (t Saturated) Temperature (°F)	200(381.86)				300(417.43)				400(444.70)			
	v	*u*	*h*	*s*	*v*	*u*	*h*	*s*	*v*	*u*	*h*	*s*
Saturated	2.289	1114.6	1199.3	1.5464	1.5442	1118.2	1203.9	1.5115	1.1620	1119.5	1205.5	1.4856
400	2.361	1123.5	1210.8	1.5600	1.4915	1108.2	1191.0	1.4967				
500	2.724	1168.0	1268.8	1.6239	1.7662	1159.5	1257.5	1.5701	1.2843	1150.1	1245.2	1.5282
600	3.058	1208.9	1322.1	1.6767	2.004	1203.2	1314.5	1.6266	1.4760	1197.3	1306.6	1.5892
700	3.379	1248.8	1373.8	1.7234	2.227	1244.6	1368.3	1.6751	1.6503	1240.4	1362.5	1.6397
800	3.693	1288.6	1425.3	1.7660	2.442	1285.4	1421.0	1.7187	1.8163	1282.1	1416.6	1.6844
900	4.003	1328.9	1477.1	1.8055	2.653	1326.3	1473.6	1.7589	1.9776	1323.7	1470.1	1.7252
1000	4.310	1369.8	1529.3	1.8425	2.860	1367.7	1526.5	1.7964	2.136	1365.5	1523.6	1.7632
1200	4.918	1453.7	1635.7	1.9109	3.270	1452.2	1633.8	1.8653	2.446	1450.7	1631.8	1.8327

Table B-3 Vapor (continued)

p (*t Saturated*)	500(467.13)				600(486.33)				700(503.23)			
Temperature (°F)	v	u	h	s	v	u	h	s	v	u	h	s
Saturated	0.9283	1119.4	1205.3	1.4645	0.7702	1118.6	1204.1	1.4464	0.6558	1117.0	1202.0	1.4305
500	0.9924	1139.7	1231.5	1.4923	0.7947	1128.0	1216.2	1.4592	0.6503	1114.5	1198.8	1.4271
600	1.1583	1191.1	1298.3	1.5585	0.9456	1184.5	1289.5	1.5320	0.7929	1177.5	1280.2	1.5081
700	1.3040	1236.0	1356.7	1.6112	1.0727	1231.5	1350.6	1.5872	0.9073	1226.9	1344.4	1.5661
800	1.4407	1278.8	1412.1	1.6571	1.1900	1275.4	1407.6	1.6343	1.0109	1272.0	1402.9	1.6145
900	1.5723	1321.0	1466.5	1.6987	1.3021	1318.4	1462.9	1.6766	1.1089	1315.6	1459.3	1.6576
1000	1.7008	1363.3	1520.7	1.7371	1.4108	1361.2	1517.8	1.7155	1.2036	1358.9	1514.9	1.6970
1200	1.9518	1449.2	1629.8	1.8072	1.6222	1447.7	1627.8	1.7861	1.3868	1446.2	1625.8	1.7682
1400	2.198	1537.6	1741.0	1.8704	1.8289	1536.5	1739.5	1.8497	1.5652	1535.3	1738.1	1.8321

p (*t Saturated*)	800(518.36)				900(532.12)				1000(544.75)			
Temperature (°F)	v	u	h	s	v	u	h	s	v	u	h	s
Saturated	0.5691	1115.0	1199.3	1.4160	0.5009	1112.6	1196.0	1.4027	0.4459	1109.9	1192.4	1.3903
600	0.6776	1170.1	1270.4	1.4861	0.5871	1162.2	1260.0	1.4652	0.5140	1153.7	1248.8	1.4450
700	0.7829	1222.1	1338.0	1.5471	0.6859	1217.1	1331.4	1.5297	0.6080	1212.0	1324.6	1.5135
800	0.8764	1268.5	1398.2	1.5969	0.7717	1264.9	1393.4	1.5810	0.6878	1261.2	1388.5	1.5664
900	0.9640	1312.9	1455.6	1.6408	0.8513	1310.1	1451.9	1.6257	0.7610	1307.3	1448.1	1.6120
1000	1.0482	1356.7	1511.9	1.6807	0.9273	1354.5	1508.9	1.6662	0.8305	1352.3	1505.9	1.6530
1200	1.2102	1444.6	1623.8	1.7526	1.0729	1443.0	1621.7	1.7386	0.9630	1441.5	1619.7	1.7061
1400	1.3674	1534.2	1736.6	1.8167	1.2135	1533.0	1735.1	1.8031	1.0905	1531.9	1733.7	1.7909
1600	1.5218	1626.2	1851.5	1.8754	1.3515	1625.3	1850.4	1.8620	1.2152	1624.4	1849.3	1.8499

p (*t Saturated*)	2000(636.00)											
Temperature (°F)	v	u	h	s								
Saturated	0.18813	1066.6	1136.3	1.2861	4000				6000			
700	0.2487	1147.7	1239.8	1.3782	0.02867	742.1	763.4	0.9345	0.02563	708.1	736.5	0.9028
800	0.3071	1220.1	1333.8	1.4562	0.10522	1095.0	1172.9	1.2740	0.03942	896.9	940.7	1.0708
900	0.3534	1276.8	1407.6	1.5126	0.14622	1201.5	1309.7	1.3789	0.07588	1102.9	1187.2	1.2599
1000	0.3945	1328.1	1474.1	1.5598	0.17520	1272.9	1402.6	1.4449	0.10207	1209.1	1322.4	1.3561
1200	0.4685	1425.2	1598.6	1.6398	0.2213	1390.1	1553.9	1.5423	0.13927	1352.7	1507.3	1.4752
1400	0.5368	1520.2	1718.8	1.7082	0.2603	1495.7	1688.4	1.6188	0.16854	1470.5	1657.6	1.5608
1600	0.6020	1615.4	1838.2	1.7692	0.2959	1597.1	1816.1	1.6841	0.19420	1578.7	1794.3	1.6307
2000	0.7284	1810.6	2080.2	1.8765	0.3625	1797.3	2065.6	1.7948	0.24087	1784.3	2051.7	1.7450

Appendix C

THERMODYNAMIC PROPERTIES OF WATER—METRIC UNITS[1]

Table C-1 Saturation: Temperatures

Temperature (°C) t	Pressure (bars) p	Specific Volume: Saturated Liquid cm^3/gm v_f	Specific Volume: Saturated Vapor cm^3/gm v_g	Internal Energy: Saturated Liquid j/gm u_f	Internal Energy: Evaporated j/gm u_{fg}	Internal Energy: Saturated Vapor j/gm u_g	Enthalpy: Saturated Liquid j/gm h_f	Enthalpy: Evaporated j/gm h_{fg}	Enthalpy: Saturated Vapor j/gm h_g	Entropy: Saturated Liquid j/gm-°K s_f	Entropy: Evaporated j/gm-°K s_{fg}	Entropy: Saturated Vapor j/gm-°K s_g
0.01	0.006113	1.0002	206 136	0.00	2375.3	2375.3	0.01	2501.3	2501.4	0.0000	9.1562	9.1562
1	0.006567	1.0002	192 577	4.15	2372.6	2376.7	4.16	2499.0	2503.2	0.0152	9.1147	9.1299
10	0.012276	1.0004	106 379	42.00	2347.2	2389.2	42.01	2477.7	2519.8	0.1510	8.7498	8.9008
20	0.02339	1.0018	57 791	83.95	2319.0	2402.9	83.96	2454.1	2538.1	0.2966	8.3706	8.6672
30	0.04246	1.0043	32 894	125.78	2290.8	2416.6	125.79	2430.5	2556.3	0.4369	8.0164	8.4533
40	0.07384	1.0078	19 523	167.56	2262.6	2430.1	167.57	2406.7	2574.3	0.5725	7.6845	8.2570
50	0.12349	1.0121	12 032	209.32	2234.2	2443.5	209.33	2382.7	2592.1	0.7038	7.3725	8.0763
60	0.19940	1.0172	7671.0	251.11	2205.5	2456.6	251.13	2358.5	2609.6	0.8312	7.0784	7.9096
70	0.3119	1.0228	5042.0	292.95	2176.6	2469.5	292.98	2333.8	2626.8	0.9549	6.8004	7.7553
80	0.4739	1.0291	3407.0	334.86	2147.4	2482.2	334.91	2308.8	2643.7	1.0753	6.5369	7.6122
90	0.7014	1.0360	2361.0	376.85	2117.7	2494.5	376.92	2283.2	2660.1	1.1925	6.2866	7.4791
100	1.0135	1.0435	1672.9	418.94	2087.6	2506.5	419.04	2257.0	2676.1	1.3069	6.0480	7.3549
110	1.4327	1.0516	1210.2	461.14	2057.0	2518.1	461.30	2230.2	2691.5	1.4185	5.8202	7.2387
120	1.9853	1.0603	891.9	503.50	2025.8	2529.3	503.71	2202.6	2706.3	1.5276	5.6020	7.1296
130	2.701	1.0697	668.5	546.02	1993.9	2539.9	546.31	2174.2	2720.5	1.6344	5.3925	7.0269
140	3.613	1.0797	508.9	588.74	1961.3	2550.0	589.13	2144.7	2733.9	1.7391	5.1908	6.9299
150	4.758	1.0905	392.8	631.68	1927.9	2559.5	632.20	2114.3	2746.5	1.8418	4.9960	6.8379
160	6.178	1.1020	307.1	674.87	1893.5	2568.4	675.55	2082.6	2758.1	1.9427	4.8075	6.7502
170	7.917	1.1143	242.8	718.33	1858.1	2576.5	719.21	2049.5	2768.7	2.0419	4.6244	6.6663
180	10.021	1.1274	194.05	762.09	1821.6	2583.7	763.22	2015.0	2778.2	2.1396	4.4461	6.5857
190	12.544	1.1414	156.54	806.19	1783.8	2590.0	807.62	1978.8	2786.4	2.2359	4.2720	6.5079
200	15.538	1.1565	127.36	850.65	1744.7	2595.3	852.45	1940.7	2793.2	2.3309	4.1014	6.4323
220	23.18	1.1900	86.19	940.87	1661.5	2602.4	943.62	1858.5	2802.1	2.5178	3.7683	6.2861
240	33.44	1.2291	59.76	1033.21	1570.8	2604.0	1037.32	1766.5	2803.8	2.7015	3.4422	6.1437
260	46.88	1.2755	42.21	1128.39	1470.6	2599.0	1134.37	1662.5	2796.9	2.8838	3.1181	6.0019
280	64.12	1.3321	30.17	1227.46	1358.7	2586.1	1235.99	1543.6	2779.6	3.0668	2.7903	5.8571
300	85.81	1.4036	21.67	1332.0	1231.0	2563.0	1344.0	1404.9	2749.0	3.2534	2.4511	5.7045
320	112.74	1.4988	15.488	1444.6	1080.9	2525.5	1461.5	1238.6	2700.1	3.4480	2.0882	5.5362
340	145.86	1.6379	10.797	1570.3	894.3	2464.6	1594.2	1027.9	2622.0	3.6594	1.6763	5.3357
360	186.51	1.8925	6.945	1625.2	626.3	2351.5	1760.5	720.5	2481.0	3.9147	1.1379	5.0526
374.136	220.9	3.155	3.155	2029.6	0	2029.6	2099.3	0	2099.3	4.4298	0	4.4298

1 bar = 1.01972 kg/cm^2; 1 joule = 1/4.1868 I.T. cal.

[1] Condensed from *Steam Tables* (International Edition) by Keenan, Keyes, Hill, and Moore (1969) by permission of authors and John Wiley & Sons, Inc., New York.

Table C-2 Saturation: Pressures

Pressure (bars) p	Temperature (°C) t	Specific Volume: Saturated Liquid cm^3/gm v_f	Specific Volume: Saturated Vapor cm^3/gm v_g	Internal Energy: Saturated Liquid j/gm u_f	Internal Energy: Evaporated j/gm u_{fg}	Internal Energy: Saturated Vapor j/gm u_g	Enthalpy: Saturated Liquid j/gm h_f	Enthalpy: Evaporated j/gm h_{fg}	Enthalpy: Saturated Vapor j/gm h_g	Entropy: Saturated Liquid j/gm-°K s_f	Entropy: Evaporated j/gm-°K s_{fg}	Entropy: Saturated Vapor j/gm-°K s_g
0.006113	0.01	1.0002	206.136	0.00	2375.3	2375.3	0.01	2501.3	2501.4	0.0000	9.1562	9.1562
0.010	6.98	1.0002	129.208	29.30	2355.7	2385.0	29.30	2484.9	2514.2	0.1509	8.8697	8.9756
0.020	17.50	1.0013	67.004	73.48	2326.0	2399.5	73.48	2460.0	2533.5	0.2607	8.4629	8.7237
0.060	36.16	1.0064	23.739	151.53	2273.4	2425.0	151.53	2415.9	2567.4	0.5210	7.8094	8.3304
0.10	45.81	1.0102	14.674	191.82	2246.1	2437.9	191.83	2392.8	2584.7	0.6493	7.5009	8.1502
0.20	60.06	1.0172	7649.0	251.38	2205.4	2456.7	251.40	2358.3	2609.7	0.8320	7.0766	7.9085
0.60	85.94	1.0331	2732.0	359.79	2129.8	2489.6	359.86	2293.6	2653.5	1.1453	6.3867	7.5320
1.00	99.63	1.0432	1694.0	417.36	2088.7	2506.1	417.46	2258.0	2675.5	1.3026	6.0568	7.3594
1.50	111.37	1.0528	1159.3	466.94	2052.7	2519.7	467.11	2226.5	2693.6	1.4336	5.7897	7.2233
2.00	120.23	1.0605	885.7	504.49	2025.0	2529.5	504.70	2201.9	2706.7	1.5301	5.5970	7.1271
2.50	127.44	1.0672	718.7	535.10	2002.1	2537.2	535.37	2181.5	2716.9	1.6072	5.4455	7.0527
3.00	133.55	1.0732	605.8	561.15	1982.4	2543.6	561.47	2163.8	2725.3	1.6718	5.3201	6.9919
3.50	138.88	1.0786	524.3	583.95	1965.0	2548.9	584.33	2148.1	2732.4	1.7275	5.2130	6.9405
4.0	143.63	1.0836	462.5	604.31	1949.3	2553.6	604.74	2133.8	2738.6	1.7766	5.1193	6.8959
6.0	158.85	1.1006	315.7	669.90	1897.5	2567.4	670.56	2086.3	2756.8	1.9312	4.8288	6.7600
8.0	170.43	1.1148	240.4	720.22	1856.6	2576.8	721.11	2048.0	2769.1	2.0462	4.6166	6.6628
10.0	179.91	1.1273	194.44	761.68	1822.0	2583.6	762.81	2015.3	2778.1	2.1387	4.4478	6.5865
12.0	187.99	1.1385	163.33	797.29	1791.5	2588.8	798.65	1986.2	2784.8	2.2166	4.3067	6.5233
14.0	195.07	1.1489	140.84	828.70	1764.1	2592.8	830.30	1959.7	2790.0	2.2842	4.1850	6.4693
16.0	201.41	1.1587	123.80	856.94	1739.0	2596.0	858.79	1935.2	2794.0	2.3442	4.0776	6.4218
18.0	207.15	1.1679	110.42	882.69	1715.7	2598.4	884.79	1912.4	2797.1	2.3981	3.9812	6.3794
20	212.42	1.1767	99.63	906.44	1693.8	2600.3	908.79	1890.7	2799.5	2.4474	3.8935	6.3409
30	233.90	1.2165	66.68	1004.78	1599.3	2604.1	1008.42	1795.7	2804.2	2.6457	3.5412	6.1869
40	250.40	1.2522	49.78	1082.31	1520.0	2602.3	1087.31	1714.1	2801.4	2.7964	3.2737	6.0701
60	275.64	1.3187	32.44	1205.44	1384.3	2589.7	1213.35	1571.0	2784.3	3.0267	2.8625	5.8892
80	295.06	1.3842	23.52	1305.57	1264.2	2569.8	1316.64	1441.3	2758.0	3.2068	2.5364	5.7432
100	311.06	1.4524	18.026	1393.04	1151.4	2544.4	1407.56	1317.1	2724.7	3.3596	2.2544	5.6141
120	324.75	1.5267	14.263	1473.0	1040.7	2513.7	1491.3	1193.6	2684.9	3.4962	1.9962	5.4924
160	347.44	1.7107	9.306	1622.7	809.0	2431.7	1650.1	930.6	2580.6	3.7461	1.4994	5.2455
200	365.81	2.036	5.834	1785.6	507.5	2293.0	1826.3	583.4	2409.7	4.0139	0.9130	4.9269
220	373.80	2.742	3.568	1961.9	125.2	2087.1	2022.2	143.4	2165.6	4.3110	0.2216	4.5327
220.9	374.14	3.155	3.155	2029.6	0	2029.6	2099.3	0	2099.3	4.4298	0	4.4298

1 bar = 1.01972 kg/cm^2; 1 joule = 1/4.1868 I.T. cal.

Table C-3 Vapor

p bars (t Sat.) t °C	cm^3/gm v	j/gm u	j/gm h	j/gm-°K s	v	u	h	s	v	u	h	s
	0.10(45.81)				0.50(81.33)				1.0(99.63)			
Saturated	14 674	2437.9	2584.7	8.1502	3240	2483.9	2645.9	7.5939	1694.0	2506.1	2675.5	7.3594
50	14 869	2443.9	2592.6	8.1749	2937	2437.0	2583.9	7.4108	1445.0	2428.2	2572.7	7.0633
100	17 196	2515.5	2687.5	8.4479	3418	2511.6	2682.5	7.6947	1695.8	2506.7	2676.2	7.3614
150	19 512	2587.9	2783.0	8.6882	3889	2585.6	2780.1	7.9401	1936.4	2582.8	2776.4	7.6134
200	21 825	2661.3	2879.5	8.9038	4356	2659.9	2877.7	8.1580	2172.0	2658.1	2875.3	7.8343
250	24 136	2736.0	2977.3	9.1002	4820	2735.0	2976.0	8.3556	2406.0	2733.7	2974.3	8.0333
300	26 445	2812.1	3076.5	9.2813	5284	2811.3	3075.5	8.5373	2639.0	2810.4	3074.3	8.2158
400	31 063	2968.9	3279.6	9.6077	6209	2968.5	3278.9	8.8642	3103.0	2967.9	3278.2	8.5435
500	35 679	3132.3	3489.1	9.8978	7134	3132.0	3488.7	9.1546	3565.0	3131.6	3488.1	8.8342

p (t Sat.) t °C	v	u	h	s	v	u	h	s	v	u	h	s
	1.5(111.37)				3.0(133.55)				5.0(151.86)			
Saturated	1159.3	2519.7	2693.6	7.2233	605.8	2543.6	2725.3	6.9919	374.9	2561.2	2748.7	6.8213
150	1285.3	2579.8	2772.6	7.4193	633.9	2570.8	2761.0	7.0778	372.9	2557.9	2744.4	6.8111
200	1444.3	2656.2	2872.9	7.6433	716.3	2650.7	2865.6	7.3115	424.9	2642.9	2855.4	7.0592
250	1601.2	2732.5	2972.7	7.8438	796.4	2728.7	2967.6	7.5166	474.4	2723.5	2960.7	7.2709
300	1757.0	2809.5	3073.1	8.0270	875.3	2806.7	3069.3	7.7022	522.6	2802.9	3064.2	7.4599
400	2067.0	2967.3	3277.4	8.3555	1031.5	2965.6	3275.0	8.0330	617.3	2963.2	3271.9	7.7938
500	2376.0	3131.2	3487.6	8.6466	1186.7	3130.0	3486.0	8.3251	710.9	3128.4	3483.9	8.0873
600	2685.0	3301.7	3704.3	8.9101	1341.4	3300.8	3703.2	8.5892	804.1	3299.6	3701.7	8.3522
700	2993.0	3479.0	3927.9	9.1524	1495.7	3478.4	3927.1	8.8319	896.9	3477.5	3925.9	8.5952

p (t Sat.) t °C	v	u	h	s	v	u	h	s	v	u	h	s
	10.0(179.91)				15.0(198.32)				20.0(212.42)			
Saturated	194.44	2583.6	2778.1	6.5865	131.77	2594.5	2792.2	6.4448	99.63	2600.3	2799.5	6.3409
200	206.0	2621.9	2827.9	6.6940	132.48	2598.1	2796.8	6.4546	95.27	2570.6	2761.1	6.2608
250	232.7	2709.9	2942.6	6.9247	151.95	2695.3	2923.3	6.7090	111.44	2679.6	2902.5	6.5453
300	257.9	2793.2	3051.2	7.1229	169.66	2783.1	3037.6	6.9179	125.47	2772.6	3023.5	6.7664
400	306.6	2957.3	3263.9	7.4651	203.0	2951.3	3255.8	7.2690	151.20	2945.2	3247.6	7.1271
500	354.1	3124.4	3478.5	7.7622	235.2	3120.3	3473.1	7.5698	175.68	3116.2	3467.6	7.4317
600	401.1	3296.8	3697.9	8.0290	266.8	3293.9	3694.0	7.8385	199.60	3290.9	3690.1	7.7024
700	447.8	3475.3	3923.1	8.2731	298.1	3473.1	3920.1	8.0837	223.2	3470.9	3917.4	7.9487
800	494.3	3660.4	4154.7	8.4996	329.2	3658.7	4152.5	8.3109	246.7	3657.0	4150.3	8.1765

1 bar = 1.01972 kg/cm^2; 1 joule = 1/4.1868 I.T. cal.

p (t Sat.) t °C	v	u	h	s	v	u	h	s	v	u	h	s
	25.0(223.99)				30.0(233.90)				40(250.40)			
Saturated	79.98	2603.1	2803.1	6.2575	66.68	2604.1	2804.2	6.1869	49.78	2602.3	2801.4	6.0701
250	87.00	2662.6	2880.1	6.4085	70.58	2644.0	2855.8	6.2872	49.69	2601.1	2799.9	6.0672
300	98.90	2761.6	3008.8	6.6438	81.14	2750.1	2993.5	6.5390	58.84	2725.3	2960.7	6.3615
400	120.10	2939.1	3239.3	7.0148	99.36	2932.8	3230.9	6.9212	73.41	2919.9	3213.6	6.7690
500	139.98	3112.1	3462.1	7.3234	116.19	3108.0	3456.5	7.2338	86.43	3099.5	3445.3	7.0901
600	159.30	3288.0	3686.3	7.5960	132.43	3285.0	3682.3	7.5085	98.85	3279.1	3674.4	7.3688
700	178.32	3468.7	3914.5	7.8435	148.38	3466.5	3911.7	7.7571	110.95	3462.1	3905.9	7.6198
800	197.16	3655.3	4148.2	8.0720	164.14	3653.5	4145.9	7.9862	122.87	3650.0	4141.5	7.8502
900	215.90	3847.9	4387.6	8.2853	179.80	3846.5	4385.9	8.1999	134.69	3843.6	4382.3	8.0647

p (t Sat.) t °C	v	u	h	s	v	u	h	s	v	u	h	s
	50(263.99)				60(275.64)				80(295.06)			
Saturated	39.44	2597.1	2794.3	5.9734	32.44	2589.7	2784.3	5.8892	23.52	2569.8	2758.0	5.7432
300	45.32	2698.0	2924.5	6.2084	36.16	2667.2	2884.2	6.0674	24.26	2590.9	2785.0	5.7906
400	57.81	2906.6	3195.7	6.6459	47.39	2892.9	3177.2	6.5408	34.32	2863.8	3138.3	6.3634
500	68.57	3091.0	3433.8	6.9759	56.65	3082.2	3422.2	6.8803	41.75	3064.3	3398.3	6.7240
600	78.69	3273.0	3666.5	7.2589	65.25	3266.9	3658.4	7.1677	48.45	3254.4	3642.0	7.0206
700	88.49	3457.6	3900.1	7.5122	73.52	3453.1	3894.2	7.4234	54.81	3443.9	3882.4	7.2812
800	98.11	3646.6	4137.1	7.7440	81.60	3643.1	4132.7	7.6566	60.97	3636.0	4123.8	7.5173
900	107.62	3840.7	4378.8	7.9593	89.58	3837.8	4375.3	7.8727	67.02	3832.1	4368.3	7.7351
1000	117.07	3040.4	4625.7	8.1612	97.49	4037.8	4622.7	8.0751	73.01	4032.8	4616.9	7.9384

p (t Sat.) t °C	v	u	h	s	v	u	h	s	v	u	h	s
	100(311.06)				200(365.81)				400			
Saturated	18.026	2544.4	2724.7	5.6141	5.834	2293.0	2409.7	4.9269	1.5710	1615.4	1678.2	3.7287
400	26.41	2832.4	3096.5	6.2120	9.942	2619.3	2818.1	5.5540	1.9077	1854.6	1930.9	4.1135
500	32.79	3045.8	3373.7	6.5966	14.768	2942.9	3238.2	6.1401	5.622	2678.4	2903.3	5.4700
600	38.37	3241.7	3625.3	6.9029	18.178	3174.0	3527.6	6.5048	8.094	3022.6	3346.4	6.0114
700	43.58	3434.7	3870.5	7.1687	21.13	3386.4	3809.0	6.7993	9.941	3283.6	3681.2	6.3750
800	48.59	3628.9	4114.4	7.4077	23.85	3592.7	4069.7	7.0544	11.523	3517.8	3978.7	6.6662
900	53.49	3826.3	4361.2	7.6272	26.45	3797.5	4326.4	7.2830	12.962	3739.4	4257.9	6.9150
1000	58.32	4027.8	4611.0	7.8315	28.97	4003.1	4582.5	7.4925	14.324	3954.6	4527.6	7.1356
1200	67.89	4444.9	5123.8	8.2055	33.91	4422.8	5101.0	7.8707	16.940	4380.1	5057.7	7.5224

1 bar = 1.01972 kg/cm^2; 1 joule = 1/4.1868 I.T. cal.

Appendix D

PROPERTIES OF FREON-12

Table D-1 Saturated Freon-12, Temperature Table[1]

		Specific Volume		Enthalpy			Entropy			
Temperature (°F) t	*Pressure* (lb/in.^{2}) p	*Saturation Liquid* v_f	*Saturation Vapor* v_g	*Saturation Liquid* h_f	*Evaporation* h_{fg}	*Saturation Vapor* h_g	*Saturation Liquid* s_f	*Evaporation* s_{fg}	*Saturation Vapor* s_g	*Temperature* (°F) t
−60	5.37	0.01036	6.516	−4.20	75.33	71.13	−0.0102	0.1681	0.1783	−60
−50	7.13	0.01047	5.012	−2.11	74.42	72.31	−0.0050	0.1717	0.1767	−50
−40	9.32	0.0106	3.911	0.00	73.50	73.50	0.00000	0.17517	0.17517	−40
−30	12.02	0.0107	3.088	2.03	72.67	74.70	0.00471	0.16916	0.17387	−30
−20	15.28	0.0108	2.474	4.07	71.80	75.87	0.00940	0.16335	0.17275	−20
−10	19.20	0.0109	2.003	6.14	70.91	77.05	0.01403	0.15772	0.17175	−10
0	23.87	0.0110	1.637	8.25	69.96	78.21	0.01869	0.15222	0.17001	0
5	26.51	0.0111	1.485	9.32	69.47	78.79	0.02097	0.14955	0.17052	5
10	29.35	0.0112	1.351	10.39	68.97	79.36	0.02328	0.14687	0.17015	10
20	35.75	0.0113	1.121	12.55	67.94	80.49	0.02783	0.14166	0.16949	20
30	43.16	0.0115	0.939	14.76	66.85	81.61	0.03233	0.13654	0.16887	30
40	51.68	0.0116	0.792	17.00	65.71	82.71	0.03680	0.13153	0.16833	40
50	61.39	0.0118	0.673	19.27	64.51	83.78	0.04126	0.12659	0.16785	50
60	72.41	0.0119	0.575	21.57	63.25	84.82	0.04568	0.12173	0.16741	60
70	84.82	0.0121	0.493	23.90	61.92	85.82	0.05009	0.11692	0.16701	70
80	98.76	0.0123	0.425	26.28	60.52	86.80	0.05446	0.11215	0.16662	80
86	107.9	0.0124	0.389	27.72	59.65	87.37	0.05708	0.10932	0.16640	86
90	114.3	0.0125	0.368	28.70	59.04	87.74	0.05882	0.10742	0.16624	90
100	131.6	0.0127	0.319	31.16	57.46	88.62	0.06316	0.10268	0.16584	100
110	150.7	0.0129	0.277	33.65	55.78	89.43	0.06749	0.09793	0.16542	110
120	171.8	0.0132	0.240	36.16	53.99	90.15	0.07180	0.09315	0.16495	120

[1] Data courtesy of E. I. du Pont de Nemours & Company.

Note: Specific volume in ft^3/lbm; enthalpy in Btu/lbm; entropy in Btu/lbm °F.

Table D-2 Saturated Freon-12, Pressure Table[1]

		Specific Volume		Enthalpy			Entropy			
Pressure (lb/in.^{2}) p	*Temperature* (°F) t	*Saturation Liquid* v_f	*Saturation Vapor* v_g	*Saturation Liquid* h_f	*Evaporation* h_{fg}	*Saturation Vapor* h_g	*Saturation Liquid* s_f	*Evaporation* s_{fg}	*Saturation Vapor* s_g	*Pressure* (lb/in.^{2}) p
5	−62.5	0.01034	6.953	−4.73	75.56	70.83	−0.0115	0.1943	0.1788	5
10	−37.3	0.0106	3.662	0.54	73.28	73.82	0.00127	0.17360	0.17487	10
15	−20.8	0.0108	2.518	3.91	71.87	75.78	0.00902	0.16381	0.17283	15
20	−8.2	0.0109	1.925	6.53	70.74	77.27	0.01488	0.15672	0.17160	20
30	11.1	0.0112	1.324	10.62	68.86	79.48	0.02410	0.14597	0.17007	30
40	25.9	0.0114	1.009	13.86	67.30	81.16	0.03049	0.13865	0.16914	40
50	38.3	0.0116	0.817	16.58	65.94	82.52	0.03597	0.13244	0.16841	50
60	48.7	0.0117	0.688	18.96	64.69	83.65	0.04065	0.12726	0.16791	60
80	66.3	0.0120	0.521	23.01	62.44	85.45	0.04844	0.11872	0.16716	80
100	80.9	0.0123	0.419	26.49	60.40	86.89	0.05483	0.11176	0.16650	100
120	93.4	0.0126	0.350	29.53	58.52	88.05	0.06030	0.10580	0.16610	120
140	104.5	0.0128	0.298	32.28	56.71	88.99	0.06513	0.10053	0.16566	140
160	114.5	0.0130	0.260	34.78	54.99	89.77	0.06958	0.09564	0.16522	160
180	123.7	0.0133	0.228	37.07	53.31	90.38	0.07337	0.09139	0.16476	180
200	132.1	0.0135	0.202	39.21	51.65	90.86	0.07694	0.08730	0.16424	200
220	139.9	0.0138	0.181	41.22	50.28	91.50	0.08021	0.08354	0.16375	220

[1] Data courtesy of E. I. du Pont de Nemours & Company.

Note: Specific volume in ft^3/lbm; enthalpy in Btu/lbm; entropy in Btu/lbm.°F.

Table D-3 Superheated Freon-12[1]

Absolute Pressure (lb/in^2) (Saturation temperature)		Temperature—°F											
		−40	−20	0	20	40	60	80	100	150	200	250	300
5 (−62.5)	v...	7.363	7.726	8.088	8.450	8.812	9.173	9.533	9.893	10.79	11.69	...	...
	h...	73.72	76.36	79.05	81.78	84.56	87.41	90.30	93.25	100.84	108.75	...	...
	s...	0.1859	0.1920	0.1979	0.2038	0.2095	0.2150	0.2205	0.2258	0.2388	0.2513	...	...
10 (−37.3)	v...	...	3.821	4.006	4.189	4.371	4.556	4.740	4.923	5.379	5.831	6.281	...
	h...	...	76.11	78.81	81.56	84.35	87.19	90.11	93.05	100.66	108.63	116.88	...
	s...	...	0.1801	0.1861	0.1919	0.1977	0.2033	0.2087	0.2141	0.2271	0.2396	0.2517	...
15 (−20.8)	v...	...	2.521	2.646	2.771	2.895	3.019	3.143	3.266	3.571	3.877	4.191	...
	h...	...	75.89	78.59	81.37	84.18	87.03	89.94	92.91	100.53	108.49	116.78	...
	s...	...	0.17307	0.17913	0.18499	0.19074	0.19635	0.20185	0.20723	0.22028	0.23282	0.24491	...
20 (−8.2)	v...	...	...	1.965	2.060	2.155	2.250	2.343	2.437	2.669	2.901	3.130	...
	h...	...	...	78.39	81.14	83.97	86.85	89.78	92.75	100.40	108.38	116.67	...
	s...	...	...	0.17407	0.17996	0.18573	0.19138	0.19688	0.20229	0.21537	0.22794	0.24005	...
25 (2.2)	v...	...	...	...	1.712	1.793	1.873	1.952	2.031	2.227	2.422	2.615	...
	h...	...	...	...	80.95	83.78	86.67	89.61	92.56	100.26	108.26	116.56	...
	s...	...	...	...	0.17637	0.18216	0.18783	0.19336	0.19748	0.21190	0.22450	0.23665	...

Table D-3 (continued)

Absolute Pressure (lb/in²) (Saturation temperature)		Temperature—°F											
		—40	—20	0	20	40	60	80	100	150	200	250	300
	v...	...	...	...	1.364	1.430	1.495	1.560	1.624	1.784	1.943	2.099	...
30	h...	...	...	...	80.75	83.59	86.49	89.43	92.42	100.12	108.13	116.45	...
(11.1)	s...	...	...	...	0.17278	0.17859	0.18429	0.18983	0.19527	0.20843	0.22105	0.23325	...
	v...	...	...	...	1.109	1.237	1.295	1.352	1.409	1.550	1.689	1.827	...
35	h...	...	...	...	80.49	83.40	86.30	89.26	92.26	99.98	108.01	116.33	...
(18.9)	s...	...	...	...	0.16963	0.17591	0.18162	0.18719	0.19266	0.20584	0.21849	0.23069	...
	v...	...	...	...	...	1.044	1.095	1.144	1.194	1.315	1.435	1.554	...
40	h...	...	...	...	...	83.20	86.11	89.09	92.09	99.83	107.88	116.21	...
(25.9)	s...	...	...	...	...	0.17322	0.17896	0.18455	0.19004	0.20325	0.21592	0.22813	...
	v...	...	...	...	...	0.821	0.863	0.904	0.944	1.044	1.142	1.239	1.332
50	h...	...	...	...	...	82.76	85.72	88.72	91.75	99.54	107.62	116.00	124.69
(38.3)	s...	...	...	...	...	0.16895	0.17475	0.19040	0.18591	0.19923	0.21196	0.22419	0.23600
	v...	...	...	...	...	...	0.708	0.743	0.778	0.863	0.946	1.028	1.108
60	h...	...	...	...	...	...	85.33	88.35	91.41	99.24	107.36	115.54	124.29
(48.7)	s...	...	...	...	...	...	0.17120	0.17689	0.18246	0.19585	0.20865	0.22094	0.23280
	v...	...	...	...	...	...	0.553	0.642	0.673	0.750	0.824	0.896	0.967
70	h...	...	...	...	...	...	84.94	87.96	91.05	98.94	107.10	115.54	124.29
(57.9)	s...	...	...	...	...	...	0.16765	0.17399	0.17961	0.19310	0.20597	0.21830	0.23020
	v...	...	...	...	...	...	...	0.540	0.568	0.636	0.701	0.764	0.826
80	h...	...	...	...	...	...	...	87.56	90.68	98.64	106.84	115.30	124.08
(66.3)	s...	...	...	...	...	...	...	0.17108	0.17675	0.19035	0.20328	0.21566	0.22760
	v...	...	...	...	...	...	...	...	0.505	0.568	0.627	0.685	0.742
90	h...	...	...	...	...	...	...	...	90.31	98.32	106.56	115.07	123.88
(73.6)	s...	...	...	...	...	...	...	...	0.17443	0.18813	0.20111	0.21356	0.22554
	v...	...	...	...	...	...	...	...	0.442	0.499	0.553	0.606	0.657
100	h...	...	...	...	...	...	...	...	89.93	97.99	106.29	114.84	123.67
(80.9)	s...	...	...	...	...	...	...	...	0.17210	0.18590	0.19894	0.21145	0.22347
	v...	...	...	...	...	...	...	...	0.357	0.407	0.454	0.500	0.543
120	h...	...	...	...	...	...	...	...	89.13	97.30	105.70	114.35	123.25
(93.4)	s...	...	...	...	...	...	...	...	0.16803	0.18207	0.19529	0.20792	0.22000
	v...	...	...	...	...	...	...	...	...	0.341	0.383	0.423	0.462
140	h...	...	...	...	...	...	...	...	...	96.65	105.14	113.85	122.85
(104.5)	s...	...	...	...	...	...	...	...	...	0.17868	0.19205	0.20479	0.21701
	v...	...	...	...	...	...	...	...	...	0.318	0.335	0.372	0.408
160	h...	...	...	...	...	...	...	...	...	95.82	104.50	113.33	122.39
(114.5)	s...	...	...	...	...	...	...	...	...	0.17561	0.18927	0.20213	0.21444
	v...	...	...	...	...	...	...	...	...	0.294	0.287	0.321	0.353
180	h...	...	...	...	...	...	...	...	...	94.99	103.85	112.81	121.92
(123.7)	s...	...	...	...	...	...	...	...	...	0.17254	0.18648	0.19947	0.21187
	v...	...	...	...	...	...	...	...	...	0.241	0.255	0.288	0.317
200	h...	...	...	...	...	...	...	...	...	94.16	103.12	112.20	121.42
(132.1)	s...	...	...	...	...	...	...	...	...	0.16970	0.18395	0.19717	0.20970
	v...	...	...	...	...	...	...	...	...	0.188	0.232	0.254	0.282
220	h...	...	...	...	...	...	...	...	...	93.32	102.39	111.59	120.91
(13.99)	s...	...	...	...	...	...	...	...	...	0.16685	0.18142	0.19387	0.20753

[1] Data courtesy of E. I. du Pont de Nemours & Company.

Note: v in ft²/lbm; h in Btu/lbm; s in Btu/lbm °F.

Appendix E

GAS TABLES

Table E-1 Air Tables[1] (For 1 lb)

T, (°F abs)	t, (°F)	h, (Btu/lb)	p_r	u, (Btu/lb)	v_r	ϕ, (Btu/lb-°F)
100	−360	23.7	0.00384	16.9	9640	0.1971
120	−340	28.5	0.00726	20.3	6120	0.2408
140	−320	33.3	0.01244	23.7	4170	0.2777
160	−300	38.1	0.01982	27.1	2990	0.3096
180	−280	42.9	0.0299	30.6	2230	0.3378
200	−260	47.7	0.0432	34.0	1715	0.3630
220	−240	52.5	0.0603	37.4	1352	0.3858
240	−220	57.2	0.0816	40.8	1089	0.4067
260	−200	62.0	0.1080	44.2	892	0.4258
280	−180	66.8	0.1399	47.6	742	0.4436
300	−160	71.6	0.1780	51.0	624	0.4601
320	−140	76.4	0.2229	54.5	532	0.4755
340	−120	81.2	0.2754	57.9	457	0.4900
360	−100	86.0	0.336	61.3	397	0.5037
380	−80	90.8	0.406	64.7	347	0.5166
400	−60	95.5	0.486	68.1	305	0.5289
420	−40	100.3	0.576	71.5	270	0.5406
440	−20	105.1	0.678	74.9	241	0.5517
460	0	109.9	0.791	78.4	215.3	0.5624
480	20	114.7	0.918	81.8	193.6	0.5726

[1] The properties given here are abridged from *Gas Tables* by J. H. Keenan and J. Kaye, by permission of John Wiley & Sons, Inc., New York, 1948.

Table E-1 (continued)

T, (°F abs)	t, (°F)	h, (Btu/lb)	p_r	u, (Btu/lb)	v_r	ϕ, (Btu/lb-°F)
500	40	119.5	1.059	85.2	174.9	0.5823
520	60	124.3	1.215	88.6	158.6	0.5917
540	80	129.1	1.386	92.0	144.3	0.6008
560	100	133.9	1.574	95.5	131.8	0.6095
580	120	138.7	1.780	98.9	120.7	0.6179
600	140	143.5	2.00	102.3	110.9	0.6261
620	160	148.3	2.25	105.8	102.1	0.6340
640	180	153.1	2.51	109.2	94.3	0.6416
660	200	157.9	2.80	112.7	87.3	0.6490
680	220	162.7	3.11	116.1	81.0	0.6562
700	240	167.6	3.45	119.6	75.2	0.6632
720	260	172.4	3.81	123.0	70.1	0.6700
740	280	177.2	4.19	126.5	65.4	0.6766
760	300	182.1	4.61	130.0	61.1	0.6831
780	320	186.9	5.05	133.5	57.2	0.6894
800	340	191.8	5.53	137.0	53.6	0.6956
820	360	196.7	6.03	140.5	50.4	0.7016
840	380	201.6	6.67	144.0	47.3	0.7075
860	400	206.5	7.15	147.5	44.6	0.7132
880	420	211.4	7.76	151.0	42.0	0.7189
900	440	216.3	8.41	154.6	39.6	0.7244
920	460	221.3	9.10	158.1	37.4	0.7298
940	480	226.1	9.83	161.7	35.4	0.7351
960	500	231.1	10.61	165.3	33.5	0.7403
980	520	236.0	11.43	168.8	31.8	0.7454
1000	540	241.0	12.30	172.4	30.1	0.7504
1020	560	246.0	13.22	176.0	28.6	0.7554
1040	580	251.0	14.18	179.7	27.2	0.7602
1060	600	256.0	15.20	183.3	25.8	0.7650
1080	620	261.0	16.28	186.9	24.6	0.7696
1100	640	266.0	17.41	190.6	23.4	0.7743
1120	660	271.0	18.60	194.2	22.3	0.7788
1140	680	287.1	19.86	197.9	21.3	0.7833
1160	700	281.1	21.2	201.6	20.29	0.7877
1180	720	286.2	22.6	205.3	19.38	0.7920
1200	740	291.3	24.0	209.0	18.51	0.7963
1220	760	296.4	25.2	212.8	17.70	0.8005
1240	780	301.5	27.1	216.5	16.93	0.8047
1260	800	306.6	28.8	220.3	16.20	0.8088
1280	820	311.8	30.6	224.0	15.52	0.8128
1300	840	316.9	32.4	227.8	14.87	0.8168
1320	860	322.1	34.3	231.6	14.25	0.8208
1340	880	327.3	36.3	235.4	13.67	0.8246
1360	900	332.5	38.4	239.2	13.12	0.8265
1380	920	337.7	40.6	243.1	12.59	0.8323
1400	940	342.9	42.9	246.9	12.10	0.8360
1420	960	348.1	45.3	250.8	11.62	0.8398
1440	980	353.4	47.8	254.7	11.17	0.8434
1460	1000	358.6	50.3	258.5	10.74	0.8470
1480	1020	363.9	53.0	262.4	10.34	0.8506

Table E-1 (continued)

T, (°F abs)	t, (°F)	h, (Btu/lb)	p_r	u, (Btu/lb)	v_r	ϕ, (Btu/lb-°F)
1500	1040	369.2	55.9	266.3	9.95	0.8542
1520	1060	374.5	58.8	270.3	9.58	0.8568
1540	1080	379.8	61.8	274.2	9.23	0.8611
1560	1100	385.1	65.0	278.1	8.89	0.8646
1580	1120	390.4	68.3	282.1	8.57	0.8679
1600	1140	395.7	71.7	286.1	8.26	0.8713
1620	1160	401.1	75.3	290.0	7.97	0.8746
1640	1180	406.4	79.0	294.0	7.69	0.8779
1660	1200	411.8	82.8	298.0	7.42	0.8812
1680	1220	417.2	86.8	302.0	7.17	0.8844
1700	1240	422.6	91.0	306.1	6.92	0.8876
1720	1260	428.0	95.2	310.1	6.69	0.8907
1740	1280	433.4	99.7	314.1	6.46	0.8939
1760	1300	438.8	104.3	318.2	6.25	0.8970
1780	1320	444.3	109.1	322.2	6.04	0.9000
1800	1340	449.7	114.0	326.3	5.85	0.9031
1820	1360	455.2	119.2	330.4	5.66	0.9061
1840	1380	460.6	124.5	334.5	5.48	0.9091
1860	1400	466.1	130.0	338.6	5.30	0.9120
1880	1420	471.6	135.6	342.7	5.13	0.9150
1900	1440	477.1	141.5	346.8	4.97	0.9179
1920	1460	482.6	147.6	351.0	4.82	0.9208
1940	1480	488.1	153.9	355.1	4.67	0.9236
1960	1500	493.6	160.4	359.3	4.53	0.9264
1980	1520	499.1	167.1	363.4	4.39	0.9293
2000	1540	504.7	174.0	367.6	4.26	0.9320
2020	1560	510.3	181.2	371.8	4.13	0.9348
2040	1580	515.8	188.5	376.0	4.01	0.9376
2060	1600	521.4	196.2	380.2	3.89	0.9403
2080	1620	527.0	204.0	384.4	3.78	0.9430
2100	1640	532.6	212	388.6	3.67	0.9456
2120	1660	538.2	220	392.8	3.56	0.9483
2140	1680	543.7	229	397.0	3.46	0.9509
2160	1700	549.4	238	401.3	3.36	0.9535
2180	1720	555.0	247	405.5	3.27	0.9561
2200	1740	560.6	257	409.8	3.18	0.9587
2220	1760	566.2	266	414.0	3.09	0.9612
2240	1780	571.9	276	418.3	3.00	0.9683
2260	1800	577.5	287	422.6	2.92	0.9663
2280	1820	583.2	297	426.9	2.84	0.9688
2300	1840	588.8	308	431.2	2.76	0.9712
2320	1860	594.5	319	435.5	2.69	0.9737
2340	1880	600.2	331	439.8	2.62	0.9761
2360	1900	605.8	343	444.1	2.55	0.9785
2380	1920	611.5	355	448.4	2.48	0.9809
2400	1940	617.2	368	452.7	2.42	0.9833
2420	1960	622.9	380	457.0	2.36	0.9857
2440	1980	628.6	394	461.4	2.30	0.9880
2460	2000	634.3	407	465.7	2.24	0.9904
2480	2020	640.0	421	470.0	2.18	0.9927
2500	2040	645.8	436	474.4	2.12	0.9950
2520	2060	651.5	450	478.8	2.07	0.9972
2540	2080	657.2	466	483.1	2.02	0.9995
2560	2100	663.0	481	487.5	1.971	1.0018
2580	2120	668.7	497	491.9	1.922	1.0040

Table E-1 (continued)

T, (°F abs)	t, (°F)	h, (Btu/lb)	p_r	u, (Btu/lb)	v_r	ϕ, (Btu/lb-°F)
2600	2140	674.5	514	496.3	1.876	1.0062
2620	2160	680.2	530	500.6	1.830	1.0084
2640	2180	686.0	548	505.0	1.786	1.0106
2660	2200	691.8	565	509.4	1.743	1.0128
2680	2220	697.6	583	513.8	1.702	1.0150
2700	2240	703.4	602	518.3	1.662	1.0171
2720	2260	709.1	621	522.7	1.623	1.0193
2740	2280	714.9	640	527.1	1.585	1.0214
2760	2300	720.7	660	531.5	1.548	1.0235
2780	2320	726.5	681	536.0	1.512	1.0256
2800	2340	723.3	702	540.4	1.478	1.0277
2820	2360	738.2	724	544.8	1.444	1.0297
2840	2380	744.0	746	549.3	1.411	1.0318
2860	2400	749.8	768	553.7	1.379	1.0338
2880	2420	755.6	791	558.2	1.348	1.0359
2900	2440	761.4	815	562.7	1.318	1.0379
2920	2460	767.3	839	567.1	1.289	1.0399
2940	2480	773.1	864	571.6	1.261	1.0419
2960	2500	779.0	889	576.1	1.233	1.0439
2980	2520	784.8	915	580.6	1.206	1.0458
3000	2540	790.7	941	585.0	1.180	1.0478
3020	2560	796.5	969	589.5	1.155	1.0497
3040	2580	802.4	996	594.0	1.130	1.0517
3060	2600	808.3	1025	598.5	1.106	1.0536
3080	2620	814.2	1054	603.0	1.083	1.0555
3100	2640	820.0	1083	607.5	1.060	1.0574
3120	2660	825.9	1141	612.0	1.038	1.0593
3140	2680	831.8	1145	616.6	1.016	1.0612
3160	2700	837.7	1176	621.1	0.995	1.0630
3180	2720	843.6	1209	625.6	0.975	1.0649
3200	2740	849.5	1242	630.1	0.955	1.0668
3220	2760	855.4	1276	634.6	0.935	1.0686
3240	2780	861.3	1310	639.2	0.916	1.0704
3260	2800	867.2	1345	643.7	0.898	1.0722
3280	2820	873.1	1381	648.3	0.880	1.0740
3300	2840	879.0	1418	652.8	0.862	1.0758
3320	2860	884.9	1455	657.4	0.845	1.0776
3340	2880	890.9	1494	661.9	0.828	1.0794
3360	2900	896.8	1533	666.5	0.812	1.0812
3380	2920	902.7	1573	671.0	0.796	1.0830
3400	2940	908.7	1613	675.6	0.781	1.0847
3420	2960	914.6	1655	680.2	0.766	1.0864
3440	2980	920.6	1697	684.8	0.751	1.0882
3460	3000	926.5	1740	689.3	0.736	1.0899
3480	3020	932.4	1784	693.9	0.722	1.0916
3500	3040	938.4	1829	698.5	0.709	1.0933
3520	3060	944.4	1875	703.1	0.695	1.0950
3540	3080	950.3	1922	707.6	0.682	1.0967
3560	3100	956.3	1970	712.2	0.670	1.0984
3580	3120	962.2	2018	716.8	0.637	1.1000
3600	3140	968.2	2068	721.4	0.645	1.1017
3620	3160	974.2	2118	726.0	0.633	1.1034
3640	3180	980.2	2170	730.6	0.621	1.1050
3660	3200	986.1	2222	735.3	0.610	1.1066
3680	3220	992.1	2276	739.9	0.599	1.1083

Table E-2 Combustion Products of 200% Air[1] (per 1 lbmole)

T	t	$\bar{h}$	p_r	$\bar{u}$	v_r	ϕ
300	−159.7	2096.7	0.16767	1500.9	19201	42.180
400	−59.7	2801.4	0.4655	2007.0	9222	44.208
500	40.3	3511.2	1.0330	2518.3	5194	45.791
600	140.3	4226.3	1.992	3034.8	3233	47.094
700	240.3	4947.7	3.487	3557.6	2154.5	48.207
800	340.3	5676.3	5.690	4087.6	1508.7	49.179
900	440.3	6413.0	8.808	4625.7	1096.4	50.047
1000	540.3	7159.8	13.089	5173.9	819.8	50.833
1100	640.3	7916.4	18.822	5731.9	627.1	51.555
1200	740.3	8683.6	26.34	6300.6	488.9	52.222
1300	840.3	9461.7	36.05	6880.1	387.0	52.845
1400	940.3	10250.7	48.38	7470.5	310.5	53.430
1500	1040.3	11050.2	63.88	8071.4	252.0	53.981
1600	1140.3	11859.6	83.10	8682.2	206.63	54.504
1700	1240.3	12678.6	106.70	9302.6	170.97	55.000
1800	1340.3	13507.0	135.43	9932.4	142.63	55.473
1900	1440.3	14344.1	170.09	10571.0	119.87	55.926
2000	1540.3	15189.3	211.6	11217.6	101.43	56.360
2100	1640.3	16042.4	260.9	11872.1	86.36	56.777
2200	1740.3	16902.5	319.2	12533.6	73.96	57.177
2300	1840.3	17769.3	387.5	13201.8	63.68	57.562
2400	1940.3	18642.1	467.4	13876.0	55.12	57.933
2500	2040.3	19520.7	559.8	14556.0	47.91	58.292
2600	2140.3	20404.6	666.6	15241.3	41.85	58.639
2700	2240.3	21293.8	789.4	15932.0	36.71	58.974
2800	2340.3	22187.5	929.8	16627.1	32.31	59.300
2900	2440.3	23086.0	1089.8	17327.0	28.56	59.615
3000	2540.3	23988.5	1271.2	18030.9	25.321	59.921

[1] Abridged from *Gas Tables* by J. H. Keenan and J. Kaye, by permission of John Wiley & Sons, Inc., New York, 1948.

Appendix F

PSYCHROMETRIC CHART

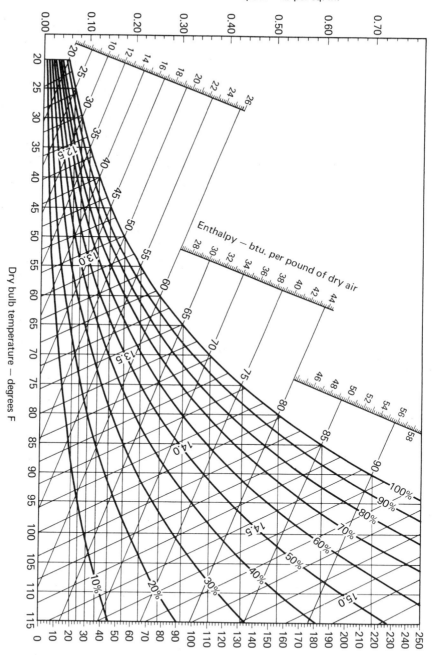

Pressure of water vapour — lb per sq. in.
0.00
0.10
0.20
0.30
0.40
0.50
0.60
0.70
Enthalpy — btu. per pound of dry air
100%
90%
80%
70%
60%
50%
40%
30%
20%
10%
12.5
13.0
13.5
14.0
14.5
15.0
Dry bulb temperature — degrees F
Weight of water vapour in one pound of dry air — grains

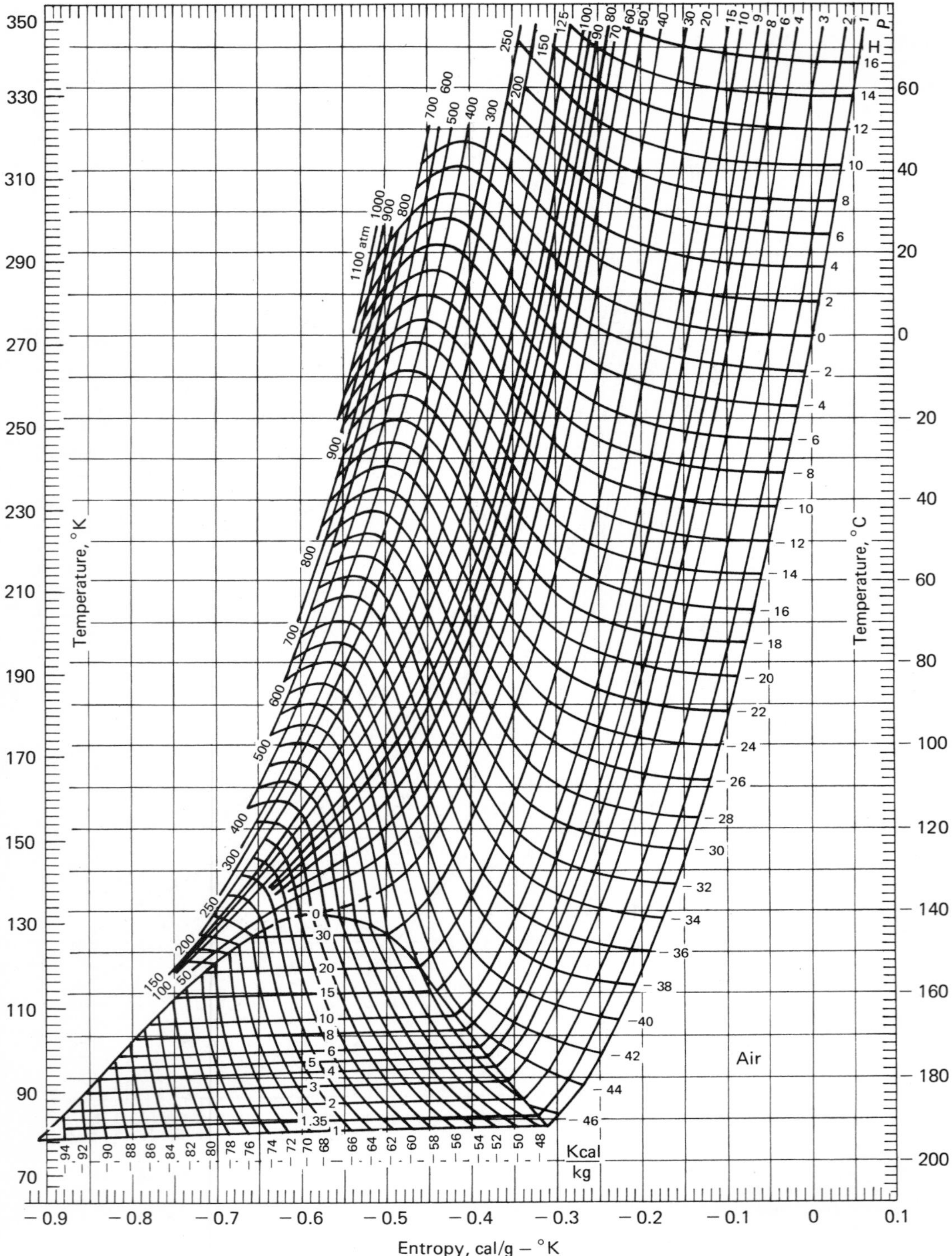
Temperature, °K
Temperature, °C
Entropy, cal/g − °K
Kcal/kg
H
P
1100 atm
Air

Appendix G

THERMODYNAMIC PROPERTIES OF AIR

Appendix H

SPECIFIC HEATS OF SUBSTANCES

Table H-1 Specific Heats of Solids[1]

$P=1$ atm *Substance*	T (°C)	c_P (cal/g-°C)
Ice	−200	0.168
	−140	0.262
	−60	0.392
	−11	0.468
	−2.6	0.500
Aluminium	−250	0.0039
	−200	0.076
	−100	0.167
	0	0.208
	+100	0.225
	300	0.248
	600	0.277
Platinum	−256	0.00123
	−152	0.0261
	0	0.0316
	+500	0.0349
	1000	0.0381
Lead	−270	0.00001
	−259	0.0073
	−100	0.0283
	0	0.0297
	+100	0.0320
	300	0.0356
Iron	20	0.107
Silver	20	0.0558
Magnesium	20	0.246
Sodium	20	0.295
Tungsten	20	0.034
Graphite	20	0.17
Wood	20	0.42
Rubber	20	0.44
Mica	20	0.21

[1] Based on values from the *Handbook of Chemistry and Physics*, American Rubber Company. Reproduced from *Thermodynamics* by Reynolds (1965) by permission of McGraw-Hill Book Co., New York.

Table H-2 Specific Heats of Liquids[1]

Substance	*State*	c_P (Btu/lbm-°F)
Water	1 atm, 32°F	1.007
	1 atm, 77°F	0.998
	1 atm, 212°F	1.007
Ammonia	sat., 0°F	1.08
	sat., 120°F	1.22
Freon-12	sat., −40°F	0.211
	sat., 0°F	0.217
	sat., 120°F	0.244
Benzene	1 atm, 60°F	0.43
	1 atm, 150°F	0.46
Light oil	1 atm, 60°F	0.43
	1 atm, 300°F	0.54
Glycerin	1 atm, 50°F	0.554
	1 atm, 120°F	0.617
Bismuth	1 atm, 800°F	0.0345
	1 atm, 1000°F	0.0369
	1 atm, 1400°F	0.0393
Mercury	1 atm, 50°F	0.033
	1 atm, 600°F	0.032
Sodium	1 atm, 200°F	0.33
	1 atm, 1000°F	0.30
n-Butane	1 atm, 32°F	0.550
Propane	1 atm, 32°F	0.576

[1] Based on values from the *Handbook of Chemistry and Physics*, American Rubber Company. Reproduced from *Thermodynamics* by Reynolds (1965) by permission of McGraw-Hill Book Co., New York.

Index

Index

A

B

C

D

E

F

G

H

I

J

K

L

M

N

O

P

Q

R

S

T

U

V

W

Z